PROBABILITY

PROBABILITY: PURE AND APPLIED

A Series of Textbooks and Reference Books

Editor

MARCEL F. NEUTS

University of Arizona
Tucson, Arizona

1. Introductory Probability Theory, *by Janos Galambos*
2. Point Processes and Their Statistical Inference, *by Alan F. Karr*
3. Advanced Probability Theory, *by Janos Galambos*
4. Statistical Reliability Theory, *by I. B. Gertsbakh*
5. Structured Stochastic Matrices of M/G/1 Type and Their Applications, *by Marcel F. Neuts*
6. Statistical Inference in Stochastic Processes, *edited by N. U. Prabhu and I. V. Basawa*
7. Point Processes and Their Statistical Inference, Second Edition, Revised and Expanded, *by Alan F. Karr*
8. Numerical Solution of Markov Chains, *edited by William J. Stewart*
9. Probability: The Mathematics of Uncertainty, *by Dorian Feldman and Martin Fox*

Other Volumes in Preparation

PROBABILITY
THE MATHEMATICS OF UNCERTAINTY

DORIAN FELDMAN
MARTIN FOX

Department of Statistics and Probability
Michigan State University
East Lansing, Michigan

Marcel Dekker, Inc.   **New York • Basel • Hong Kong**

ISBN 0-8247-8452-9

This book is printed on acid-free paper.

MARCEL DEKKER, INC.
270 Madison Avenue, New York, New York 10016

Current printing (last digit):
10 9 8 7 6 5 4 3 2 1

PRINTED IN THE UNITED STATES OF AMERICA

To David Blackwell,
our teacher and friend

Preface

Probability provides a formal basis for reasoning about and quantifying uncertainty. The uncertainty to which we refer may center on such frivolous items as roulette wheels, dice, slot machines, and horse races; or around weightier matters such as guilt or innocence, investment policies, scientific theories, election forecasts, and the likelihood of nuclear war. With this breadth of application in mind, we have chosen an approach to probability that we believe to be particularly flexible and conducive to a variety of interpretations and settings. We believe it to be a more natural approach than the usual one based on sets.

We begin with the idea that probability measures (on a 0–1 scale) the "likelihood" that a given statement is true. No particular form or structure is imposed on the nature of statements or the determination of truth or falsity. The ensuing development is, at least at the beginning, considerably different from the traditional sample space–set theoretic foundation expounded in most textbooks on this subject. Ultimately, however, after a thorough examination of the logic of occurrence, experiments and sample spaces emerge as important applications of general concepts. It is worth noting that in Sections 1 to 7 and 11 to 14 all of the usual results of probability for finite collections of events, including those related to random variables and their expectations, are derived without any reference to sets. They are based solely on event relations and the rules of probability. We have found it advantageous, pedagogically, that the absence of sets as a framework forces a greater concentration on the logical relations between statements.

In addition to the theoretical development, the text deals with a variety of applications. Chapter 3, for example, is entirely devoted to an excursion into some of them.

The text contains more than enough material for a semester, and we view its contents as preparation for a follow-up course in mathematical statistics for which we hope to write a text. The text by M. H. DeGroot, *Probability and Statistics*, 2nd ed. (Addison-Wesley, 1986), has worked well as a continuation. To provide additional source material, references are given at the end of each chapter.

The course for which this text has been used is a senior-graduate level course for students who have taken (or are currently enrolled in) a course in advanced calculus. Only a smattering of advanced calculus material is actually used in this text, but familiarity with and ability to follow detailed mathematical arguments is definitely required of the reader. Advanced calculus results as well as some results from linear algebra are summarized in the section on notation and results.

The following outline shows how this text might be used in one semester.

Chapter 1: Sections 1 to 7 and 9 are essential to the later development. The results in Sections 3 to 5 are particularly critical and must be adequately covered even if the proofs are omitted or skimmed. Section 9 relates the classical approach to ours. For readers who are not familiar with the basics of combinatorics, that material is included in an appendix (Section 10) which, we hope, is sufficient.

Chapter 2: The development of conditional probability and independence in Sections 11 to 15 is essential. On the assumption that most readers have had some exposure to finite dimensional Cartesian products, that topic is relegated to an appendix (Section 16).

Chapter 3: The applications here can be covered only briefly or left to the reader.

Chapter 4: Sections 21 to 24 contain the basics on random variables and random vectors. Some deletion of coverage is possible in later portions of Sections 22 to 24. Section 25 need not be covered.

Chapter 5: Conditional distributions and independence for random variables and random vectors are discussed in Sections 26 to 28. The examples in Section 29 may be left to the reader.

Chapter 6: Sections 30 to 35 contain the essential material on moments and conditional moments. The last part of Section 30 contains a completely general definition of expectation and can be omitted. Section 36 provides essential preparation for statistics and for the Central Limit Theorem. Section 37 contains the Weak Law of Large Numbers.

Chapter 7: Sections 39 to 44 contain all the essential development of special distributions including applications and relations between them. Section 41 on multivariate frequencies can be skipped as can the last subsections of Sections 42 and 43. The *t*- and *F*-distributions are deferred to Chapter 8.

Chapter 8: Sections 47 and 48 as well as the portion of Section 46 on convolutions are essential for statistics and should be covered.

Chapter 9: Coverage of this chapter on generating functions is optional.

Our motivation for delaying the introduction of sample spaces centers on several considerations including: (i) The concept of sample space is part of a mathematical structure that identifies events with sets. We have chosen, initially, to focus attention on the logical relationships between events. (ii) The choice of a workable sample space often hinges on technical issues that we feel are best raised after some experience with probabilistic reasoning (in particular, the concept of a partition). (iii) In many instances, such as the subjective evaluation of probabilities, sample spaces are of secondary interest.

We are grateful to Professor Dennis Gilliland and to the late Professor V. Susarla for testing an early version of our manuscript in classes. We thank Professor R. V. Ramamoorthi, who read and made useful comments on major portions of the manuscript. We also thank the students who put up with all the deficiencies of the manuscript and helped us to improve it. Anonymous referees also contributed valuable comments.

Since this book was first conceived, several typists have helped us with various versions of the manuscript: Noralee Burkehardt, Joann Peterson, Carol Case, Sharon Carson, and Loretta Ferguson, all of whom we thank. Michelle Mikos provided valuable assistance in preparing the figures.

Dorian Feldman
Martin Fox

Contents

ix

Notation and Useful Mathematical Results

REFERENCE NOTATION

The numbering system for formulae, theorems, and examples begins with 1 in any section. References to formula (1) in Section 3 will be as follows:

1. It will be called (1) everywhere in Section 3.
2. It will be called (3.1) in all other sections.

The same referencing system will be used for theorems and examples. General Problem 2 in Chapter 5 is, for example, referenced as GP5.2 Section numbering runs consecutively throughout the text. For example, Chapter 1 ends with Section 10 and Chapter 2 begins with Section 11. The reader should refer to the Table of Contents to locate sections.

A dagger (†) is used to tag examples and problems referred to in later sections.

INTERVALS AND PRODUCTS

The following standard mathematical notation is used:

$\Re$ = the real line;
$\Re^n$ = n-dimensional Euclidean space;

For intervals in $\mathcal{R}$,

$$[a, b] = \{x: a \leq x \leq b\}; \qquad (-\infty, b] = \{x: x \leq b\};$$
$$[a, b) = \{x: a \leq x < b\}; \qquad (a, \infty) = \{x: x > a\}$$

and others are denoted analogously.

Parallel to the usual use of Σ for summations, we use Π for products as below:

$$\prod_{k=1}^{n} a_k = a_1 \cdot a_2 \cdots \cdot a_n;$$

$$\prod_{k=1}^{\infty} a_k = a_1 \cdot a_2 \cdots;$$

$$\prod_{a \in \mathcal{a}} a = \text{the product of all members of the set } \mathcal{a}.$$

RESULTS FROM ADVANCED CALCULUS

1. If $F(x) = \int_a^x f(t)\, dt < \infty$ for all $x \in (a, b)$, then F is continuous on (a, b) and $F'(x) = f(x)$ for all $x \in (a, b)$ at which f is continuous. Here $a = -\infty$ or $b = \infty$ are possible.

Let f be a function on $A \subset \mathcal{R}^2$. Let (x, y) denote a point in A. Result (2.) requires $\int_a^b [\int_c^d |f(x, y)|\, dy]\, dx < \infty$ and (3.) and (4.) involve analogous conditions.

2. *Interchange of order of integration.* When integrating over A, the order of integration is immaterial. For example, $\int_a^b [\int_c^d f(x, y)\, dy]\, dx = \int_c^d [\int_a^b f(x, y)\, dx]\, dy$.

3. *Interchange of order of summation.* Suppose that A is a countable set. Then, $\Sigma_x [\Sigma_y f(x, y)] = \Sigma_y [\Sigma_x f(x, y)]$ where the sums are over $(x, y) \in A$.

4. *Interchange of integral and sum.* Suppose that A is given by $x \in (a, b)$ and y in some countable set. Then, $\int_a^b [\Sigma_y f(x, y)]\, dx = \Sigma_y [\int_a^b f(x, y)\, dx]$.

5. *Interchange of integral and derivative.*

$$\frac{d}{dx} \int_a^b f(x, y)\, dy \, \bigg|_{x=x_0} = \int_a^b \frac{\partial}{\partial x} f(x, y) \, \bigg|_{x=x_0} dy$$

provided that (a.) $\int_a^b f(x, y)\, dy$ is defined for x sufficiently close to x_0; (b.) $\partial/\partial x[f(x, y)]|_{x=x_0}$ exists for every fixed y such that $(x_0, y) \in A$; (c.) the derivative in (b.) is integrable on (a, b); and (d.) there exists an integrable function g such that, for $|h|$ sufficiently small,

$$\left| \frac{f(x_0 + h, y) - f(x_0, y)}{h} \right| \leq g(y).$$

6. *Interchange of sum and derivative.*

$$\frac{d}{dx} \sum_y f(x, y) \bigg|_{x=x_0} = \sum_y \frac{\partial}{\partial x} f(x, y) \bigg|_{x=x_0}$$

under the same conditions as in (5) with integrals replaced by sums.

RESULTS FROM MATRIX ALGEBRA

A few portions of this text are designated as requiring familiarity with matrix algebra. Although some results in these portions are used later, detailed reading of them is unnecessary for understanding of the rest of the text.

If $\mathbf{A}$ is a positive definite, symmetric matrix, then there exists a unique nonsingular, symmetric matrix $\mathbf{B}$ such that $\mathbf{A} = \mathbf{B}^2$. We denote $\mathbf{B}$ by $\mathbf{A}^{1/2}$, called the *square root of* $\mathbf{A}$.

If $\mathbf{A}$ is a nonnegative definite, symmetric matrix, then there exists a matrix $\mathbf{B}$ such that $\mathbf{A} = \mathbf{BB}'$.

If $\mathbf{A}$ is $n \times n$ and rank $(\mathbf{A}) = k < n$, then there are $n - k$ linear combinations of columns (or of rows) of $\mathbf{A}$ which are $\mathbf{0}$.

If $\mathbf{A}$ is a nonsingular matrix, then $\det(\mathbf{A}^{-1}) = 1/\det(\mathbf{A})$.

If $\mathbf{A}$ is a symmetric matrix, then there exists an orthogonal matrix $\mathbf{P}$ $(\mathbf{PP}' = \mathbf{I})$ such that $\mathbf{PAP}'$ is a diagonal matrix.

PROBABILITY

1.
Probability Models

1. INTRODUCTION

Probability is used for coping with problems related to uncertainty and variability. The mathematical development of this basic subject did not begin until the sixteenth and seventeenth centuries, and only then in connection with gambling problems—to which it is still closely linked. Nevertheless, its methodology has become indispensable to those engaged in such activities as research, decision making, planning, or prediction. Mathematicians, scientists, philosophers, and scholars in general have been intrigued and stimulated by the richness of its concepts and theorems and the multitude of applications and interpretations they evoke.

The basic objects of study in this theory are *event*, *occurrence*, and *probability*. As in other branches of mathematics, the basic entities are not specifically defined but are assumed to satisfy certain axioms or rules from which a body of theorems can be derived. To apply the theory to concrete problems, a correspondence must be established between the real elements of the problem and the abstractions of the theory. The process evolves through intuition, compromise, and approximation. Ultimately, the conclusions are evaluated on the basis of their usefulness—a matter that is not always free of controversy.

A probability is a number attached to an event and represents a measurement roughly analogous to physical size measurements, such as area, weight, height, and width. What the probability of an event measures we call the *likelihood that the event occurs*. The type of "measurement" we have in mind yields a number between 0 and 1 and is illustrated by the following three cases, which also provide possible interpretations of the phrase *likelihood of occurrence*.

Case I: Population Proportions—Prevalence

Consider a population π with N members (e.g., people, animals, objects). An *event* in this context is a property that members of π may or may not possess. Let A denote such a property and let $n(A)$ be the number of members with property A. Then the probability of A is defined as the proportion: $P(A) = n(A)/N$. The numerical value of $P(A)$ is a descriptive measure of the prevalence of A among the members of π.

Example 1.   In a certain district there are $N = 20{,}000$ registered voters of whom 9000 are registered Democrats (D), 8000 are registered Republicans (R), and the rest are classified as independents (I). For this population we therefore have $P(D) = .45$, $P(R) = .40$, and $P(I) = .15$.

Example 2.†   A Nevada roulette wheel is a device consisting of 38 numbered and colored slots. There are two green slots numbered 0 and 00, respectively. The remaining slots are red and black and numbered 1 through 36. Let G, R, and B denote the color and let O and E denote the properties odd and even, respectively. For this purpose 0 and 00 are neither odd nor even.

Look closely at such a wheel and, by direct count, you will find that $n(G) = 2$, $n(R) = n(B) = n(O) = n(E) = 18$. Thus $P(G) = 1/19$ and $P(R) = P(B) = P(O) = P(E) = 9/19$. If we let RO, RE, BO, and BE denote slots with *both* indicated properties, it turns out (perhaps surprisingly) that $P(RO) \neq P(RE)$ and $P(BO) \neq P(BE)$. In fact, $P(RO) = P(BE) = 5/19$ and $P(RE) = P(BO) = 4/19$. Of course, $P(GE) = P(GO) = 0$.

Example 3.   In an ordinary deck of cards there are four suits, each with 13 representatives distinguished by face value. In the population of 52 cards, then $P(\text{spades}) = P(\text{hearts}) = P(\text{diamonds}) = P(\text{clubs}) = 1/4$. Now, consider the population of all pairs of distinct cards that could be formed from such a deck. Some of these pairs will consist of cards from the same suit (let A denote this property) and the rest will not. To determine $P(A)$ we could write down every pair (of which there are 1326) and simply count those with A. If we did, we would find that $n(A) = 312$, so that

$P(A) = 312/1326 = .24$, a number we actually obtained from simple combinatorial calculations of the type frequently used in probability.

Example 4. We want to describe the way a measured value varies over a population. Here, for example, is a summary of the grades assigned in a course consisting of 37 students:

GP	0.0	1.0	1.5	2.0	2.5	3.0	3.5	4.0
Proportion	.108	.081	.081	.270	.135	.162	.027	.135

From this information, additional characteristics of the grade distribution, such as the average or percentiles, can be determined for this class.

When the only information about an object is that it belongs to a particular population, the uncertainty concerning the object's membership in a subgroup (i.e., those with a specified property) is expressed by giving the probability (i.e., prevalence) of the subgroup rather than stating, as a fact, that it does/does not belong. A probability near 1 (near 0) indicates substantial certainty of membership (nonmembership) in the subgroup.

When a population is very large or widely scattered geographically, it may be quite difficult, if not impossible, to obtain the necessary counts to establish precise proportions. A considerable portion of statistical methodology is devoted to the problem of making inferences about population proportions based on samples, that is, on subsets of the population selected in an appropriate manner.

Case II: Relative Frequencies in Repeated Trials—
Rate of Occurrence

An experiment is repeated N times. Let A be some result that may or may not occur on any given trial, and let $n(A)$ be the number of trials on which A does occur. The relative frequency of occurrence of A is $P_N(A) = n(A)/N$. It depends on N and also on the results of the sequence of trials performed. It is a descriptive measure of the rate (per trial) of the occurrence of A in the N trials.

Example 5. On Monday a certain machine produced 30 items, of which 2 were defective. The relative frequency of defectives for Monday was therefore 2/30 or .067. On Tuesday the same machine produced 50 items, 6 of which were defective. The relative frequency of defectives for Tuesday was .12, and for the first two days of the week was $8/80 = .1$. Here each item produced by the machine constitutes one trial of the experiment.

Our concern here is with the quality of the machine's output. It seems natural to evaluate a machine in terms of the rate at which it produces defective items and to prefer machines with lower rate (unfortunately, a rate of zero is rarely obtainable). A machine's rates are measured by monitoring its output—as the example suggests. If total monitoring is not feasible, occasional samples of the output are taken, from which estimates of the rate may be obtained. The fact that the rate differs from one day to the next (or from sample to sample) is indicative of the fluctuating nature of relative frequencies.

The theory of probability predicts that there will be fluctuations within reasonable limits. It also predicts that the range of fluctuations will narrow as the number of trials on which the relative frequencies are based increases. Empirical observation of this effect has led many theorists, researchers, and other practitioners to postulate a property called the *stability of long-run relative frequencies*, and to define or interpret probability as a "limit" of relative frequencies based on an unending sequence of repeated trials. This is by far the most widely advocated version of the "true" meaning of probability and is perfectly acceptable in the context of repeated trials. However, we believe that the mathematical model can be usefully adapted to a wider class of applications.

Example 6. A roulette wheel (Example 2) is operated by spinning the wheel with a ball that comes to rest in one of the slots. Gamblers are allowed to bet (against the "house") at specified odds on various properties of the selected slot (e.g., odd, even, red, black, etc.). While the spin-to-spin results are unpredictable, it is generally understood by the participants that in a "large" number of spins the relative frequency of a property mirrors its prevalence. That is what is meant by such terms as *fair wheel* or *random selection*. Thus after many spins we would expect a green slot roughly 1/19 of the time, a red slot 9/19, and a black–even slot 5/19. The extent to which results depart from expectations is a major topic in statistical analysis.

Example 7. Relative frequencies play an important role in the operations of insurance firms. In particular, premiums for life insurance are established through analysis of mortality tables. These are tables giving the observed relative frequency of survival beyond any age for selected groups of individuals. Let P_k, $k = 0, 1, 2, \ldots, T$, denote the observed relative frequencies of survival beyond age k years (T is a large integer). The ratio

P_{k+m}/P_k is the relative frequency that an age k survivor lives another m years. This number contributes to the determination of a suitable premium for a person aged k who wishes to purchase an m-year life insurance policy.

Case III: Opinions; Beliefs

In certain situations, some people are willing (and able) to express the extent of their belief in the truth of statement A as a number, $P(A)$, between 0 and 1 (or as an equivalent percentage). People also express belief by the choices they make when faced by various alternatives. They may of course, be guided in these matters by considerations of the type discussed in Cases I and II, but they are not bound to do so. In many cases there are no relative frequencies or populations that bear on the issue. Football games, horse races, and similar occasions are unique confrontations and may give rise to widely varying opinions. The wagering that takes place in this environment is an expression of the different probabilities that people assign to a given event. This can also be observed with regard to the stock market or with regard to business strategies in general. Speculative statements regarding the past or the future may well lead to different assessments by different individuals. Opinions, even expert opinions, are often stated as personal probabilities.

There are those who reject subjective probabilities as unscientific or impractical. In their view the question of what probabilities to assign to certain events is meaningless. Others believe that all probabilities are, to some extent, personal.

Example 8.   During the 1988 Republican convention we asked five of our colleagues the question, "What probability do you assign to the event that the Republicans will retain the presidency?" Their responses were .43, .45, .50, .40, and .58, respectively. These are personal, subjective opinions and, not surprisingly, vary.

Numerical assignments do not necessarily constitute probabilities just because they lie between 0 and 1—other properties, to be introduced later, must be satisfied. Adherence to these rules is ultimately based on the agreement that they are reasonable and intuitively correct in terms of the logic of occurrence. To help individuals quantify their opinions according to these rules, various schemes have been devised for eliciting information from them which can be used to measure *their* probabilities of events. One such scheme is to make a person choose the monetary worth of a fictitious gamble.

```
┌─────────────────────────────┐
│    Pay Bearer  $50          │
│   if Mt. St. Helen's        │
│  erupts within 3 months     │
│        Date_____         │
└─────────────────────────────┘
```

Figure 1

Example 9. We want *your* probability that Mt. St. Helen's has a major eruption within the next three months (call this event A). You, of course, have no idea of how to measure that since you do not know what we mean by probability. Imagine therefore that we offer you a ticket, such as the one shown in Figure 1. We ask you: What is the largest amount you would pay for this ticket? If you answer \$$m$, *your* assignment is $P(A) = m/50$.

We assume that the m you choose will be somewhere between 0 and 50, thus producing a value of $P(A)$ between 0 and 1. We also assume other things about your behavior that will guarantee the validity of the procedure. For example, if the prize is \$100 instead of \$50, we would assume that your response would be \$$2m$. In principle, we can carry out this inquiry for any event whatever and for as large a collection of events as patience and time will allow. If doing so yielded a collection of numbers satisfying all the rules of probability, we would be prepared to call you "rational," or as some probabilists would say, "coherent."

Now that the empirical basis for our subject is outlined, we should admit that this volume will be devoted almost entirely to the mathematical assignment of probabilities using what we call "probability models." In the next volume our concern will center on the relation between theory and observation—namely, statistical analysis.

Meanwhile, we hope to show that probability theory is a useful mathematical model for a wide variety of practical problems, many of which will deal with such measurements as are outlined in Cases I, II, and III. To begin with, however, we will simply describe its mathematical structure.

Various rules governing the behavior of these measurements (numbers) will be prescribed. The rules serve a dual purpose. First, they serve to define what specific numerical measurements will be called probabilities. Second, they provide a basis for the *calculus of probability*, which deals with the problem of how one uses the known probabilities of *some* events to deduce the probabilities of others.

2. EVENTS AND INDICATORS

Event

We use the word *event* to designate anything that admits of two, and only two, distinguishable possibilities: It either "occurs" or it "doesn't occur." We allow the cases, described below in the definitions of impossible and sure events, in which one of these is, in fact, not really possible. The choice of names for these possibilities is arbitrary. Other names for them, suggestive of the wide range of applicability of the notion, are "yes" and "no," "true" and "false," "on" and "off," or "success" and "failure."

Tense is irrelevant. "Oswald acted alone in the assassination of Kennedy" is an event that occurred if the statement is true, did not occur if the statement is false. "This light bulb will burn for 500 hours" is an event that may or may not occur (in the future).

Statements are events. There may be uncertainty about the truth (occurrence) or falsity (nonoccurrence) of any statement.

No specific mechanism for determining whether the event in question occurs or does not occur is implied. In fact, for some events, we never will know if they occur. In the analysis of random experiments the occurrence or nonoccurrence of an event will be determined by observing experimental results.

In many applications, events are represented by subsets of a specified set, called a *space* or *population*. A point is selected from the space in some prescribed fashion. Occurrence of a specified event will then mean that the selected point belongs to the subset determined by the event.

Example 1. There are 10 applicants for a job, 3 of which are women. The event "a woman is hired" refers to the subset of 3 women on the list of applicants. The event occurs if the selected applicant belongs to this subset.

Events will generally be denoted by the letters A, B, C, . . . with or without subscripts.

Impossible and Sure Events

An event is *impossible* or *empty* if it cannot (either logically or by assumption) occur; an event is *sure* or *certain* if it must occur. Impossible events are denoted by ϕ and sure events are denoted by Ω.

Example 2. An individual is selected from some population. Let A be the event that the person chosen is at least 18 years old. If the population consists of those who voted in the 1984 presidential election, A is a sure event. If the population consists of preschool children, A is an empty event.

Much of the analysis of occurrence and probability is facilitated by identifying an event with what is called its indicator. This is a convenient mathematical device that reduces all event operations and relations to simple arithmetic with the numbers 0 and 1.

Indicator

The *indicator* of an event, A, is denoted by I_A and is identified with A by the relation

$$I_A = 1 \quad \text{if } A \text{ occurs}$$
$$ = 0 \quad \text{if } A \text{ does not occur.}$$

For the sure and impossible events, we obtain

$$I_\phi \equiv 0, \qquad I_\Omega \equiv 1.$$

(The notation $\equiv$ is used to indicate that no other values are possible.)

Because indicators assume numerical values, we can combine them in various ways (addition, multiplication, etc.) to produce functional expressions. If the functional expression takes only values 0 and 1, it too is an indicator.

Example 3. Let A and B be events. The expressions $1 - I_A$, $1 - I_B$, $I_A I_B$, and $I_A + I_B - I_A I_B$ are indicators. The expression $I_A - I_B$ may or may not be an indicator, but $(I_A - I_B)^2$ is always an indicator. If the occurrence of B implies the occurrence of A, then $I_A - I_B$ is an indicator since the value -1 is precluded.

Two rules worth mentioning in connection with algebraic expressions are the following:

1. Any product of indicators is itself an indicator.
2. If I_A is any indicator, so is $1 - I_A$.

Summary

Event: Anything admitting two possibilities (occurs or does not occur).
Sure event: Certain to occur (denoted by Ω).
Impossible event: Cannot occur (denoted by ϕ).
Indicator of event A: $I_A = 1$ if A occurs, $I_A = 0$ if A fails to occur.

Products of indicators are indicators and if I_A is an indicator, then so is $1 - I_A$. Furthermore, $I_\Omega \equiv 1$ and $I_\phi \equiv 0$.

Problems

Which of the expressions in Problems 1 to 9 are necessarily indicators?

1. $I_A + I_B$
2. I_A^2
3. $I_A^2 + I_B^2 - 2I_A I_B$

4. $I_A(1 - I_A)$
5. $I_A I_B + I_B I_C + I_C I_D - I_A I_B I_C - I_B I_C I_D$
6. $1/(1 - I_A)$
7. $I_A + I_B + I_C - 3I_A I_B I_C$
8. $I_A + I_B - 3I_A I_B$
9. $I_A + I_B - 2I_A I_B I_C$

Which of the expressions in Problems 10 to 13 are indicators of sure events? Of impossible events?

10. $I_A + I_B - I_A I_B + (1 - I_A)(1 - I_B)$
11. $I_A I_B + I_A I_C + I_B I_C - 2I_A I_B I_C$
12. $(I_A^3 + I_B^2 - 2I_A I_B)(I_A I_B)^{3/2}$
13. $I_A + I_B(1 - I_A) + I_C(1 - I_A)(1 - I_B) + (1 - I_A)(1 - I_B)(1 - I_C)$

3. EVENT OPERATIONS AND RELATIONS

A. Event Operations

Given a collection of events, we may be interested in combining them in various ways or expressing relations between them. In this section we define operations on events that enable us to do this formally. Corresponding algebraic expressions for indicators are given.

Here is a simple illustration.

Example 1. Suppose that a device is made up of two components, C_1 and C_2, either of which may or may not be functioning. We say that the device is hooked up in *series* if proper functioning of the device requires *both* components to function. We say that the device is hooked up in *parallel* if proper functioning of the device requires only that *at least one* of the components functions. An example of a series hookup is a monaural system consisting of a receiver and a speaker. For this system to make music, both the receiver *and* the speaker must function properly. An example of a parallel hookup is a sailboat with auxiliary engine. The boat will move if either the wind blows *or* the engine functions. Series and parallel hookups are illustrated in Figure 1.

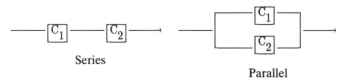

Series   Parallel

Figure 1

What we mean by an *event operation* is illustrated by the relation between the functioning of the device and the functioning of the individual components, in either the "series" or "parallel" case. Let A denote the event that the device functions, let B_1 denote the event that component C_1 functions, and let B_2 denote the event that component C_2 functions. For a series hookup "A occurs" means that "B_1 and B_2 *both* occur." For a parallel hookup "A occurs" means that "B_1 occurs or B_2 occurs or both occur." In terms of indicators we can represent these relations as follows:

Series: $I_A = I_{B_1} I_{B_2}$
Parallel: $I_A = I_{B_1} + I_{B_2} - I_{B_1} I_{B_2}.$

In the series case we require both components to function; therefore, I_A must be 1 if, and only if, both I_{B_1} and I_{B_2} are 1. In the parallel case I_A will be 1 if, and only if, at least one of the indicators, I_{B_1} or I_{B_2}, is 1.

An alternative way of describing the operation of a device is by giving conditions under which it *fails* to function. The event "device fails to function," call it F, is related to the event $A = $ "the device functions" in the following obvious way: "F occurs" means "A fails to occur." The indicator of F, then, is

$$I_F = 1 - I_A.$$

We see from this example that the notion of "occurrence of an event" gives rise to certain operations with events. Although couched in terms of occurrence, these operations are directly analogous to the operations of set algebra and those of common sentential logic ("not," "and," and "or"). Because of their assumed familiarity we will use the terminology and notation of set algebra and, on occasion, Venn diagrams.

One common mathematical model represents events as subsets of a specified universe, $\mathcal{S}$, called a *sample space*. The indicator of a set (event) A is the function on $\mathcal{S}$ given by

$$I_A(x) = 1 \quad \text{if } x \in A$$
$$= 0 \quad \text{if } x \notin A.$$

The basic operations are as follows:

Complement

To any event A there corresponds an event A^c, called the *complement* of A, which occurs if A does not occur. Then

$$I_{A^c} = 1 - I_A.$$

Intersection

To any pair of events A, B there corresponds an event $A \cap B$, called the *intersection* of A and B, which occurs if both A and B occur. Then $I_{A \cap B} = I_A I_B$.

Union

To any pair of events A, B there corresponds an event $A \cup B$, called the *union* of A and B, which occurs if A occurs or B occurs or both occur. An equivalent way of referring to the union is "at least one of the events A or B occurs." Then $I_{A \cup B} = I_A + I_B - I_{A \cap B}$.

An alternative indicator expression for the union is $I_{A \cup B} = 1 - (1 - I_A)(1 - I_B)$. This expresses the fact that $A \cup B$ fails to occur if, and only if, *both* A and B fail to occur. Carrying out the indicated algebra yields the first expression given for $I_{A \cup B}$.

To simplify notation we shall adopt the convention that when events are written side by side with no operation symbol (and no possibility of confusion), it will be understood that we mean intersection. This convention is analogous to that of numerical algebra where multiplication of the numbers a and b is written ab. The expression AB, then, denotes the intersection of A and B (i.e., $AB = A \cap B$). Indicator expressions are treated accordingly:

$$I_{AB} = I_A I_B$$

$$I_{A \cup B} = I_A + I_B - I_{AB} = I_A + I_B - I_A I_B.$$

In logic, the term *complementation* is replaced by *negation* or simply referred to as "not A"; the term *intersection* is replaced by *conjunction* or referred to as "A and B"; the term *union* is replaced by *disjunction* or referred to as "A or B." We will often use the words *not*, *and*, and *or* as substitutes for *complement*, *intersection*, and *union*, respectively.

The following are simple conclusions from the definitions, which the reader is invited to verify:

(a) $I_{\phi^c} = I_\Omega$ (d) $I_{A^c \cup A} = I_\Omega$

(b) $I_{(A^c)^c} = I_A$ (e) $I_{(A \cup B)^c} = I_{A^c B^c}$ (1)

(c) $I_{A^c A} = I_\phi$ (f) $I_{(AB)^c} = I_{A^c \cup B^c}$

In words, (a) says that if an event is impossible, its complement must occur; (b) says that an event occurs if, and only if, its complement does not occur; (c) says that A and A^c cannot both occur; (d) says that at least one of the events A, A^c must occur; (e) says that the event $A \cup B$ fails to occur if, and only if, both A^c and B^c occur; (f) says AB fails to occur if, and only if, A^c or B^c or both occur.

The operations we have defined for pairs of events can be extended in obvious ways to any number of events. Let $\mathcal{C}$ denote any collection of events. The *intersection of the events in* $\mathcal{C}$ occurs if, and only if, every event in $\mathcal{C}$ occurs. The *union of the events in* $\mathcal{C}$ occurs if, and only if, at least one of the events in $\mathcal{C}$ occurs.

Example 2. Consider a set of 10 individuals and let A_i denote the event that the ith individual smokes one or more packs of cigarettes every day. $\mathcal{C}$ is the collection of events $A_1, \ldots, A_{10}$. The union of these events occurs if *at least one* of the individuals in the group smokes one or more packs a day. The intersection occurs if *every* member of the group smokes one or more packs a day.

The notation we use for multiple unions will generally be self-explanatory. If the class $\mathcal{C}$ consists of an indexed sequence of events [e.g., $\mathcal{C} = \{A_1, A_2, \ldots\}$], we will use the symbol $\cup$ in an analogous manner to the symbol Σ for summation (e.g., $\cup_{i=1}^n A_i$ or $\cup_{i=1}^\infty A_i$). If the events are not indexed, we may simply write $\cup_{A \in \mathcal{C}} A$ to denote the union of the events that are members of $\mathcal{C}$. Similarly, we use the symbol $\cap$ for multiple intersections: for example, $\cap_{i=1}^n A_i$ or $\cap_{i=1}^\infty A_i$ and, in the unindexed case, $\cap_{A \in \mathcal{C}} A$.

The indicator for multiple intersections presents no difficulty since products of indicators are themselves indicators. We use the symbol Π to represent product and if $B = \cap_{A \in \mathcal{C}} A$ we write $I_B = \Pi_{A \in \mathcal{C}} I_A$. If $B = \cap_{i=1}^n A_i$, we write $I_B = \Pi_{i=1}^n I_{A_i}$. A product of indicators (no matter how many) is 1 if, and only if, every indicator in the product has value 1.

The indicator of a multiple union is less obvious. Nevertheless, it has a simple form based on its complement. The event "at least one" is complementary to the event "none" and the latter is represented by intersection. Therefore, the indicator of an arbitrary union can be given as follows: If $B = \cup_{A \in \mathcal{C}} A$, then

$$I_B = 1 - \prod_{A \in \mathcal{C}} I_{Ac} = 1 - \prod_{A \in \mathcal{C}} (1 - I_A)$$

If $B = \cup_{i=1}^n A_i$, then

$$I_B = 1 - \prod_{i=1}^n I_{A_i^c} = 1 - \prod_{i=1}^n (1 - I_{A_i}). \tag{2}$$

We thus reduce all indicator expressions for event operations to products and subtraction from 1.

Exercise 1. Verify the following useful expressions:

(a) If $B = \bigcap_{A \in \mathscr{C}} A$, then $I_B = \min_{A \in \mathscr{C}} I_A$.
(b) If $B = \bigcup_{A \in \mathscr{C}} A$, then $I_B = \max_{A \in \mathscr{C}} I_A$.

B. Event Relations

The operations we have defined for events can be used to define various important relations among them.

Implication

We say of two events A and B that A *implies* B if the occurrence of A requires the occurrence of B. This is denoted by $A \subset B$ (or $B \supset A$) and can be expressed as "$A \subset B$ if, and only if, AB^c is an impossible event" (i.e., A without B is impossible).

Exercise 2. Show that $A \subset B$ if, and only if, $I_A \leq I_B$.

Referring to Example 1, let A be the event that the device functions and B_i be the event that component C_i functions. For a series hookup we have $A \subset B_1$ and $A \subset B_2$ since functioning of the device requires that both components function. In the parallel hookup we have $B_1 \subset A$ and $B_2 \subset A$ since if either component functions, the device functions.

Implication provides a way of comparing events since $A \subset B$ means that B is, in a sense, larger than A. Some events, of course, cannot be compared in this way. For example, in the two-component device it is not true that $B_1 \subset B_2$ or $B_2 \subset B_1$ when one component's operation is unrelated to the operation of the other.

If $A \subset B$, then A is called a *subevent* of B.

Equality

The events A and B are *equal* if $A \subset B$ and $B \subset A$. In terms of indicators this becomes $I_A \equiv I_B$.

As a consequence of the indicator equations (1), we have the relations

$$
\begin{array}{ll}
\text{(a) } \phi^c = \Omega; & \text{(d) } A^c \cup A = \Omega; \\
\text{(b) } (A^c)^c = A; & \text{(e) } (A \cup B)^c = A^c B^c; \\
\text{(c) } A^c A = \phi; & \text{(f) } (AB)^c = A^c \cup B^c.
\end{array}
\tag{3}
$$

Exercise 3. Show that $A \subset B$ if, and only if, $AB = A$.

Exercise 4.   Show that $A \subset B$ if, and only if, $A \cup B = B$.

Exercise 5.   Show that $A \subset B$ if, and only if, $A^c \supset B^c$.

Disjoint

Two events A and B are called *disjoint* or *incompatible* if the occurrence of one precludes the occurrence of the other (i.e., $AB = \phi$). Disjoint events are sometimes called *mutually exclusive*.

An alternative expression for $AB = \phi$ is $A \subset B^c$. Another is $B \subset A^c$.

In terms of indicators we have $AB = \phi$ if, and only if, $I_{AB} = I_A I_B \equiv 0$.

Exercise 6.   Show that A and B are disjoint if, and only if, $I_A + I_B \leq 1$.

Exhaustive

A class $\mathcal{C}$ of events is *exhaustive* if $\cup_{A \in \mathcal{C}} A = \Omega$. Thus if $\mathcal{C}$ is exhaustive, at least one of its members must occur.

Exercise 7.   Show that $\mathcal{C}$ is exhaustive if, and only if, $\Pi_{A \in \mathcal{C}} (1 - I_A) \equiv 0$.

Summary

Complement: A^c occurs if, and only if, A does not occur.
Intersection: $\cap_{A \in \mathcal{C}} A$ occurs if, and only if, every event in the class $\mathcal{C}$ occurs.
Union: $\cup_{A \in \mathcal{C}} A$ occurs if, and only if, at least one event in the class $\mathcal{C}$ occurs.
Implication: $A \subset B$ if the occurrence of A implies the occurrence of B.
Equality: $A = B$ if $A \subset B$ and $B \subset A$.
Disjoint: $AB = \phi$.
Exhaustive class, $\mathcal{C}$: $\cup_{A \in \mathcal{C}} A = \Omega$.

Indicator relations:

1. $I_{A^c} = 1 - I_A$.
2. $I_{\cap_{A \in \mathcal{C}} A} = \Pi_{A \in \mathcal{C}} I_A = \min_{A \in \mathcal{C}} I_A$.
3. $I_{A \cup B} = I_A + I_B - I_{AB}$.
4. $I_{\cup_{A \in \mathcal{C}} A} = 1 - \Pi_{A \in \mathcal{C}} (1 - I_A) = \max_{A \in \mathcal{C}} I_A$.
5. $A \subset B$ if, and only if, $I_A \leq I_B$.
6. $A = B$ if, and only if, $I_A = I_B$.
7. A and B are disjoint if, and only if, $I_{AB} \equiv 0$.
8. $\mathcal{C}$ is exhaustive if, and only if, $\Pi_{A \in \mathcal{C}} (1 - I_A) \equiv 0$.

Problems

1. Let A, B, and C be events. Express the indicators of the given events in terms of I_A, I_B, and I_C.
   (a) $A \cup B \cup C$
   (b) $A \cup (BC)$
   (c) $A(B \cup C)$
   (d) $(A \cup B \cup C)^c$
   (e) $(ABC)^c$
   (f) The event that at most one of A, B, and C occurs.
   (g) The event that at least two of A, B, and C occur.

For Problems 2 to 5, consider a system with two subsystems s_1 and s_2, hooked up either in series, case (a), or in parallel, case (b). Subsystem s_1 has three components, c_1, c_2, and c_3, and subsystem s_2 has two components, c_4 and c_5. Let B_i be the event that c_i functions and A be the event that the system functions. In each problem express I_A in terms of the I_{B_i} for both cases (a) and (b).

2.† Assume that c_1, c_2, and c_3 are in series in s_1 and that c_4 and c_5 are in series in s_2.
3.† Assume that c_1, c_2, and c_3 are in series in s_1 and that c_4 and c_5 are in parallel in s_2.
4.† Assume that c_1, c_2, and c_3 are in parallel in s_1 and that c_4 and c_5 are in parallel in s_2.
5.† Assume that s_1 operates if any two of its three components operate and that c_4 and c_5 are in parallel in s_2.
6. Let A, B, and C be three arbitrary events. Find expressions in terms of event operations and indicators for each of the following:
   (a) Only A occurs.
   (b) At least two of the events A, B, C occur.
   (c) Exactly two of the events A, B, C occur.
   (d) Not more than two of the events A, B, C occur.
7. A deck containing five numbered cards is dealt to five numbered positions. Let A_i be the event that card i is in position i (*match* in position i). For each event below, give the indicator in terms of the I_{A_i} and the event in terms of event operations on the A_i.
   (a) No match occurs.
   (b) Exactly one match occurs.
   (c) At least two matches occur.
   (d) Exactly four matches occur.
8. A hand of $n < 52$ cards is dealt from an ordinary deck. Let A_i be the event that the ace of spades is the ith card dealt $(i = 1, \ldots, n)$. Let

B be the event that the ace of spades is in the hand. Using the simplest possible form, express B in terms of the A_i and I_B in terms of the I_{A_i}.

9. Consider a collection $x_1, x_2, \ldots$ of numbers. Define the events:

$$A: \quad \max x_i \le t \qquad B: \quad \min x_i > t$$
$$C: \quad \text{all } x_i \le t \qquad D: \quad \text{all } x_i > t.$$

Show that $A = C \subset D^c$ and $B = D \subset C^c$.

10. Prove the following event relations using indicators.
    (a) $A \cup (BC) = (A \cup B)(A \cup C)$
    (b) $A(B \cup C) = (AB) \cup (AC)$
    (c) $(A \cup B \cup C)^c = A^c B^c C^c$
    (d) $(ABC)^c = A^c \cup B^c \cup C^c$

11. Let d be the distance from Detroit to Lansing. Let A be the event that $|d - 100| \ge 10$ and B be the event that $d \le 90$. Establish an implication relation between A and B.

12. Let d be as in Problem 11. Let A be the event that $|d - a| \ge \delta$ and B be the event that $d \le t$. For fixed a, what values of δ and t yield an implication relation between A and B. Specify the relation.

13. Let $A_1, \ldots, A_n$ be any events. Show that $\max I_{A_i} \le \sum_{i=1}^{n} I_{A_i}$. When is equality achieved?

4. PARTITIONS; LINEAR COMBINATIONS OF INDICATORS

The analysis of a complicated event can often be simplified by breaking it up into "small," disjoint pieces (i.e., partitioning it) and analyzing the pieces separately. This process of partitioning, as will be seen in Section 5, is an absolutely fundamental and essential device for the calculation of probabilities. In this section we examine the formal structure of partitions, standard methods of obtaining and generating them, and some of their uses.

A. Basics

Partition

A class $\mathcal{D}$ of events is called a *partition of an event B* if the following two conditions are satisfied:

(a) All members of $\mathcal{D}$ are disjoint. That is, if D_1 and D_2 are distinct members of $\mathcal{D}$, then $D_1 D_2 = \phi$.

(b) $\bigcup_{D \in \mathcal{D}} D = B$.

If B is a sure event, $\mathcal{D}$ will be called a *partition* (without specifying the event). Thus a partition is an exhaustive collection of disjoint events.

Example 1. Let A, B be arbitrary events and let $\mathcal{D}_0 = \{A, A^c\}$ and $\mathcal{D}_B = \{AB, A^cB\}$. Then $\mathcal{D}_0$ is a partition and $\mathcal{D}_B$ is a partition of B.

We call a partition *finite*, *countable*, or *uncountable*, depending on the number of nonempty events it contains. In Example 1, both partitions are obviously finite. The following examples illustrate the various possibilities.

Example 2. Every day the state lottery selects a three-digit number and there may be interest in considering the following events:

E_i = "first occurrence of 777 (your favorite number) takes place i days from now" ($i = 0$ means today).

B_r = "777 is drawn within the next r days."

C = "777 is drawn eventually."

The following partitions might prove useful in analyzing the occurrence of 777:

$\mathcal{D}_1 = \{E_0, E_0^c\}$ is a finite partition (of Ω).

$\mathcal{D}_2 = \{E_0, E_1, \ldots\}$ is a countable partition of C.

$\mathcal{D}_3 = \{E_0, E_1, \ldots, E_r\}$ is a finite partition of B_r.

$\mathcal{D}_4 = \{E_{r+1}, E_{r+2}, \ldots\}$ is a countable partition of $B_r^c C$.

Example 3.† The length of time between the next two arrivals of cars at a toll booth is to be measured. Let D_t denote the event that the measured time is exactly t and let $\mathcal{D} = \{D_t : t \geq 0\}$. If the measuring device has perfect accuracy, $\mathcal{D}$ is an uncountable partition. If, however, time can only be measured to the nearest, say, hundredth of a second, $\mathcal{D}$ is a countable partition. Any event relating to the time in question is a union of (i.e., would be partitioned by) members of $\mathcal{D}$.

We conclude this subsection with two examples that illustrate the construction of partitions for special purposes. Because of their general utility, we will refer to these often in later sections.

Example 4.† (Generalization of Example 1) Let $\mathcal{D}$ be a partition (of Ω), let B be an arbitrary event, and let $\mathcal{D}_B = \{C = BD: D \in \mathcal{D}\}$. Distinct members of $\mathcal{D}_B$ are clearly disjoint events since they are subevents of disjoint D's. Furthermore,

$$\bigcup_{D \in \mathcal{D}} BD = B\left(\bigcup_{D \in \mathcal{D}} D\right) = B\Omega = B, \tag{1}$$

so that $\mathcal{D}_B$ is a partition of B. We are free to include or exclude from $\mathcal{D}_B$ those BD that are empty. This simple construction is illustrated in the Venn diagram in Figure 1.

Here is a typical way in which such an approach arises. City block i contains b_i individuals, of which r_i are employed ($i = 1, 2, 3, 4$). An experiment consists of first choosing a block and, then, selecting a person from the block chosen. Let D_i be the event that block i is chosen, and let B be the event that the individual selected is employed. Then $\mathcal{D} = \{D_1, D_2, D_3, D_4\}$ is a partition and $\mathcal{D}_B = \{BD_1, BD_2, BD_3, BD_4\}$ is a partition of B, each member of which refers to a different block from which an employed person might be selected. The decomposition arises naturally since it is much easier to analyze the selection of an employed person from each block separately than from the conglomeration of blocks simultaneously. Two (or multi)-stage experiments (e.g., Problems 3 and 4) are usually analyzed in this manner.

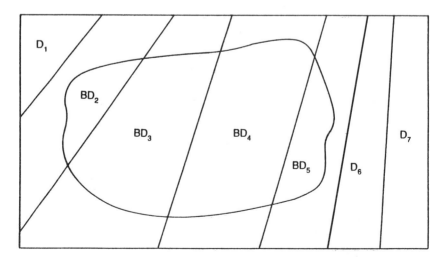

Figure 1 Event B partitioned by $\mathcal{D}_B = \{BD_2, BD_3, BD_4, BD_5\}$. $BD_i = \phi$ for $i = 1, 6, 7$.

Example 5.†   (Partitioning a union)   There are various ways to represent a countable union as a union of (related) disjoint events. For example,

$$E \cup F = E \cup E^c F$$

or, in terms of indicators,

$$I_E + I_{E^c F} = I_E + I_F(1 - I_E) = I_{E \cup F}.$$

In general, if $A_1, A_2, \ldots$ is any sequence of events whose union is B, then

$$B = A_1 \cup A_1^c A_2 \cup A_1^c A_2^c A_3 \cup \cdots = \bigcup_k D_k, \qquad (2)$$

where $D_k = A_1^c \cdots A_{k-1}^c A_k$ is the unique contribution of A_k to B among the first k events. The indicator of B is

$$I_B = I_{A_1} + (1 - I_{A_1})I_{A_2} + (1 - I_{A_1})(1 - I_{A_2})I_{A_3} + \cdots, \qquad (3)$$

the kth term of the expression being the indicator of D_k and

$$I_{D_k} = I_{A_k} \prod_{j=1}^{k-1} (1 - I_{A_j}). \qquad (4)$$

The class $\mathcal{D} = \{D_1, D_2, \ldots\}$ is a partition of B that is often useful. Figure 2 indicates the procedure.

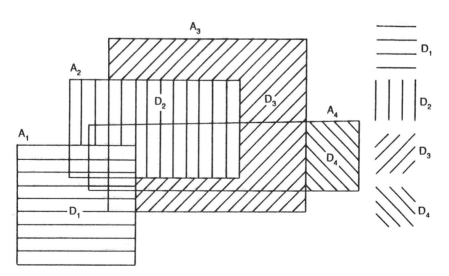

Figure 2   Partitioning a union: $\bigcup_{i=1}^{4} A_i = \bigcup_{i=1}^{4} D_i$, D_i's disjoint, $D_i = A_i \cap_{j<i} A_j^c$.

B. Partition Generated by a Class

Let $\mathcal{C} = \{A_1, A_2, \ldots\}$ be an arbitrary countable (possibly finite) collection of events, and let G_k stand symbolically for either A_k or A_k^c. Thus $G_1 G_2$ may be any of $A_1 A_2$, $A_1^c A_2$, $A_1 A_2^c$, or $A_1^c A_2^c$. The partition we will examine is the class $\mathcal{D}$, consisting of all events of the form $D = \cap_k G_k$ where, as above, for each k, $G_k = A_k$ or $G_k = A_k^c$. Each member of $\mathcal{D}$ represents the simultaneous occurrence of certain members of $\mathcal{C}$ and the nonoccurrence of the remaining members of $\mathcal{C}$.

Partition Generated by $\mathcal{C}$

We call $\mathcal{D}$, described above, the *partition generated by* $\mathcal{C}$.

The fact that $\mathcal{D}$ is a partition is easily verified (Problem 12). Various observations regarding this definition are contained in the following remarks and examples.

Remark 1.   One or more of the D's defined above may (and often will) be empty (i.e., impossible). Only the nonempty ones are of any consequence, but when $\mathcal{C}$ is finite (say, it contains n events), it will be convenient to treat all 2^n expressions of the form $\cap_{k=1}^n G_k$ as distinct members of $\mathcal{D}$.

Remark 2.   The fact that $\mathcal{D}$ is a partition implies (see Example 4) that A_i is partitioned by $\mathcal{D}_i = \{A_i D: D \in \mathcal{D}\}$. By construction, each $D \in \mathcal{D}$ is either a subevent of A_i (if $G_i = A_i$) or a subevent of A_i^c (if $G_i = A_i^c$). Hence if $D \neq \phi$, then $A_i D$ is either D itself (if $G_i = A_i$) or ϕ (if $G_i = A_i^c$). Therefore,

$$\mathcal{D}_i = \{D: D \in \mathcal{D}, D \subset A_i\} \tag{5}$$

and $A_i = \cup_{D \in \mathcal{D}_i} D$. Thus each of the original events A_i is the union of members of $\mathcal{D}$.

C. The Finite Case

We now assume finiteness of $\mathcal{C}$ and present a more detailed analysis of the partition it generates. If $\mathcal{C}$ is a finite class, the nonempty members of $\mathcal{D}$ can be identified, in principle, from the Venn diagram of the collection. For small n they are usually easy to identify. Figure 3 gives several variations for classes, $\mathcal{C}$, which contain two, three, or four events. The nonempty D_k's are labeled.

Remark 3.   For each $\cap_{k=1}^n G_k$, there is a unique subset, J, of the set of indices $\{1, 2, \ldots, n\}$ for which $G_k = A_k$ if $k \in J$ and $G_k = A_k^c$ if $k \notin J$.

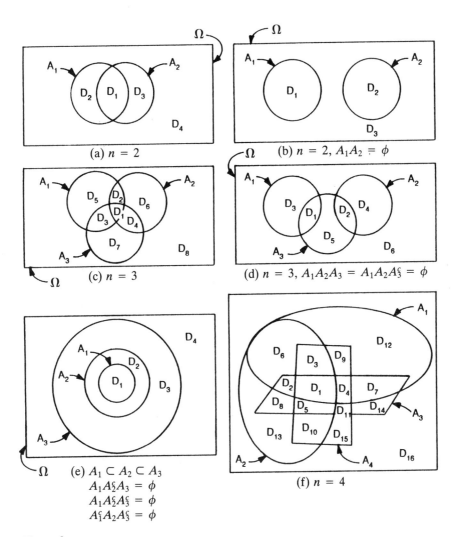

(a) $n = 2$

(b) $n = 2$, $A_1 A_2 = \phi$

(c) $n = 3$

(d) $n = 3$, $A_1 A_2 A_3 = A_1 A_2 A_3^c = \phi$

(e) $A_1 \subset A_2 \subset A_3$
$A_1 A_2^c A_3 = \phi$
$A_1 A_2^c A_3^c = \phi$
$A_1^c A_2 A_3^c = \phi$

(f) $n = 4$

Figure 3

Identify $\bigcap_{k=1}^{n} G_k$ with J and note that this establishes a one-to-one correspondence between expressions and subsets of $\{1, 2, \ldots, n\}$ (of which there are 2^n). The set J identifies those A_k's that occur and, by omission, those that do not.

Example 6. Suppose that $n = 10$ and consider $D_1 = \bigcap_{i=1}^{10} A_i^c$, $D_2 = A_1 A_2^c A_3 A_4 A_5^c A_6 A_7^c A_8 A_9 A_{10}^c$, and $D_3 = A_1 A_2 A_3 \bigcap_{i=4}^{10} A_i^c$. Then, to D_i

there corresponds J_i where $J_1 = \phi$, $J_2 = \{1, 3, 4, 6, 8, 9\}$, and $J_3 = \{1, 2, 3\}$. This illustrates just 3 of the 1024 expressions/subsets available.

In the lemma below, we use the following notation:

D is an arbitrary nonempty member of $\mathcal{D}$.
J_D is the subset of $\{1, 2, \ldots, n\}$ corresponding to D.
$\mathcal{D}_i$ (see Remark 2, above) is the subclass of $\mathcal{D}$ that partitions A_i.

Lemma 1. With D, J_D, $\mathcal{D}_i$ defined as above, we have

$$J_D = \{i: D \in \mathcal{D}_i\} \tag{6}$$

and

$$\mathcal{D}_i = \{D: i \in J_D\}. \tag{7}$$

Proof. If $i \in J_D$ and $D = \cap_{k=1}^{n} G_k$, then $G_i = A_i$ by definition of J_D. Hence, from (5), $D \in \mathcal{D}_i$ and (6) follows.

Since $D \in \mathcal{D}_i$, then $D = \cap_{k=1}^{n} G_k$ and $G_i = A_i$. This implies that $i \in J_D$ and hence proves (7).

Example 7. Let $\mathcal{C} = \{A_1, A_2, A_3\}$ and designate members of $\mathcal{D}$ as in the Venn diagram, Figure 3(c). Table 1 exhibits the various entities defined above.

Example 8. Consider a three-component device and let A_i denote the event that component i functions ($i = 1, 2, 3$). Each member of $\mathcal{D}$ (generated by $\{A_1, A_2, A_3\}$) describes the functioning or malfunctioning of each of the components. Any event described in terms of the A_i's can be ex-

Table 1

i	D_i	J_i	$D_i \in \mathcal{D}_t$	$\mathcal{D}_1$	$\mathcal{D}_2$	$\mathcal{D}_3$
1	$A_1 A_2 A_3$	$\{1, 2, 3\}$	$\mathcal{D}_1, \mathcal{D}_2, \mathcal{D}_3$	$\in$	$\in$	$\in$
2	$A_1 A_2 A_3^c$	$\{1, 2\}$	$\mathcal{D}_1, \mathcal{D}_2$	$\in$	$\in$	
3	$A_1 A_2^c A_3$	$\{1, 3\}$	$\mathcal{D}_1, \mathcal{D}_3$	$\in$		$\in$
4	$A_1^c A_2 A_3$	$\{2, 3\}$	$\mathcal{D}_2, \mathcal{D}_3$		$\in$	$\in$
5	$A_1 A_2^c A_3^c$	$\{1\}$	$\mathcal{D}_1$	$\in$		
6	$A_1^c A_2 A_3^c$	$\{2\}$	$\mathcal{D}_2$		$\in$	
7	$A_1^c A_2^c A_3$	$\{3\}$	$\mathcal{D}_3$			$\in$
8	$A_1^c A_2^c A_3^c$	ϕ	none			

pressed as a union of members of this partition. If the device is hooked up in series, we are, of course, interested in the event $B = A_1A_2A_3$, which is itself a member of the partition. If the device is hooked up in parallel, we would be interested in the event $A_1 \cup A_2 \cup A_3$, which is the union of all members of the partition except $A_1^c A_2^c A_3^c$.

Exercise 1. Use Figure 3(f) to produce a table analogous to Table 1 for a class containing four events. Let B be the event that exactly two A_i's occur. Write B as a union of D_k's.

D. Linear Combinations of Indicators

By a (finite) linear combination of indicators we mean any expression of the form $L = \sum_{i=1}^{n} a_i I_{A_i}$, where the a_i's are arbitrary numbers and the A_i's are arbitrary events. Think of the a_i's as, perhaps, monetary consequences (positive or negative) associated with the occurrence of each event. It is useful to express L in terms of disjoint events as follows. For each nonempty member of the partition generated by $A_1, A_2, \ldots, A_n$, the value of L is determined. For example, if $A_1 A_2^c A_3^c \cdots A_n^c$ occurs, the value of L is a_1 since A_1 occurs but none of the others do; if $A_1 A_2 \cdots A_n$ occurs, the value of L is $\sum_{i=1}^{n} a_i$; and so on. Each nonempty member of the partition determines a value of L because it identifies those I_{A_i} that are 1 and those that are 0. If we label the formal expressions for members of the partition by D_k $(k = 1, 2, \ldots, 2^n)$, we may therefore write

$$L = \sum_{k=1}^{2^n} d_k I_{D_k},$$

where d_k is the value of L determined by the occurrence of D_k. In writing this expression for L, we recognize that several of the D_k may equal ϕ.

The value of d_k is obtainable from the expression corresponding to D_k or from the corresponding subset of $\{1, 2, \ldots, n\}$. Thus if $D_k = \cap_{i=1}^{n} G_i$, then [see (6)]

$$d_k = \sum_{i:G_i=A_i} a_i = \sum_{i \in J_k} a_i, \tag{8}$$

where J_k is the subset of $\{1, 2, \ldots, n\}$ corresponding to D_k. From (7) we have

$$I_{A_i} = \sum_{k:i \in J_k} I_{D_k}. \tag{9}$$

The decomposition is completely described by the diagram of Figure 4 for a linear combination of three events.

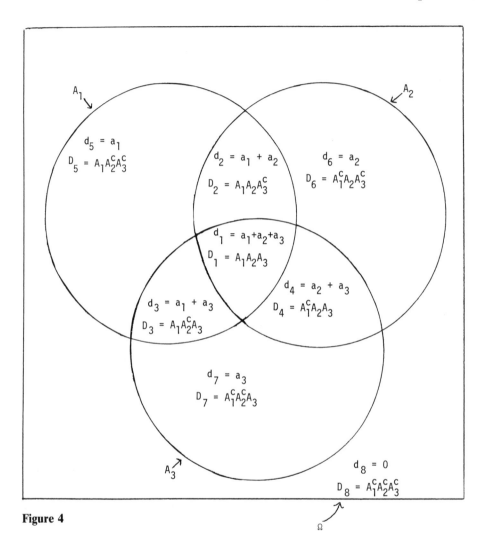

Figure 4

Example 9. A shop has two machines, m_1 and m_2, which are used to produce a certain item. Each morning, m_1 produces 10 items and m_2 produces 15 items. A decision is then made (depending on demand) to stop or continue production on either or both machines. If production continues on m_1, it produces 10 more items in the afternoon. If production is continued on m_2, it produces 15 more items in the afternoon. Every so often an inspection takes place at the end of the day. When this occurs, 5 items are chosen from the day's output and (destructively) tested. Let L denote the total number of items available for sale at the end of a day. Then

$$L = 25 + 10I_{A_1} + 15I_{A_2} - 5I_{A_3}, \tag{10}$$

where A_1 is the event "m_1 is used in the afternoon," A_2 is the event "m_2 is used in the afternoon," and A_3 is the event "inspection."

Let $L' = L - 25$. With D_k as in Table 1 and Figure 4, we have $a_1 = 10$, $a_2 = 15$, $a_3 = -5$, and hence

$$
\begin{aligned}
d_1 &= a_1 + a_2 + a_3 = 20 & d_5 &= a_1 = 10 \\
d_2 &= a_1 + a_2 = 25 & d_6 &= a_2 = 15 \\
d_3 &= a_1 + a_3 = 5 & d_7 &= a_3 = -5 \\
d_4 &= a_2 + a_3 = 10 & d_8 &= 0.
\end{aligned}
$$

Note that $\mathcal{D}_1 = \{D_1, D_2, D_3, D_5\}$ is the resulting partition of A_1. Thus

$$L' = 20I_{D_1} + 25I_{D_2} + 5I_{D_3} + 10I_{D_4} + 10I_{D_5}$$
$$+ 15I_{D_6} - 5I_{D_7} + 0I_{D_8}.$$

Hence, since $25 = \sum_{k=1}^{8} 25I_{D_k}$,

$$L = L' + 25$$
$$= 45I_{D_1} + 50I_{D_2} + 30I_{D_3} + 35I_{D_4} + 30I_{D_5}$$
$$+ 40I_{D_6} + 20I_{D_7} + 25I_{D_8}.$$

Example 10. Consider a room with lights controlled by two switches. The lights are on if, and only if, the switches are in opposite positions. Let B denote the event the light is on and let A_i denote the event switch i is up. Then

$$I_B = I_{A_1} + I_{A_2} - 2I_{A_1A_2} \tag{11}$$

expresses the fact that the bulb will be on if either, but not both, of the switches is in the up position. Alternatively,

$$I_B = I_{A_1A_2^c} + I_{A_2A_1^c}$$

expresses I_B in terms of the partition generated by A_1 and A_2.

Example 10 demonstrates that a linear combination of indicators may be an indicator even if the coefficients are not all integers. Solving (10) for $I_{A_1A_2}$ yields

$$I_{A_1A_2} = \frac{I_{A_1} + I_{A_2} - I_B}{2}.$$

Summary

Partition of B: $D_iD_j = \phi$ for $i \neq j$, $\cup_n D_n = B$.
Partition generated by $\mathcal{C} = \{A_1, A_2, \ldots\}$: See page 20.

Let $\{D_1, \ldots, D_m\}$ $(m \leq 2^n)$ be the partition generated by $\{A_1, \ldots, A_n\}$. Then $L = \Sigma_{i=1} a_i I_{A_i}$ implies that there exist $d_1, \ldots, d_m$ such that $L = \Sigma_{k=1}^m d_k I_{D_k}$.

Problems

1. Show that $\mathcal{D} = \{D_1, D_2, \ldots\}$ is a countable partition of B if, and only if, $I_B = \Sigma_i I_{D_i}$.
2. The members of a club belong to three disjoint groups. Three belong to group A, two to B, and one to C. Three different members are chosen sequentially. Let A_i be the event the ith member chosen belongs to A and define B_i and C_i similarly.
   (a) Give a partition in terms of the A_i, B_i, and C_i of the event that at least two of the three belong to A.
   (b) Give a partition in terms of the A_i, B_i, and C_i of the event that at least as many B's are chosen as A's.
3. ABC Manufacturing Co. has five factories making widgets. The ith factory produces n_i widgets with a defective rate p_i. A factory is chosen, and from that factory, a widget is chosen. Let B be the event that the chosen widget is defective. What would you choose as the partition $\mathcal{D}$? What is the resulting $\mathcal{D}_B$?
4. A city consists of 100 square blocks and the kth block has n_k households. Two blocks are chosen, and from each chosen block, a sample of $r < \min n_k$ households is chosen. All adults in the sampled households are interviewed. Let B be the event that a majority of those interviewed favors a sewage bond proposal. What would you choose as the partition $\mathcal{D}$? What is the resulting $\mathcal{D}_B$?
5. A coin is tossed until at least one head and at least one tail result. Let B be the event that the last toss is a head. Given a countably infinite partition $\mathcal{D}$ and the resulting $\mathcal{D}_B$. Is $\mathcal{D}_B$ infinite?
6. A poker hand (five cards) is dealt from an ordinary deck. Let A_i be the event that the ith card dealt is an ace.
   (a) Give a verbal interpretation of $B = A_1 \cup A_2 \cup A_3 \cup A_4 \cup A_5$.
   (b) Partition B and give a verbal interpretation of each member of your partition of B.
7. Let $\mathcal{C}$ be a countable class. Verify that "the partition generated by $\mathcal{C}$," given in Subsection B, is, in fact, a partition as defined in Subsection A.

8. Refer to the lottery of Example 2. For each of the following classes, (1) identify the partition generated, (2) describe, in words, what membership in the partition means, and (3) give its cardinality.
   (a) $\mathcal{C}_0 = \{B_0, B_1, \ldots\}$.
   (b) $\mathcal{C} = \{A_1, A_2, \ldots\}$, where A_i is the event that 777 is drawn on day i.

9. A poker hand is a five-card subset of an ordinary deck. For a poker hand, consider the following events:

   A_1: The cards are of the same suit.
   A_2: The cards have consecutive values.
   A_3: The high card is a 10.
   A_4: The low card is a 3.

   (a) List the nonempty members of the partition generated by A_1, A_2, A_3, and A_4.
   (b) Given an expression in terms of the events in part (a) for the event "straight flush" (cards are of the same suit and have consecutive values).

10. A yard has five trees, some of which may be diseased. Let C be the event that three trees are diseased. Let A_i be the event that the ith tree is diseased. Write C as the union of members of the partition generated by the A_i.

11. Let $\mathcal{C}_n = \{A_1, \ldots, A_n\}$ for $n = 1, 2, \ldots$ and $\mathcal{D}_n$ be the partition generated by $\mathcal{C}_n$. Express an arbitrary member of $\mathcal{D}_{n-1}$ in terms of members of $\mathcal{D}_n$.

12. The partition $\mathcal{D}$ is a *refinement* of the partition $\mathcal{D}_0$ if, given $D \in \mathcal{D}_0$, there exists a partition of D that consists of members of $\mathcal{D}$. Show that the partition generated by $\{A_1, \ldots, A_n\}$ is a refinement of the partition generated by $\{A_1, \ldots, A_{n-1}\}$.

13. Let $\mathcal{D}_1$ and $\mathcal{D}_2$, be partitions and let $\mathcal{D} = \{D_1 D_2 : D_1 \in \mathcal{D}_1, D_2 \in \mathcal{D}_2\}$. Show that $\mathcal{D}$ is a refinement of both $\mathcal{D}_1$ and $\mathcal{D}_2$. ($\mathcal{D}$ is called the *common refinement* of $\mathcal{D}_1$ and $\mathcal{D}_2$.)

14. A lottery awards one $500 first prize and one $100 second prize. You hold three tickets. Let L be the total amount you win and A_{ij} be the event that your ith ticket wins the jth prize. Assume that no ticket wins more than one prize.
   (a) Express L in terms of the $I_{A_{ij}}$.
   (b) Express L in terms of the indicators of the events in the partition generated by the A_{ij}.

15. At roulette, you bet $1 on each of the events (see Example 1.2):

   A: Odd

B: Black
C: 1 through 12

You win \$1 if A occurs, \$1 if B occurs, and \$2 if C occurs. Let L be your net gain.
(a) Express L as a linear combination of I_A, I_B, and I_C.
(b) Give the value of L for each event in the partition generated by $\{A, B, C\}$.

16. You have four torpedoes. You will fire one at a time at the enemy ship until you either hit the ship or have no more torpedoes left. Let A_i be the event that the ith torpedo hits the ship.
(a) Write the partition generated by the A_i.
(b) If you hit the ship, the Admiralty will award you \$1000 plus \$100 for each torpedo left. Let L be your total award and write L as a linear combination of indicators of the events in part (a).

5. THE RULES OF PROBABILITY

A. Basic Rules

As stated previously, probabilities are numbers attached to events. The mathematical theory does not provide a rule for establishing what these numbers should be any more than Euclidean geometry tells us what a point is or how to evaluate distances or angles.

The theory is based on three rules that (1) define the relations required of a probability assignment and (2) provide a basis for *probability calculus*, that is, for using the assigned probabilities of *some* events to calculate the probabilities of others.

From now on $P(A)$ will denote the "probability of the event A."

The first rule establishes the scale of measurement for probabilities.

Rule 1. $0 \le P(A) \le 1$.

Since probability measures likelihood of occurrence, no event will be assigned a probability greater than that assigned to a sure event, Ω. Similarly, no event will be assigned a probability smaller than that of an impossible event, ϕ. The second rule locates those events at the extreme boundaries of the scale.

Rule 2. $P(\Omega) = 1$ and $P(\phi) = 0$.

We have placed both of these statements under Rule 2, though, as will become clear, one can be deduced from the other (using Rule 3).

While to some extent the scale of measurement embodied in Rules 1 and 2 is arbitrary, Rule 3 establishes the most basic structural feature of the theory. It is referred to as the *additivity property*.

Rule 3. If $\{A_1, A_2, \ldots, A_n\}$ is any finite partition of the event A, then

$$P(A) = P(A_1) + P(A_2) + \cdots + P(A_n).$$

The theory that follows from these rules is adequate for dealing with all problems in which only finitely many events are being considered. When dealing with problems concerning infinitely many events, the following countable additivity property will be imposed.

Rule 3′. If $\{A_1, A_2, \ldots, A_n, \ldots\}$ is any finite or countably infinite partition of the event A, then

$$P(A) = \sum_i P(A_i).$$

Rule 3′ cannot be derived from Rules 1 to 3, but Rule 3′ together with Rule 2 can be used to prove Rule 3.

Note: The rules we have given specify the probabilities of only two events: $P(\Omega) = 1$, $P(\phi) = 0$. Furthermore, the rules do not specify that every event must have a probability attached to it. The rules (1, 2, 3′) are to be understood as conditional statements: *If* the events alluded to have probabilities associated with them, either by initial assignment or by computation, *then* the particular rule must apply. The events (other than Ω and ϕ) to which probabilities are assigned will depend on the context, the particular interpretation of these numbers, and the relevance of the events to the issue at hand.

The starting point for the probabilistic description of any phenomenon is the assignment of probabilities to a class, $\mathcal{C}$, of events, from which, using the rules given, the probabilities of other events (outside $\mathcal{C}$) may be deduced. The initial assignment is generally based on intuition, assumption, or experience with the particular phenomenon in question. Section 1 provides some of the considerations that might be involved in making this initial assignment. For example, unless we have reason to believe otherwise, symmetry suggests that the faces of an ordinary die will have equal probability of occurring when the die is tossed. Experience with the relative frequencies produced by a particular die might lead to another model.

The primary tool for extending probabilities from $\mathcal{C}$ to a larger class is Rule 3 (or 3′). Suppose that A, B, and AB have specified probabilities (i.e., they belong to $\mathcal{C}$) but that A^c, A^cB, and $A \cup B$ do not. Since $A \cup A^c = \Omega$, it follows from Rules 2 and 3 that

$$P(A) + P(A^c) = P(\Omega) = 1.$$

Similarly, $B = AB \cup A^cB$ implies that

$$P(B) = P(AB) + P(A^cB)$$

and $A \cup B = A \cup A^cB$ implies that

$$P(A \cup B) = P(A) + P(A^cB).$$

Hence the probabilities of A^c, A^cB, and $A \cup B$ must be assigned accordingly.

In general, if $\mathcal{C}$ is any collection of events with specified probabilities, we can augment $\mathcal{C}$ by assigning probabilities to additional events as follows:

1. For every $A \in \mathcal{C}$ we assign to A^c the value $P(A^c) = 1 - P(A)$.
2. If $\{A_1, A_2, \ldots\}$ is a countable collection of disjoint members of $\mathcal{C}$, we assign to $\cup_i A_i$ the value $P(\cup_i A_i) = \Sigma_i P(A_i)$.

A question of considerable mathematical complexity is: *How do we know that when we extend probabilities outside the initial class $\mathcal{C}$, none of the rules of probability will be violated?*

The following example illustrates one unfortunate possibility. Suppose that $\mathcal{C} = \{A, B, AB\}$ and that A is assigned .9, B is assigned .9, and AB is assigned .5. None of the stated rules of probability are violated by this assignment. However, if we proceed to try to expand this class according to the methods given above, we find that A^cB is assigned the value .9 − .5 = .4 and $A \cup B$ is assigned the value .9 + .4 = 1.3, which violates Rule 1. Our conclusion is that the initial assignment was improper.

To guarantee that no violations can occur we require that $\mathcal{C}$, the initial class of events, satisfies certain simple conditions. In the remainder of this book $\mathcal{C}$ will satisfy:

(i) If A and B are members of $\mathcal{C}$, so is their intersection, AB.
(ii) If A belongs to $\mathcal{C}$, there is a countable partition of A^c, each of whose members belong to $\mathcal{C}$.

It is beyond the scope of this book to show that, in fact, conditions (i) and (ii) guarantee that no violations of the Rules 1, 2, and 3′ can ever occur (assuming, of course, that none are violated in the initial assignment to members of $\mathcal{C}$).

Probability Model

A *probability model* consists of a class $\mathcal{C}$ satisfying (i) and (ii) above together with an assignment, P, of probabilities to members of $\mathcal{C}$.

Example 1.† There is one special case in which extension of the initial probabilities is particularly simple. This is the case in which the probability model specifies the probability of each member of a countable partition. If $\mathcal{C} = \{A_1, A_2, \ldots\}$ is a countable partition, the class of events to which probabilities can be assigned is simply the class of all unions of a finite or countably infinite subcollection of members of $\mathcal{C}$. We refer to this as the *discrete case*. Note that it satisfies conditions (i) and (ii) above. If J is any set of positive integers and $B = \cup_{j \in J} A_j$, then

$$P(B) = \sum_{j \in J} P(A_j).$$

Rules 2 and 3′ require that $\Sigma_i P(A_i) = 1$.

Example 2.† If $\mathcal{C} = \{A_1, \ldots, A_n\}$ is a finite partition, one possible assignment is $P(A_i) = 1/n$ for $i = 1, \ldots, n$. This is the equally likely case. If B is any union of k members of $\mathcal{C}$, then $P(B) = k/n$.

Example 3.† Consider a random experiment (e.g., rolling a die, measuring a temperature, tomorrow's closing Dow Jones industrial average) that produces a real number. Let $\Omega = \mathcal{R}$, let $\mathcal{C}$ be the class of all intervals, and define *occurrence of interval A* to mean that the experiment produces a number belonging to A. The term *interval* includes ϕ, all single element subsets of $\mathcal{R}$, and all variations of endpoint inclusion and exclusion. If A and B are intervals, then (i) AB is an interval and (ii) A^c is either an interval or the union of two disjoint intervals [e.g., $(a, b)^c = (-\infty, a] \cup [b, \infty)$]. Figure 1 illustrates three possibilities.

It should be emphasized that the probabilities assigned to intervals must satisfy Rules 1, 2, and 3′. Thus these probabilities must be between 0 and 1 with $P(\mathcal{R}) = 1$ and $P(\phi) = 0$. Furthermore, for any choice of $a_1, a_2, \ldots$ such that $-\infty < a_0 < a_1 < \cdots < a_n < \cdots < a < \infty$ with $\lim_{n \to \infty} a_n = a$, the intervals $(a_{n-1}, a_n]$ $(n = 1, 2, \ldots)$ form a countable partition of (a_0, a). Hence any probability assignment satisfies

$$P((a_0, a)) = \sum_{n=1}^{\infty} P((a_{n-1}, a_n]).$$

B. Linearity Properties

Consider a probability model given by $\mathcal{C}$ and P. Let $\{A_1, A_2, \ldots, A_n\}$ be any finite collection of members of $\mathcal{C}$. The probabilities assigned to members

Figure 1 $[a_3, b_3][a_4, b_4] = \{b_3\} = \{a_4\}$, $[a_1, b_1)(a_2, b_2) = (a_2, b_1)$, and $[a_1, b_1)[a_4, b_4] = \phi$.

of $\mathcal{C}$ determine the probabilities of any event in the partition induced by $\{A_1, A_2, \ldots, A_n\}$. (Recall from Section 4 that this is the collection of all n-fold intersections that involve the A_i's or their complements.) Furthermore, we have shown that any expression of the form $L = \sum_{i=1}^{n} a_i I_{A_i}$ can be expressed in terms of the events in this partition. We will prove a theorem that summarizes the essential implication of Rule 3.

The following example illustrates the proof of Theorem 1.

Example 4. Let A_1, A_2, A_3 be any events and assume that a_1, a_2, a_3 are numbers such that $L = a_1 I_{A_1} + a_2 I_{A_2} + a_3 I_{A_3} \equiv 0$. The constraints imposed by this identity are described below. A consequence to be demonstrated is that $a_1 P(A_1) + a_2 P(A_2) + a_3 P(A_3) = 0$. In Section 4 it was shown that we can write $L = \sum_{k=1}^{8} d_k I_{D_k}$ where the D_k belong to the partition generated by A_1, A_2, A_3 and the d_k are partial sums of the a_i. Table 1 gives a list of the D_k and the d_k. Figure 4.4 illustrates schematically the partition generated by three events, A_1, A_2, A_3, and the values of L on each member of the partition.

Since the D_k are disjoint, no two of the I_{D_k} may be simultaneously equal to 1, so that for each k, either $d_k = 0$ or $D_k = \phi$. Hence $d_k P(D_k) = 0$ for each $k = 1, \ldots, 8$ and $\sum_{k=1}^{8} d_k P(D_k) = 0$. Referring to Table 1 and rearranging the terms yields

$$a_1[P(D_1) + P(D_2) + P(D_3) + P(D_5)]$$
$$+ a_2[P(D_1) + P(D_2) + P(D_4) + P(D_6)]$$
$$+ a_3[P(D_1) + P(D_3) + P(D_4) + P(D_7)] = 0.$$

But the bracketed expressions are $P(A_1)$, $P(A_2)$, and $P(A_3)$, respectively. Hence $a_1 P(A_1) + a_2 P(A_2) + a_3 P(A_3) = 0$.

Theorem 1. *Let* $L = \sum_{i=1}^{n} a_i I_{A_i} \equiv 0$. *Then* $\sum_{i=1}^{n} a_i P(A_i) = 0$.

Table 1

k	D_k	d_k	k	D_k	d_k
1	$A_1 A_2 A_3$	$a_1 + a_2 + a_3$	5	$A_1 A_2^c A_3^c$	a_1
2	$A_1 A_2 A_3^c$	$a_1 + a_2$	6	$A_1^c A_2 A_3^c$	a_2
3	$A_1 A_2^c A_3$	$a_1 + a_3$	7	$A_1^c A_2^c A_3$	a_3
4	$A_1^c A_2 A_3$	$a_2 + a_3$	8	$A_1^c A_2^c A_3^c$	0

Proof. Using (4.9) followed by (4.8) gives

$$L = \sum_{i=1}^{n} a_i \sum_{k:i\in D_k} I_{D_k} = \sum_{k=1}^{2^n} \sum_{i\in J_k} a_i I_{D_k}.$$

Since the D_k are disjoint, $L \equiv 0$ requires that for each k, either $D_k = \phi$ or the coefficient of I_{D_k} is zero; that is,

$$d_k = \sum_{i\in J_k} a_i = 0.$$

This proves that

$$\sum_{k=1}^{2^n} d_k P(D_k) = \sum_{k=1}^{2^n} \left(\sum_{i\in J_k} a_i \right) P(D_k) = 0. \tag{1}$$

The left side of (1) can be rearranged (as in Example 4) to read

$$\sum_{i=1}^{n} a_i \left(\sum_{k:i\in J_k} P(D_k) \right) = \sum_{i=1}^{n} a_i P(A_i) \tag{2}$$

since the D_k with $i \in J_k$ form a partition of A_i.

The conclusion of Theorem 1 is not true in complete generality when L is a countably infinite linear combination of indicators.

Corollary 1. *Let $\sum_{i=1}^{n} a_i I_{A_i} = I_B$. Then*

$$P(B) = \sum_{i=1}^{n} a_i P(A_i). \tag{3}$$

Proof. Since $\sum_{i=1}^{n} a_i I_{A_i} = I_B$, then $\sum_{i=1}^{n+1} a_i I_{A_i} \equiv 0$, where $A_{n+1} = B$ and $a_{n+1} = -1$. The result follows from Theorem 1.

Example 5. Let $P(A_i) = 1/5$, $P(A_i A_j) = 1/20$ $(i \neq j)$, and $P(A_1 A_2 A_3) = 1/60$. If B is the event that exactly one of A_1, A_2, A_3 occurs, then

$$I_B = I_{A_1} + I_{A_2} + I_{A_3} - 2I_{A_1 A_2} - 2I_{A_1 A_3} - 2I_{A_2 A_3} + 3I_{A_1 A_2 A_3},$$

so that $P(B) = (3)(1/5) - (3)(2)(1/20) + (3)(1/60) = 7/20$.

Summary

Rules for probability:

(a) $0 \leq P(A) \leq 1$ for any event A. (4)
(b) $P(\Omega) = 1$ and $P(\phi) = 0$.
(c) If $\{A_1, A_2, \ldots\}$ is a countable partition of A, then $P(A) = \sum_{n=1}^{\infty} P(A_n)$.

Extension of probabilities: Probabilities defined on an appropriate class
$\mathcal{C}$ of events can be extended beyond $\mathcal{C}$ to countable unions and complements
of members of $\mathcal{C}$ using (4) provided:

1. $A, B \in \mathcal{C}$ implies that $AB \in \mathcal{C}$.
2. $A \in \mathcal{C}$ implies that there exists a countable partition, $\{B_1, B_2, \ldots\}$ of
    A^c such that each $B_n \in \mathcal{C}$.

Probability model: A class $\mathcal{C}$ satisfying 1 and 2 above together with an
assignment, P, of probabilities to members of $\mathcal{C}$.

Theorem 1.   If $\Sigma_{i=1}^n a_i I_{A_i} = 0$, then $\Sigma_{i=1}^n a_i P(A_i) = 0$.
Corollary 1.   If $I_B = \Sigma_{i=1}^n a_i I_{A_i}$, then $P(B) = \Sigma_{i=1}^n a_i P(A_i)$.

These results are useful for computing probabilities when linear relations
exist among indicators.

Problems

1. Let A and B be events for which $P(A) = .6$, $P(B) = .7$, and
    $P(AB) = .5$. determine the probabilities of all members of the partition
    generated by A and B. Hence verify that this initial assignment is proper.
2. For Problems 3.2 to 3.5, let each $P(B_i) = .9$ and the probability of any
    intersection of k of the B_i have probability $(.9)^k$. For each problem, in
    each of cases (a) and (b), find $P(A)$.
3. A device produces numbers in $[0, 1]$. Let $\mathcal{C}$ denote the collection of all
    subintervals of $[0, 1]$. Each interval may include or exclude either or
    both endpoints. For any interval J in $\mathcal{C}$ we assign $P(J) = $ length of J.
    (a) Determine $P(A)$ where A is the event that the number is at least
        1/3 or between 1/4 and 3/4.
    (b) Determine $P(B)$ where B is the event that the number is less than
        1/4 or greater than 2/3.
    (c) Determine $P(C)$ where C is the event that the first significant digit
        in the decimal expansion of the number is divisible by 3.
4. There are five trees in a row, two of which are diseased. Number the
    trees 1, 2, 3, 4, and 5. Consider two models:

    A. (Noncontagious)   All arrangements of diseased and healthy trees
        are equally likely.
    B. (Contagious)   Let B be the event that the diseased trees are ad-
        jacent. Then $P(B) = 2P(B^c)$. All arrangements consistent with B

are equally likely, as are those consistent with B^c. Let A_i be the event that the ith tree is diseased.

(a) List the nonempty members of the partition generated by $(A_1, A_2, A_3, A_4, A_5)$.
(b) Give the probabilities of the events in part (a) under both models A and B.
(c) Suppose that the trees are arranged in a circle so that tree 1 and tree 5 are adjacent. Repeat part (b).

5. Delete from (4) the assumption $P(A) \leq 1$ and $P(\phi) = 0$. Use only the remaining assumptions to show that $P(A) \leq 1$ and $P(\phi) = 0$.

6. In Example 1 show that probability cannot be extended beyond events of the form of B.

7. Suppose that $P(A_i) = p_i$ $(i = 1, \ldots, n)$ and, for any $K \subset \{1, \ldots, n\}$, that $P(\cap_{i \in K} A_i) = \Pi_{i \in K} P(A_i)$. Find the probabilities of all events in the partition generated by $A_1, \ldots, A_n$. *Hint*: Solve the problem for $n = 2, 3, \ldots$ to obtain the pattern. Begin with $K = \{1, \ldots, k\}$.

8. Determine strict upper and lower bounds for $P(AB)$ when $P(A)$ and $P(B)$ are specified.

9. Let $\mathcal{C} = \{A_1, A_2, \ldots\}$ where the A_i's are disjoint, but $\cup_{i=1}^{\infty} A_i \neq \Omega$. Show that an initial probability assignment to $\mathcal{C}$ with $\Sigma_{i=1}^{\infty} P(A_i) \leq 1$ can be extended.

6. ELEMENTARY CONSEQUENCES

We list here, for easy reference, several elementary, sometimes obvious, consequences of the rules (5.4) and Theorem 5.1. Some have already been proved and used in Section 5. Some can be proved in a variety of ways. Most of the proofs are left as problems.

Theorem 1. *For events A and B,*

(a) $P(A^c) = 1 - P(A)$.
(b) $P(AB) = P(A) - P(AB^c)$.
(c) *If $A \subset B$, then $P(B) = P(A) + P(A^cB) \geq P(A)$.*
(d) $P(A \cup B) = P(A) + P(B) - P(AB) = P(B) + P(AB^c) \leq P(A) + P(B)$.

Proof. Statements (a) and (b) both follow from Corollary 5.1 and the indicator expressions

$$I_\Omega = I_A + I_{A^c}$$

and

$$I_{A \cup B} = I_A + I_B - I_{AB} = I_B + I_{AB^c}.$$

The remainder of the proof is left as a problem.

Theorem 1(a), although quite obvious, provides an important technique. To calculate $P(A)$ we can either go about it directly or calculate $P(A^c)$ and subtract from 1. The latter is particularly attractive when $P(A^c)$ is readily evaluated, but direct evaluation of $P(A)$ is difficult (as in Example 2).

Example 1. Suppose that the probability of rain in East Lansing is .1 for any given day in July. What is the probability that either July 3 or July 4 is dry (no rain)?

Let A be the event that it is dry on July 3 and B be the event that it is dry on July 4. By complementation, $P(A) = P(B) = .9$. Furthermore, $P(A \cup B) = P(A) + P(B) - P(AB)$. Suppose that examination of rainfall data shows that 85% of the time both July 3 and 4 have been dry, so that $P(AB) = .85$. Then $P(A \cup B) = .9 + .9 - .85 = .95$. Also, $P(AB^c) = P(A) - P(AB) = .9 - .85 = .05$.

If such detailed rainfall data are not available, what we know enables us to conclude that $.8 \le P(AB) \le .9$. The lower bound results from $P(A \cup B) \le 1$. The upper bound is a consequence of the fact that $AB \subset A$.

The following example makes rudimentary use of combinatorics. The reader unfamiliar with combinatorics should pause here to read Section 10, the appendix to this chapter.

Example 2. A fair die is rolled five times so there are 6^5 possible sequences of results. The sequences constitute a partition. Fairness of the die means that we have the equally likely case of Example 5.2, so each member of this partition has probability $1/6^5$. Let A be the event that at least one six is rolled. Now, $A = \cup_{i=1}^5 B_i$, where B_i is the event that the ith roll results in a six. The indicator, I_A, is a complicated function of the I_{B_i}. However, A^c is the event that no six is rolled. There are 5^5 sequences of results which contain no sixes, so that $P(A^c) = (5/6)^5$ and $P(A) = 1 - (5/6)^5 = .598$.

Theorem 2 deals with n events, extending some parts of Theorem 1.

Theorem 2. *Let $\mathcal{C} = \{A_1, \ldots, A_n\}$, let $U = \cup_{i=1}^n A_i$, and let B be any event. Then*

(a) $P(U) = \sum_{k=1}^n (-1)^{k+1} \Lambda_k$, *where $\Lambda_k = \sum_{J \in N_k} P(\cap_{i \in J} A_i)$ and N_k is the class of all subsets of size k from $\{1, \ldots, n\}$.*

(b) $P(U) = P(A_1) = \sum_{k=2}^{n} P(A_k \cap_{j=1}^{k-1} A_j^c) \le \sum_{i=1}^{n} P(A_i)$.
(c) *If $\mathcal{C}$ is a partition, then $\sum_{i=1}^{n} P(A_i) = 1$ and $P(B) = \sum_{i=1}^{n} P(A_i B)$.*

Proof. To prove (a), expand the expression (3.2) for I_U. This yields $I_U = \sum_{k=1}^{n} (-1)^{k+1} L_k$, where $L_k = \sum_{J \in N_k} \prod_{i \in J} I_{A_i}$. The proof of (b) uses Example 4.5, the partitioning of a union. The remaining details are left as a problem.

Example 3. The way to calculate the probability of a union depends on the available information regarding intersections. Suppose, for example, that a certain studio has three films (call them α, β, and γ) showing in a certain city of size 20,000 and we want the probability (proportion in this case) that any given resident has seen at least one of these films. If we denote by A, B, and C, respectively, the events corresponding to having seen α, β, and γ, then what we want is $P(A \cup B \cup C)$.

We might be informed that:

(i)   2000 residents saw α;
(ii) of those who did not see α, 1000 saw β; and
(iii) of those who saw neither α nor β, 500 saw γ.

Theorem 2(c) yields

$$P(A \cup B \cup C) = P(A) + P(A^c B) + P(A^c B^c C)$$
$$= \frac{2000}{20,000} + \frac{1000}{20,000} + \frac{500}{20,000} = .175.$$

Alternatively, we might be informed that

(i)   2000 saw α, 2000 saw β, 1100 saw γ;
(ii) 1000 saw α and β, 500 saw α and γ, 400 saw β and γ; and
(iii) 300 saw all three (α, β, and γ).

Then, from Theorem 3(b),

$$P(A \cup B \cup C) = P(A) + P(B) + P(C) - P(AB)$$
$$- P(AC) - P(BC) + P(ABC)$$
$$= .175.$$

The second set of figures is more informative than the first; in fact, it provides enough to determine the probabilities of all members of the partition generated by A, B, and C. Nevertheless, if all we want is the probability of the union, the first set of figures is adequate. Finally, the simple fact (if we are so informed) that 16,500 residents saw none of the films in question would also suffice.

Monotone Sequences of Events and Their Limits

The sequence of events $A_1, A_2, \ldots$ is *increasing* if $A_1 \subset A_2 \subset \cdots$ and is *decreasing* if $A_1 \supset A_2 \supset \cdots$; the sequence is called *monotone* if it is increasing or decreasing. For a monotone sequence of events, $A_1, A_2, \ldots$, we define the *limit* by $\lim_{n\to\infty} A_n = \bigcup_{n=1}^{\infty} A_n$ if the sequence is increasing and by $\lim_{n\to\infty} A_n = \bigcap_{n=1}^{\infty} A_n$ if the sequence is decreasing.

Theorem 3. *For a monotone sequence of events,* $P(\lim_{n\to\infty} A_n) = \lim_{n\to\infty} P(A_n)$.

Proof. Assume that $A_1, A_2, \ldots$ is increasing. Set $B_1 = A_1$ and, for $j = 2, 3, \ldots$, set $B_j = A_{j-1}^c A_j$. Then $\{B_1, \ldots, B_n\}$ is a partition of A_n and $\{B_1, B_2, \ldots\}$ is a partition of $\lim_{n\to\infty} A_n$. Hence $P(A_n) = \sum_{i=1}^{n} P(B_i)$ and $P(\lim_{n\to\infty} A_n) = \sum_{n=1}^{\infty} P(B_n) = \lim_{n\to\infty} P(A_n)$.

For $A_1, A_2, \ldots$ decreasing, $A_1^c, A_2^c, \ldots$ is increasing and $(\lim_{n\to\infty} A_n)^c = \lim_{n\to\infty} A_n^c$, so the proof follows from the preceding case.

Example 4.† Let P assign a probability to each interval in $\mathcal{R}$, let $A_x = (-\infty, x]$, let $B_x = (-\infty, x)$, and let $F(x) = P(A_x)$. Consider any sequence $a_1 > a_2 > \cdots > 0$ with $\lim_{n\to\infty} a_n = 0$. Then $A_{x-a_1}, A_{x-a_2}, \ldots$ is an increasing sequence of events and $\lim_{n\to\infty} A_{x-a_n} = B_x$. Thus $P(B_x) = \lim_{n\to\infty} P(A_{x-a_n}) = \lim_{n\to\infty} F(x - a_n)$. It follows that $P(B_x) = \lim_{t\uparrow x} F(t)$, where this notation stands for the limit taken with $t < x$, increasing to x. Furthermore, $A_x = \bigcap_{n=1}^{\infty} A_{x+a_n} = \lim_{n\to\infty} A_{x+a_n}$, so that $F(x) = \lim_{n\to\infty} F(x + a_n)$.

Corollary 1. *For any events* $A_1, A_2, \ldots$, (i) $P(\bigcup_{i=1}^{\infty} A_i) = \lim_{n\to\infty} P(\bigcup_{i=1}^{n} A_i)$ *and* (ii) $P(\bigcap_{i=1}^{\infty} A_i) = \lim_{n\to\infty} P(\bigcap_{i=1}^{n} A_i)$. *It follows from* (i) *and Theorem 2(b) that*

$$P\left(\bigcup_{i=1}^{\infty} A_i\right) \leq \sum_{i=1}^{\infty} P(A_i). \tag{1}$$

Proof. Left as a problem.

Summary

Increasing sequence of events: $A_1 \subset A_2 \subset \cdots$ ($\lim_{n\to\infty} A_n = \bigcup_{n=1}^{\infty} A_n$).
Decreasing sequence of events: $A_1 \supset A_2 \supset \cdots$ ($\lim_{n\to\infty} A_n = \bigcap_{n=1}^{\infty} A_n$).
Monotone sequence of events: Increasing or decreasing sequence.

Some important consequences of the rules (5.4) are:

(i) $P(A^c) = 1 - P(A)$.
(ii) $P(AB) = P(A) - P(AB^c)$.
(iii) If $A \subset B$, then $P(A) \leq P(B)$.
(iv) $P(A \cup B) = P(A) + P(B) - P(AB)$.
(v) $P(\cup_{i=1}^{\infty} A_i) \leq \Sigma_{i=1}^{\infty} P(A_i)$.
(vi) If $A_1, A_2, \ldots$ is a monotone sequence of events, then $P(\lim_{n\to\infty} A_n)$ $= \lim_{n\to\infty} P(A_n)$.

Additional consequences may be found in Theorem 2.

Problems

1. In Example 1 find $P(A^c \cup B^c)$ and $P(A^c B^c)$. Explain, in words, what the events $A^c \cup B^c$, AB^c, and $A^c B^c$ represent.
2. A city has three newspapers; call them A, B, and C. Of the population, 40% read A, 25% read B, 10% read C, 10% read both A and B, 6% read both A and C, 8% read both B and C, and 5% read all three. What percentage read none of the newspapers?

Problems 3 to 8 require familiarity with combinatorics in addition to material from this section. See Section 10.

3. A box contains five good and five defective mousetraps. Three mousetraps are selected without replacement. Let A be the event that at least one selected mousetrap is defective. Suppose that each possible choice of three mousetraps out of the 10 has the same probability. Compute $P(A)$ from this model.
4. Find the probability that a bridge hand (13 cards) contains
   (a) At least one spade.
   (b) At least two spades.
5. For a poker hand (five cards) determine the probabilities of the events below in the order given that
   (a) The hand has at least one spade.
   (b) The hand has at least one ace.
   (c) The hand has at least one spade or at least one ace.
   (d) The hand has at least one spade and at least one ace.
6. For a bridge hand, find the probabilities that
   (a) The hand contains five spades.
   (b) The hand contains five spades and five hearts.
   (c) The hand contains at least one five card suit.
7. A club has 50 members of whom 30 are Whigs and 20 are Federalists. The club will have three officers and a council consisting of these and five additional members. All council members are chosen at random.

Find the probabilities that
   (a) All officers are of the same party.
   (b) Each party has at least two representatives on the council.
   (c) The council has a Whig majority.
8. A fair die is rolled n times. What is the probability that not all rolls result in the same face?
9. Let $A_1, A_2, \ldots$ be disjoint events with $P(A_i) = (1/4)(3/4)^{i-1}$.
   (a) Verify that $\sum_{i=1}^{\infty} P(A_i) = 1$.
   (b) Find the largest n such that $P(\cup_{i=n}^{\infty} A_i) \geq 1/2$.
10. (a) Prove that $1 - \sum_{i=1}^{n} P(A_i^c) \leq P(\cap_{i=1}^{n} A_i) \leq \min P(A_i)$.
    (b) Give conditions on the A_i so that the lower bound in part (a) is attained.
    (c) Give conditions on the A_i so that the upper bound in part (a) is attained.
11. Obtain a formula for $P(\cap_{i=1}^{n} A_i^c)$ in terms of probabilities of intersections of A_i's.
12. Let $A_1, A_2, \ldots$ be any infinite sequence of disjoint events.
    (a) Show that $\lim_{n \to \infty} P(A_n) = 0$.
    (b) Let B_n be the event that none of the events $A_n, A_{n+1}, \ldots$ occurs. Show that $\lim_{n \to \infty} P(B_n) = 1$.
13.† Let P and F be as in Example 4. Show that F has the following properties:
    (a) $\lim_{x \to -\infty} F(x) = 0$.
    (b) $\lim_{x \to \infty} F(x) = 1$.
    (c) $x \geq y$ implies that $F(x) \geq F(y)$.
    (d) If $t \geq x$ converges to x, then $F(t) \to F(x)$.
14. Suppose that (5.4) is satisfied except that the additivity property (5.4c) is only assumed for finite partitions. Assume, further, for any decreasing sequence $\{B_1, B_2, \ldots\}$ of events with $\lim_{n \to \infty} B_n = \phi$ that $\lim_{n \to \infty} P(B_n) = 0$. Show that (5.4c) is valid for countable partitions.
15. Complete the proof of Theorem 1.
16. Complete the proof of Theorem 2.
17. Prove Corollary 1.

7. SIMPLE RANDOM VARIABLES; EXPECTED VALUE

Many of the concepts of probability are inspired by and applied to numerical phenomena. We have in mind such diverse quantities as physical measurements, demand for commodities, SAT scores, the number of stars in the Milky Way, and/or the gains and losses of a gambler playing a sequence of games. We call such quantities *random variables* (rv's) and, in later chapters, we shall give an extensive and systematic account of various properties, applications, and techniques involving them. At this stage, however, we will deal only with a special case. The rv's of this section are characterized by the fact that there are only finitely many possible numerical values that they can attain. We call such rv's *simple*.

Simple Random Variable

A quantity, X, is called a *simple random variable* (simple rv) if it can be expressed as a linear combination of a finite collection of indicators, that is, if there exist events $A_1, \ldots, A_n$ and numbers $a_1, \ldots, a_n$ such that

$$X = \sum_{i=1}^{n} a_i I_{A_i}.$$

The class of simple rv's obviously contains all indicators and all constants since $X = c I_\Omega$ is identically equal to c. We first introduced linear combinations of indicators in Section 4. Example 4.9, where we used L instead of X, serves as a typical illustration.

We called attention to the fact that any such X can be expressed in terms of the partition generated by the A_i's. There may, in fact, be many ways of expressing the same X as a linear function of indicators.

Example 1.† Teams I and II play a three-game series. No ties are possible. Let A_i denote the event that team I wins the ith game, and let X denote the total number of games won by team I. Then

$$X = I_{A_1} + I_{A_2} + I_{A_3}. \tag{1}$$

Another expression for X can be obtained (as in Section 4) by considering the partition generated by $\{A_1, A_2, A_3\}$:

$$
\begin{aligned}
D_1 &= A_1 A_2 A_3 (X = 3) & D_5 &= A_1 A_2^c A_3^c (X = 1) \\
D_2 &= A_1 A_2 A_3^c (X = 2) & D_6 &= A_1^c A_2 A_3^c (X = 1) \\
D_3 &= A_1 A_2^c A_3 (X = 2) & D_7 &= A_1^c A_2^c A_3 (X = 1) \\
D_4 &= A_1^c A_2 A_3 (X = 2) & D_8 &= A_1^c A_2^c A_3^c (X = 0).
\end{aligned}
$$

Then

$$X = 3 I_{D_1} + 2 I_{D_2} + 2 I_{D_3} + 2 I_{D_4} + I_{D_5} + I_{D_6} + I_{D_7} \tag{2}$$

is another way of expressing the same simple rv. Still another can be obtained by combining all B_i's on which X has the same value. Let $C_1 =$

$D_5 \cup D_6 \cup D_7$, $C_2 = D_2 \cup D_3 \cup D_4$, $C_3 = D_1$, and, for the sake of completeness, $C_0 = D_8$. Then

$$X = \sum_{i=0}^{3} iI_{C_i} = 0 \cdot I_{C_0} + 1 \cdot I_{C_1} + 2 \cdot I_{C_2} + 3 \cdot I_{C_3}. \qquad (3)$$

Note that $\{C_0, C_1, C_2, C_3\}$ is a partition, each member of which represents a different value of X. We generally use the notation $[X = j]$ to refer to the event C_j. This bracket notation will be used in a more general way. For example, we write $[X > k]$ for the event that $X > k$ and $[a \le X \le b]$ for the event that $a \le X \le b$. In general, [a statement about X] will denote the event "the statement about X is true."

To the three representations for X that have been given, we add a (final) fourth. The events $[X > k]$, $k = 0, 1, 2, 3$, are not disjoint. Nevertheless, we have

$$X = I_{[X>0]} + I_{[X>1]} + I_{[X>2]} + I_{[X>3]}. \qquad (4)$$

Note that $I_{[X>k]} = 0$ implies that $I_{[X>k+1]} = 0$ and hence $I_{[X>k]} = 1$ implies that $I_{[X>k-1]} = 1$. Then the proof of (4) is an expansion of the following argument:

$$X = 0 \quad \text{if, and only if, } I_{[X>0]} = 0$$
$$X = 1 \quad \text{if, and only if, } I_{[X>0]} = 1 \quad \text{and} \quad I_{[X>1]} = 0.$$

The remainder of the proof of (4) is left to the reader.

Expressions such as (3) and (4) can be used with more general rv's.
Let X be a simple rv and let $\{v_1, \ldots, v_m\}$ be the finite set of possible values attainable by X.

Partition Induced

The collection of the m events of the form $[X = v_j]$ is called the *partition induced by* X.
For X a simple rv,

$$X = \sum_{j=1}^{m} v_j I_{[X=v_j]}. \qquad (5)$$

Many properties of a simple rv, X (and more general discrete rv's) can be expressed in terms of the partition induced by X. More details on this will be given later.

Tail Events

Let t be any real number. The events $[X > t]$, $[X \geq t]$, $[X < t]$, and $[X \leq t]$ are called *tail events*. Tail events are sometimes distinguished by such adjectives as *upper*, *lower*, *open*, and *closed*.

Expressions generalizing (4) can be obtained for arbitrary simple rv's using linear combination of tail events. The following theorem covers the nonnegative, integer-valued case.

Theorem 1. *Let X be a simple rv that takes only nonnegative, integer values. Then*

$$X = \sum_{i=0}^{m} I_{[X>i]} = \sum_{i=1}^{m} I_{[X \geq i]} \tag{6}$$

for any integer m such that $P[X \leq m] = 1$.

Proof. Left as a problem.

Theorem 5.1 gives a useful identity for evaluating probabilities of events. It also provides the basis for the assignment of an important descriptive number to each simple rv.

Expected Value

Let $X = \sum_{i=1}^{n} a_i I_{A_i}$ be a simple rv. The number $E(X)$ determined by

$$E(X) = \sum_{i=1}^{n} a_i P(A_i)$$

is called the *expected value* of X. [There are many other names for $E(X)$, among which are *expectation*, *average*, and *mean*.]

The definition of expected value provides for the assignment of a number, $E(X)$, to any simple rv, X. However, $E(X)$ has been expressed in terms of a specific representation of X, and it has been shown that using different collections of events, a given X can have many representations. Theorem 2, which follows, is a simple consequence of Theorem 5.1 and guarantees that no matter what events are used to represent X, the value of $E(X)$ will be the same.

Theorem 2. *Let X be a simple rv and suppose that X has two representations,*

$$X = \sum_{i=1}^{n} a_i I_{A_i} = \sum_{j=1}^{m} b_j I_{B_j}.$$

Then

$$E(X) = \sum_{i=1}^{n} a_i P(A_i) = \sum_{j=1}^{m} b_j P(B_j).$$

Proof. The theorem follows immediately by noting that $\sum_{i=1}^{n} a_i I_{A_i} - \sum_{j=1}^{m} b_j I_{B_j} \equiv 0$ and applying Theorem 5.1 to obtain $\sum_{i=1}^{n} a_i P(A_i) - \sum_{j=1}^{m} b_j P(B_j) = 0$.

Since $E(I_A) = P(A)$, expectation is a generalization of probability.

Example 2.   In Example 1 we gave four expressions for a specified simple rv, X. Theorem 2 assures that whichever expression is used to compute $E(X)$, its value will be the same. However, the information required for the calculations differs. To obtain $E(X)$ from (1), we merely need $P(A_i)$, $i = 1, 2, 3$. That is, we need to know the probability that I wins game i ($i = 1, 2, 3$). Often this would be assumed constant, say, $P(A_i) = p$, and thus $E(X) = 3p$. To obtain $E(X)$ from any of the other expressions (2) to (4) requires some knowledge of the probabilities of members of the partition generated by the events A_1, A_2, A_3. In Example 1 these were denoted by $D_1, \ldots, D_8$. Suppose, for example, that $P(D_j) = 1/8$ ($j = 1, \ldots, 8$). Then (2) yields

$$E(X) = 3 \cdot 1/8 + 2 \cdot 1/8 + 2 \cdot 1/8 + 2 \cdot 1/8 + 1/8$$
$$+ 1/8 + 1/8 = 3/2.$$

Note that $P(D_j) = 1/8$ ($j = 1, \ldots, 8$) implies that $P(A_i) = 1/2$ since each A_i is the union of four (disjoint) D_j's.

The representations of X given by (5) and (6) yield the following immediate corollaries of Theorem 2.

Corollary 1.   *Let $\{[X = v_j]: j = 1, \ldots, m\}$ be the partition induced by the simple rv X. Then*

$$E(X) = \sum_{j=1}^{m} v_j P[X = v_j]. \tag{7}$$

Corollary 2.   *Let X be a nonnegative integer-valued simple rv with $P[X \leq r + 1] = 1$. Then $E(X) = \sum_{k=0}^{r} P[X > k]$.*

Problem 2 concerns a generalized version of Theorem 1.
The meaning we associate with $E(X)$ varies with the context, as does the meaning of $P(A)$. We illustrate three possibilities. Let $X = 10I_A -$

$5I_{A^c}$ represent monetary consequences, in dollars, of the occurrence or nonoccurrence of some event, A, with $P(A) = .4$. Then $E(X) = 10(.4) - 5(.6) = \$1$.

Case 1. A group consists of two types of persons, A's and non-A's. Each A is given \$10 and each non-A is charged \$5. The proportion of A's in the group is .4. Then $X = 10I_A - 5I_{A^c}$ represents the change in monetary value of one group member, and $E(X) = \$1$ represents the average (per capita) change in monetary value. If, for example, the group has 500 members, its total change in monetary value is \$500.

Case 2. Adam and A^cbel play a sequence of games. If Adam wins, A^cbel pays him \$10; if A^cbel wins, Adam pays him \$5. The rv X is the payoff to Adam and $I_A = 1 - I_{A^c}$ is the indicator of a win for Adam. On the basis of past performance, Adam wins 40% of the time. (A^cbel wins 60%.) Here $E(X) = \$1$ is the average per game that Adam can expect to earn.

Case 3. You just paid \$5 for a ticket on a horse to win. If the horse wins, you get \$10 (plus your \$5 back). In your opinion there is a 40% chance that your horse will win. Your "gain" is $X = 10I_A - 5I_{A^c}$, where A is the event that your horse wins. Here $E(X) = \$1$ is your evaluation of how much you expect to win with your ticket.

Example 3. You have paid an entry fee to play the game Matrix. You will choose a row and your opponent will choose a column of the matrix:

$$\begin{pmatrix} 1 & 0 \\ 0 & 2 \end{pmatrix}.$$

Neither player knows his or her opponent's choice and you will win back, from your opponent, the amount in the row and column chosen. If your opponent knows your choice, he or she can choose a column to guarantee that you will win 0. To defeat this, you may use a randomized strategy. That is, you choose row 1 with probability p and row 2 with probability $q = 1 - p$.

Let X_i be the amount you win if your opponent chooses column i and A be the event that you choose row 1. Then $X_1 = I_A$ and $X_2 = 2I_{A^c}$ so that $E(X_1) = P(A) = p$ and $E(X_2) = 2P(A^c) = 2q$.

If you choose $p = 1/2$ and your opponent knows this, he or she should choose column 1 since $E(X_1) = 1/2$, which is less than $E(X_2) = 1$. Thus your opponent guarantees that your expected winnings will be 1/2. If you choose $p = 2/3$, then $E(X_1) = E(X_2) = 2/3$, so that, regardless of which column your opponent chooses, your expected winnings will be 2/3.

Let X be a simple rv and suppose that we wish to consider a quantity $Z = g(X)$ (i.e., a quantity whose value is determined by the value of X). Then Z is also a simple rv. In Example 1, for instance, suppose that each time team I wins, team II pays I \$5 and each time II wins, I pays II \$3. If Z denotes I's net winnings, then

$$Z = 5X - 3(3 - X) = 8X - 9.$$

This is an example in which a rv Z is a function $g(X)$ of another rv X. In general, the expression $Z = g(X)$ means that if $X = v_j$, then $Z = g(v_j)$. Hence $Z = \sum_{j=1}^n g(v_j) I_{[X=v_j]}$ so that Z is a simple rv and the corollary below follows easily from Theorem 2.

Corollary 3. *If X is a simple rv and $Z = g(X)$, then Z is also simple and*

$$E(Z) = \sum_{j=1}^m g(v_j) P[X = v_j]. \tag{8}$$

Thus the partition induced by X can be used to determine the expected value of any function of X.

Example 4. Suppose that X, B_j, and C_i are as in Examples 1 and 2 and let $Z = g(X) = |X - 1|$. Then

$$Z = I_{C_0} + I_{C_2} + 2I_{C_3}$$

and if each B_j has probability $1/8$, then

$$P(C_0) = \frac{1}{8}, \qquad P(C_2) = \frac{3}{8}, \qquad P(C_3) = \frac{1}{8},$$

so that

$$E(Z) = \frac{1}{8} + \frac{3}{8} + 2\left(\frac{1}{8}\right) = \frac{3}{4}.$$

Finally, if $X_1, \ldots, X_n$ are simple rv's, so is any quantity of the form $Z = f(X_1, \ldots, X_n)$.

Theorem 3. *If $X_1, \ldots, X_n$ are simple rv's, $c_1, \ldots, c_n, K$ are numbers, and $Z = \sum_{i=1}^n c_i X_i + K$, then*

$$E(Z) = \sum_{i=1}^n c_i E(X_i) + K. \tag{9}$$

Proof. Left as a problem.

Example 5. Acme Products, Inc. (API) has five salespeople. Salesperson i will contact r_i customers today, each contact has probability p_i of producing a sale, and API's profit on each sale is c_i. Let X_i denote the number of sales today for salesperson i. Table 1 gives the values of these quantities and the $E(X_i)$.

The daily overhead cost for these salespeople is $500, so set $K = -5$. The expected net profit in units of $100 is

$$E(Z) = (5)(2.0) + (3)(3.5) + (7)(1.5) + (6)(1.6)$$
$$+ (3)(4.8) - 5 = 50$$

Since a simple rv, X, takes only finitely many values, there must exist numbers $a \leq b$ such that X is always at least a and at most b. We write $a \leq X \leq b$.

Theorem 4. *If X is a simple rv and $a \leq X \leq b$, then $a \leq E(X) \leq b$.*

Proof. Let $v_1, \ldots, v_m$ be the possible values of X. Then $a \leq X \leq b$ implies that $a \leq v_j \leq b$ for all $j = 1, \ldots, m$. Since $E(X) = \sum_{j=1}^{m} v_j P[X = v_j]$ and $\sum_{j=1}^{m} P[X = v_j] = 1$, the theorem follows.

Let X and Y be simple rv's such that $Y - X \geq 0$. Then we write $X \leq Y$ or $Y \geq X$.

Corollary 4. *If X and Y are simple rv's and $X \leq Y$, then $E(X) \leq E(Y)$.*

Proof. Since $Y - X \geq 0$, the result is immediate from Theorem 4 and (9).

Theorem 4 may be paraphrased as follows: If a set of numbers is between a and b, their average is between a and b. In this form, the statement may appear more obvious. The corollary generalizes the inequality in Theorem 6.1(c).

Table 1

i	r_i	p_i	c_i (units of $100)	$E(X_i)$
1	5	.4	5	2.0
2	7	.5	3	3.5
3	3	.5	7	1.5
4	4	.4	6	1.6
5	8	.6	3	4.8

Summary

Simple random variable: $X = \sum_{i=1}^{n} a_i I_{A_i}$.
Expected value: $E(X) = \sum_{i=1}^{n} a_i P(A_i)$.

The value of $E(X)$ does not depend on the representation of X that is used. In general, $E(X) = \sum_j v_j P[X = v_j]$, and if X takes values 0, 1, $\ldots$, $r + 1$, then $E(X) = \sum_{k=0}^{r} P[X > k]$. If $Z = g(X)$, then $E(Z) = \sum_j g(v_j) P[X = v_j]$.

Problems

1. Consider a population with n individuals. The ith person has annual income a_i. One person is selected at random from the population. By this we mean that each person has probability $1/n$ of being selected. Let X be the annual income of the person selected.
   (a) Why is X a simple rv?
   (b) Show that $E(X) = \sum_{i=1}^{n} a_i/n$, the average annual income in the population.

2. An urn contains six numbered balls (numbers 1, 2, 3, 4, 5, 6). Two balls are drawn at random and without replacement. Let X be the sum of the numbers on the two balls drawn. Let A_i and B_j be the events that the first and the second balls, respectively, are numbered i and j. The assumption is that $P(A_i B_j) = 1/30$ for every pair $i \neq j$.
   (a) Write X as a linear combination of the indicators of the A_i and the B_j.
   (b) Write X as a linear combination of indicators of disjoint events. Write these disjoint events in terms of the A_i and B_j.
   (c) Compute $E(X)$.

3. In Example 1, suppose that for each game, the loser pays the winner $1. Let Y be the net amount won by team I.
   (a) Express Y in terms of X.
   (b) Express Y as a linear combination of the indicators of the A_i and as a linear combination of the indicators of the B_j.
   (c) Setting $P(A_i) = p$, compute $E(Y)$.
   (d) Suppose that if the third game is a tie breaker (if each team wins one of the first two games), the payment for the third game is $5. Otherwise, the payments remain $1 per game. Let Z be the net amount won by team I. Express Z as a linear combination of the indicators of the B_j. Can Z be expressed similarly in terms of the indicators of the A_i? In terms of X? Why or why not?

4. Teams A and B have m and n runners, respectively. All $m + n$ of them run a race and we define the following rv's: R_i = the order of

finish for A's ith runner ($R_i = 1, \ldots, m + n$); and U_{ij} is the indicator of the event that team A's ith runner beats team B's jth runner. Express $\Sigma_{i=1}^m R_i$ in terms of the U_{ij}.

5. Consider Example 3.

   (a) Verify that $p = 2/3$ is your best choice against an opponent who discovers your randomized strategy.

   (b) What is your opponent's best choice of a randomized strategy to protect against your discovery of his or her randomized strategy? Show that your opponent, using this strategy, guarantees that you cannot expect to win more than $2/3$.

   (c) What is a fair entry fee for you in this game?

6. A deck contains n consecutively numbered cards (numbers 1 to n). The deck is dealt one card at a time. If the ith card dealt is number i, a *match* occurs. Let X be the number of matches and find $E(X)$. *Hint*: Express X in terms of the indicators of "match for card i."

7. A population consists of m individuals. Consider a random sample of size n drawn with replacement.

   (a) Let X be the number of people in the population who are not in the sample and find $E(X)$.

   (b) Let Y be the number of people in the population who are in the sample and find $E(Y)$.

8. Items are selected one at a time and without replacement from a lot of 1000 items, 80 of which are defective. Let Δ_r denote the number of good items between the $(r - 1)$st and the rth defectives in the sequence selected. Let T_r denote the waiting time (number of trials) until r defectives appear in the sequence.

   (a) Express T_r as a function of the Δ_i ($i = 1, \ldots, r$).

   (b) Let Δ_{81} denote the number of good items left when the 80th defective is chosen. What is $\Sigma_{i=1}^{81} \Delta_i$?

   (c) Express T_r as a function of the Δ_i ($i = r + 1, \ldots, 81$).

9. At time 0, we turn on n new bulbs and allow them to burn out. Let T_i denote the first time at which a total of i bulbs have burned out. (T_n marks the end of the experiment.) Let Δ_i denote the time elapsed between the $(i - 1)$st burnout and the ith burnout ($\Delta_1 = T_1$). Finally, let $Z = \Sigma_{i=1}^n T_i$. Express Z as a linear combination of the Δ_i's.

10. Prove Theorem 1.

11. Let X be an integer-valued simple rv with $P[-m \leq X \leq m] = 1$. Show that $X = \Sigma_{j=1}^m I_{A_j} - \Sigma_{j=1}^m I_{B_j}$ where $A_j = [X \geq j]$ and $B_j = [X \leq -j]$. This extends Theorem 1.

12.† Prove Theorem 3.

8. SCOPE OF THE APPLICATIONS

Section 1 discusses three interpretations of $P(A)$. These are (with the same notation):

  (i)   *Population proportions (prevalence)*: $P(A)$ is the population proportion, $n(A)/N$.
 (ii)   *Relative frequency (rate of occurrence)*: $P(A)$ is the relative frequency, $n(A)/N$, of occurrences of A in N trials of an experiment.
(iii)   *Subjective probability (personal opinions, beliefs)*:   Let A be a statement and let $P(A)$ be a number between 0 and 1 that expresses the degree of belief that you (for instance) have in the truth of statement A. Then $P(A)$ is *your* probability that A is true.

It remains to be shown that the rules (5.4) that characterize mathematical probability are applicable to these measured quantities.

In cases (i) and (ii), $0 \le n(A) \le N$, so $0 \le n(A)/N \le 1$.

In case (i), if A and B are two properties such that no person has both, that is, if in the population A and B are mutually exclusive, then $n(A \cup B) = n(A) + n(B)$ so that $n(A \cup B)/N = n(A)/N + n(B)/N$.

In case (ii), if it is impossible for both A and B to occur on the same trial, $n(A \cup B) = n(A) + n(B)$, so that, once more, $n(A \cup B)/N = n(A)/N + n(B)/N$.

What has been shown is that in cases (i) and (ii), $n(A)/N$ satisfies the additivity property (5.4c) only for finite partitions. However, in developing theoretical probability models it is often necessary to contemplate infinite populations or experiments with an infinite number of possible results. Then it is natural to assume that the additivity property holds for countably infinite partitions. This is also required to obtain continuity results of the type in Theorem 6.3.

The remainder of this section is not required for later sections. We will make a set of assumptions about "rational" behavior that lead to an assignment of personal probabilities satisfying (5.4) for finite partitions.

To use probabilities to represent a person's opinion, we need (1) a method for measuring opinion quantitatively and (2) reasonably simple, yet realistic assumptions from which the rules (5.4) of probability can be deduced. What follows is just one among many methods that have been proposed for accomplishing these goals.

The "you" referred to below is the person whose opinions are to be measured.

Let A be any event (statement) and imagine that you may purchase, for any amount whatever, a lottery ticket that pays an amount $\$M \ne 0$ if A occurs and $\$0$ if A fails to occur. The interpretation of $M < 0$ is that you pay \$|M| if A occurs. Call this ticket (A, M). You are then asked to state an amount, \$m, which represents the *largest* sum you would be willing to pay for the possession of this ticket. "Your" probability for A is then taken to be the number $P(A) = m/M$. Call m the *worth* you place on the ticket (A, M). The interpretation of $m < 0$ is that you insist on receiving at least \$|m| if you are to take the ticket. Conceivably, we can do this for as many events, A, as we wish and thus produce a numerical assignment, P, defined on events.

Although we use the symbol P and call this probability, there is no guarantee that the numbers thus assigned to events obey the required rules. What is needed at this point is a set of assumptions about your behavior that will enable us to deduce that P is actually a probability in the sense of the rules (5.4). A person whose opinions satisfy such assumptions will be called *coherent* or *rational*. Understand, however, that (1) different people will have, perhaps, different notions of what constitutes "rational" judgment, and (2) even if a criterion for rationality is agreed to, there may be no person whose judgments are rational in this particular sense. Therefore, any such assumptions are, to some extent, idealized and any application of the resulting model hinges on the extent to which the user believes the assumptions are satisfied.

Because we have used monetary values in measuring opinion, the following rationality assumptions are conditions on your attitude toward money.[1]

If the worth you place on ticket (A, M) is m, the worth you
place on ticket (A, tM) is tm.   (1)

You will never assign a worth to a ticket or set of tickets that will
lead to losing money with absolute certainty.   (2)

The first assumption guarantees that the measurement, P, is independent of the potential prize, M. Thus if the payoff on a ticket is \$10 and you assign it a worth of \$5 ($P = 1/2$), you would assign it a worth \$20 if a ticket for the same event paid \$40. No restriction is placed in (1) on the sign of t.

We will show that assumptions (1) and (2) guarantee that our assignments lead to a set of P's that satisfy (5.4). It is obvious that the worths of (ϕ, M) and (Ω, M) are 0 and M, respectively.

1. $0 \leq P(A) \leq 1$ for every A: Suppose that m is your worth of the ticket (A, M). The net gain is given in Table 1 for two cases when the

---

[1]Analogous assumptions replacing money by utility for money (nonlinear) lead to the same results.

Table 1

A occurs	$M - m$
A fails	$-m$

ticket is bought for m. The two entries cannot both be negative, since in that case, you will lose money with certainty. The two entries cannot both be positive since we can always consider $(A, -M)$.

We now consider two cases.

(a) $M > 0$. Then $P(A) = m/M > 1$ implies that both entries are negative, while $P(A) < 0$ implies that both entries are positive.

(b) $M < 0$. The situation is the reverse of that in case (a).

2. If A_1 and A_2 are disjoint events, then $P(A_1 \cup A_2) = P(A_1) + P(A_2)$: Consider three lottery tickets, (A_1, M_1), (A_2, M_2), and (C, M_3) where $C = A_1 \cup A_2$. Assume that the worths you place on these tickets are m_1 on (A_1, M_1), m_2 on (A_2, M_2), and m_3 on (C, M_3). The total worth of these tickets is therefore $m_1 + m_2 + m_3 = P(A_1)M_1 + P(A_2)M_2 + P(C)M_3$. Because A_1 and A_2 are disjoint, the following can happen:

(i) A_1 occurs, you get $M_1 + M_3$.
(ii) A_2 occurs, you get $M_2 + M_3$.
(iii)   Neither occurs, you get 0.

The difference between what you pay and what you get is

$$L_1 = P(A_1)M_1 + P(A_2)M_2 + P(C)M_3 - (M_1 + M_3) \quad \text{if } A_1 \text{ occurs;}$$
$$L_2 = P(A_1)M_1 + P(A_2)M_2 + P(C)M_3 - (M_2 + M_3) \quad \text{if } A_2 \text{ occurs;}$$
$$L_3 = P(A_1)M_1 + P(A_2)M_2 + P(C)M_3 \quad \text{if neither occurs.}$$

Collecting terms,

$$[P(A_1) - 1]M_1 + P(A_2)M_2 + [P(C) - 1]M_3 = L_1;$$
$$P(A_1)M_1 + [P(A_2) - 1]M_2 + [P(C) - 1]M_3 = L_2; \qquad (3)$$
$$P(A_1)M_1 + P(A_2)M_2 + P(C)M_3 = L_3.$$

Assumption (2) requires that no matter how M_1, M_2, and M_3 are chosen, L_1, L_2, and L_3 should never all be negative. Thus we wish to guarantee that there are no solutions M_1, M_2, and M_3 for (3) when $L_1, L_2, L_3 < 0$.

However, for any choice of negative (or positive) values for L_1, L_2, and L_3, there will be a solution for M_1, M_2, and M_3 in the system of equations

$$a_1 M_1 + b_1 M_2 + c_1 M_3 = L_1;$$

$$a_2 M_1 + b_2 M_2 + c_2 M_3 = L_2;$$

$$a_3 M_1 + b_3 M_2 + c_3 M_3 = L_3.$$

unless the determinant of the coefficients is 0.

The determinant in (3) reduces, after some algebra, to $P(C) - P(A_1) - P(A_2)$, so that $P(C) - P(A_1) - P(A_2) = 0$ is a necessary and sufficient condition for (1) and (2).

We have thus shown, for any disjoint events A_1 and A_2, that $P(A_1 \cup A_2) = P(A_1) + P(A_2)$, which implies that for any collection $\{A_1, \ldots, A_n\}$ of disjoint events, $P(\cup_{i=1}^n A_i) = \Sigma_{i=1}^n P(A_i)$.

The preceding argument shows that under assumptions (1) and (2), P satisfies the additivity rule for finite partitions. We will again assume, when necessary, that subjective probability satisfies the additivity rule for countably infinite partitions, just as in the previous (objective) cases.

Suppose that you purchase the ticket (A, M) for the worth, m, you place on the ticket. Your net gain, X, is a simple rv given by $X = MI_A - mI_\Omega = (M - m)I_A - mI_{A^c}$. The relation between probability and worth guarantees that $E(X) = 0$.

Consider the three lottery tickets (A_1, M_1), (A_2, M_2), and (C, M_3) used to prove additivity for personal probability. If you have purchased each ticket for the worth you place on it, your net loss is the simple rv.

$$Y = (m_1 + m_2 + m_3)I_\Omega - (M_1 + M_3)I_{A_1} - (M_2 + M_3)I_{A_3}$$

$$= L_1 I_{A_1} + L_2 I_{A_2} + L_3 I_{A_1^c A_2^c}.$$

From this it follows that $E(Y) = 0$.

Consider any gamble with prospective payoffs $v_1, \ldots, v_n$ to which you assign respective probabilities $p_1, \ldots, p_n$. The worth, m (to you), to this gamble is $m = E(X) = \Sigma_{i=1}^n v_i p_i$.

Example 1. This example is taken from an actual consulting experience of one of us. A geneticist was interested in knowing if a certain human gene leading to an abnormality is dominant or recessive. While we were discussing the problem, it became clear that she was quite willing to assign personal probabilities and then modify these in light of observations. She was asked to consider a lottery ticket paying $1 if the gene is dominant. The following is a digest of the conversation.

Q. Would you pay 50 cents for this ticket?
A. No.

Q. Would you pay 25 cents for this ticket?
A. No.
Q. Would you pay 10 cents for this ticket?
A. Yes.
Q. Would you pay 15 cents for this ticket?
A. No.
Q. Would you pay 12 cents for this ticket?
A. No.
Q. Would you pay 11 cents for this ticket?
A. No.

From this conversation we adopted the probabilities $P(\text{dominant}) = .1$ and $P(\text{recessive}) = .9$.

Summary

In this section we reviewed the application of probability to:

(i)   Population proportions.
(ii) Relative frequency.
(iii) Personal opinions.

Problem

1.† Five horses are entered in a race. A bookmaker offers the following odds for each horse (to win):

H_1: 1–2 H_4: 1–6
H_2: 1–4 H_5: 1–10.
H_3: 1–4

A \$2 bet on H_1, for example, will yield \$4 if H_1 wins. (You also get your \$2 back.) Suppose that you have \$300 and that you can bet (in integer multiples of \$2) on any number of the horses (at the quoted odds). Is it possible to allocate the \$300 among the horses in such a way that you will make money no matter which horse wins? If not, explain. If yes, produce such an allocation.

9. EXPERIMENTS AND SAMPLE SPACES

The terminology of probability theory is drawn from a variety of sources: gambling, science, statistics, philosophy, and so on. The structure of the probability models for experiments developed in this section using the terminology most commonly used for that purpose. These form the most important class of probability models.

A. Mathematical Structure

A great many (all?) situations involving uncertainty can be reduced to the following form. There is a set $\mathcal{S}$ and a "random device" for selecting a member of $\mathcal{S}$. The uncertainty concerning which member will be selected is expressed through a probability assignment of subsets of $\mathcal{S}$. If A is a subset of $\mathcal{S}$, then $P(A)$ is the probability that the device selects a member of A.

With respect to the foregoing description we use the following terminology:

Sample Space, Outcome, Experiment

The set $\mathcal{S}$ is called a *sample space*, members of $\mathcal{S}$ are called *outcomes*, and the random device is called an *experiment*.

Probability Model for an Experiment

A *probability model for an experiment* is a probability model (as defined in Section 5) for which each member of $\mathcal{C}$ is a subset of a specified set, $\mathcal{S}$.

In this context, an event A, is a statement specifying a property of the outcome of the experiment. The subset of $\mathcal{S}$ consisting of all outcomes for which A is true represents (in fact, identifies) this event and we make no distinction, notational or otherwise, between A and the subset of $\mathcal{S}$ to which it corresponds. *Occurrence* is, then, identified with "membership" and what we mean by "*A occurs*" is that the outcome of the experiment is a member of A. Clearly, then, all event operations are the usual operations of set algebra. The indicator I_A of the event A is the function on $\mathcal{S}$ defined by

$$I_A(s) = 1 \qquad \text{if } s \in A$$
$$= 0 \qquad \text{if } s \in A^c.$$

Accordingly, a simple rv is a function on $\mathcal{S}$ with finitely many possible values.

Let $D_s = \{s\}$ be the event "the outcome of the experiment is s." Then $\mathcal{D} = \{D_s : s \in \mathcal{S}\}$ is a partition whose members are the singleton subsets of $\mathcal{S}$. Thus a sample space is a sure event and its members, considered as singleton subsets, are disjoint and exhaustive.

To identify a probability model for an experiment we need only stipulate the set $\mathcal{S}$ and the probabilities of some class $\mathcal{C}$ (satisfying the usual conditions) of subsets of $\mathcal{S}$. From the mathematical point of view it makes no difference whether or not the specific model is attainable by any "real" physical process.

Example 1. Consider the selection of a point in the unit interval $[0, 1]$ and let A be any of the subintervals (s, t), $(s, t]$, $[s, t)$, or $[s, t]$ with $0 \le s \le t \le 1$. Suppose that the selection can be done in such a way that

$$P(A) = t^2 - s^2 \tag{1}$$

for any such A. For this experiment the sample space is $\mathcal{S} = [0, 1]$, the class $\mathcal{C}$ consists of all subintervals of $[0,1]$, and the probability assignment P is given by (1).

Example 2. Let $\mathcal{S} = \{1, 01, 001, \ldots\}$ be the countable set whose $(k + 1)$st member, s_{k+1}, is a sequence consisting of k 0's followed by a 1. Suppose, for each integer $k \ge 0$, that $P(s_{k+1}) = 1/2^{k+1}$ is the probability that the outcome of the experiment is s_{k+1}. Then $\mathcal{S}$ is the sample space and $\mathcal{C}$ consists of all singleton subsets of $\mathcal{S}$. Here, as in the sequel, we use the notation $P(s)$ for the probability of the event $D_s = \{s\}$.

Although these two examples provide theoretical models for fictitious experiments, each will be found to be useful for describing "real" experiments to be discussed later.

B. Building a Model: Choice of a Sample Space

We began the exposition of the preceding subsection with the assertion that many situations involving uncertainty can be reduced to a particular formal structure. The first step in achieving this reduction is the selection of an appropriate sample space, the set $\mathcal{S}$. Once the choice of $\mathcal{S}$ has been made, the only events (concerning the phenomenon in question) that can be addressed are those expressible as subsets of $\mathcal{S}$. We usually want $\mathcal{S}$ to be as simple (and small) as possible, but it needs to be detailed enough (and adequately large) to allow for the analysis of any event with which we may be concerned. The essential requirement is that the elements of $\mathcal{S}$, considered as singleton subsets, form a partition.

Example 3. A coin is tossed once. The usual sample space is $\{H, T\}$. Note the relationship between this sample space and the partition $\{A, A^c\}$, where A is the event that a head is tossed. In the 1939 movie *Mr Smith Goes to Washington*, a key scene involves a coin standing on end. Hence you may wish to use the sample space $\{H, T, E\}$. If the coin is tossed in a street, another possibility is that the coin may disappear down a storm drain, so we may wish to use the sample space $\{H, T, E, D\}$.

Example 4. Seven runners (label them 1 to 7) are entered in a race and we wish to construct a sample space for the result. The sample space we choose can be very simple or very complex, depending in the minutiae of detail that motivates our concern. Here are several possibilities.

(a) Who won?: $\mathcal{S} = \{1, \ldots, 7\}$;
(b) How did runner 1 do?: $\mathcal{S} = \{1, \ldots, 7\}$;
(c) What is the order of finish?: $\mathcal{S}$ = the set of all permutations of the integers 1 to 7;
(d) Who were the first three finishers in order?: $\mathcal{S}$ = the set of all permutations of the integers 1 to 7 taken three at a time;
(e) Who was the winner and what was the time?: $\mathcal{S} = \{(i, x): i = 1, \ldots, 7; x > 0\}$
(f) What were the times of the participants?: $\mathcal{S} = \{(x_1, \ldots, x_7): x_i > 0\}$

The last sample space, the most detailed record of what happened, can be used to obtain information contained in any of the others, but not conversely.

Example 5. A rat is run through a T-maze five times. Each time the direction, left or right, that the rat turns is recorded. If the purpose of the experiment is simply to study the rat's preference for left or right, then $\mathcal{S} = \{0, 1, 2, 3, 4, 5\}$ is an adequate sample space. Suppose, instead, that the left arm contains a reward and that the rat's learning behavior is of primary interest. Then a sample space identifying the direction for each trial (in order) would be required: $\mathcal{S}' = \{(x_1, x_2, x_3, x_4, x_5): x_i = L, R$ for $i = 1, 2, 3, 4, 5\}$. This sample space will also suffice if interest is confined to the number of left turns. The possible values of the x_i are irrelevant as long as exactly two values are allowed. For example, we could take $x_i = 0$ or 1 or $x_i = $ y(es) or n(o).

Let A_i be the event "left turn on the ith trial." Then $\mathcal{S}'$ represents the partition $\{D_1, \ldots, D_{32}\}$ generated by the A_i and $\mathcal{S}$ represents the partition induced by $X = \Sigma_{i=1}^5 I_{A_i}$. Let E be the event "the rat turns left twice." In $\mathcal{S}$, then, $E = \{2\}$, while in $\mathcal{S}'$ there are 10 outcomes in E: (L, L, R, R, R), (L, R, L, R, R), and so on.

Suppose that this experiment is modified so that the number of left turns is 100. This experiment must terminate in a finite number of trials. However, the number of trials is unbounded and the rat may have to be nearly immortal to complete the experiment. Let T_i be the event that the experiment terminates on the ith trial ($i = 1, 2, \ldots$). Clearly $T_i = \phi$ for $i \leq$ 99. The event that the number of trials required is divisible by 3 is $\cup_{i=1}^{\infty}$ T_{3i}.

As a practical matter, even the fictitious rat of this example is mortal. Allowing time for food, drink, sleep, and recreation, we may suppose that the rat can only run the maze 1 million times during its adult life. Do we gain any simplicity or accuracy by replacing $\cup_{i=1}^{\infty} T_{3i}$ by $\cup_{i=1}^{333,333} T_{3i}$?

In fact, the model is simpler and reasonably realistic if no upper bound is placed on the number of trials. The differences between probabilities of typical events will be small when we compare the results of imposing and not imposing such bounds.

The choice of a sample space may also be influenced by the ease with which the probability assignment can be made or described. Even if we only require a small sample space, we may need a larger one to establish appropriate probabilities on the smaller one.

Example 6. Consider two tosses of a fair coin and suppose that the only concern is with the number of heads (H's). To describe the result, $\mathbb{S} = \{0, 1, 2\}$ could be used. In assigning probabilities to these outcomes, however, it is more convenient to use $\mathbb{S}' = \{HH, TH, HT, TT\}$, a sample space that identifies the result of each toss separately. "Fairness" of the coin translates into "equally likely outcomes" for $\mathbb{S}'$, but *not* for $\mathbb{S}$. Thus each member of $\mathbb{S}'$ is assigned probability 1/4 and hence, in $\mathbb{S}$, we have $P(0) = P(2) = 1/4$, and $P(1) = 1/2$.

C. Assigning Probabilities: Completion of the Model

We devote this subsection to a description of rather broad categories of models and how probability assignments might be made in special cases. In Section 5 we mentioned that probability assignments are generally based on intuition, assumptions, or experience. Experience may, for example, consist of a historical account of repeated trials of some experiment.

Of course, an important consideration in assigning probabilities is the nature of the uncertainty in question.

Example 7. You are about to dip a piece of litmus paper into a liquid. An appropriate sample space for this experiment would seem to be $\mathbb{S} = \{red, blue\}$. Various meanings may be attributed to $P(red) = 1 - P(blue)$. First, it may be that the liquid comes from a population of liquids of which a proportion $P(red)$ are acids while $P(blue)$ are bases. Second, you may be dealing with an unstable liquid that fluctuates between acidic and basic states, with $P(\text{red})$ then as the proportion of time it exhibits acidic behavior. Finally, the number $P(\text{red})$ may simply reflect prior subjective uncertainty regarding the appropriate classification for this liquid. If you know that the liquid is hydrochloric acid, then $P(\text{red}) = 1$ for you.

In general, we approach the task of assigning probabilities to an appropriate class of subsets of $\mathcal{S}$ in the manner suggested by Examples 5.1, 5.2, and 5.3. There are two basic categories of models.

The Discrete Case

$\mathcal{S} = \{s_1, s_2, \ldots\}$ is countable, $\mathcal{C}$ consists of all singleton subsets of $\mathcal{S}$, and probabilities are assigned to the points of $\mathcal{S}$ so that $0 \le P(s_i) \le 1$ and $\Sigma_i P(s_i) = 1$. If A is any subset of $\mathcal{S}$, then $P(A) = \Sigma_{s_i \in A} P(s_i)$.

The Nondiscrete Case

$\mathcal{S}$ is uncountable, $\mathcal{C}$ consists of a distinguished class of subsets of $\mathcal{S}$, and probabilities (or a rule for determining them) are assigned to members of $\mathcal{C}$. Other events to which probabilities can be assigned are those that can be obtained from $\mathcal{C}$ by countable set operations.

Example 1 illustrates a nondiscrete model, and Example 2 illustrates a discrete model. There are several other classes of models that need to be identified and we conclude this section by introducing a few of them.

Finite Case

This is the special subclass of the discrete case for which the sample space is finite (i.e., $\mathcal{S} = \{s_1, \ldots, s_N\}$).

Uniform Case

A uniform probability model is one for which the probability of an event (set) is proportional to its "size" (measured appropriately). It may be discrete or nondiscrete, but in either case, it can only apply when the "size" of $\mathcal{S}$ is finite. For any uniform model (and appropriate A), $P(A) = (\text{size of } A)/(\text{size of } \mathcal{S})$. We use this terminology in two cases.

In the discrete case, the size of a set is the number of elements it contains and therefore a discrete uniform model is possible only when $\mathcal{S}$ is a finite set. Then $P(s_i) = 1/N$ for $i = 1, \ldots, N$ and $P(A) = n(A)/N$, where $n(A)$ is the cardinality of A. Other names for the finite $\mathcal{S}$ uniform model are *equally likely model* and *random selection model*.

In the nondiscrete case, $\mathbb{S}$ will be a subset of $\mathscr{R}^n$ and the size of a subset is its "volume" (positive for $\mathbb{S}$). Such a model is also called a *continuous uniform* model. An illustration is given in Example 10.

Empirical and Theoretical Models

A probability model is *empirical* if the probabilities assigned to events are obtained from observations on the phenomenon in question. In the case of a population, these are the actual values of proportions in the population. In the case of repeated trials of an experiment, these are the observed relative frequencies. In case of personal probabilities, these are the stated assessments of the particular individual.

A probability model is *theoretical* if the probabilities assigned to events are deduced from assumptions about the phenomenon in question. We often have a multiplicity of possible theoretical models and wish to pick the model that "best" fits. This is done by comparison with an empirical model based on actual observations. This comparison is one of the topics in statistics.

A finite model may be either theoretical or empirical. If the $P(s_i)$ are the result of some assumptions about the mechanism by which an s_i is chosen, they are theoretical. If, however, the $P(s_i)$ are observed relative frequencies of the s_i in repeated trials, they are empirical.

If $\mathbb{S}$ consists of the members of some population, the equally likely model yields population proportions, an empirical model. The same model serves as a theoretical model for the result of selecting member at random from the population based on the assumption that the selection is made in a particular fashion. For example, if $\mathbb{S}$ consists of the faces of an N-faced die, the equally likely assignment is descriptive of the symmetry among faces, a fact that can be established by inspection. Whether or not the same model applies to the results of rolling the die is another matter.

Example 8.[2] (Random sampling) One of the most important applications of the equally likely case is to sampling from a finite population. Let Π be a population of size M and consider two types of sampling.

(a) *Sampling with replacement*: n items are sequentially chosen at random from Π. Each item chosen is replaced before the next is drawn. Thus an individual may be drawn more than once. In this case, $N = M^n$ is the number of ordered sequences (permitting repetitions) of length n from Π.

---

[2]From time to time, some familiarity with combinations as discussed in Section 10 will be required.

(b) *Sampling without replacement*: Again, n items are sequentially drawn at random from Π. Each item chosen is withdrawn from the population and the next is randomly selected from what remains. Thus no item can be drawn twice. In this case, $N = (M)_n = M!/(M - n)!$ is the number of ordered sequences (without repetitions) of length n from Π. If order is irrelevant, this procedure yields a randomly chosen subset of size n from Π. The number of subsets of size n from Π is $\binom{M}{n} = (M)_n/n!$.

In sampling without replacement, the size of the subset obtained is fixed and equal to the sample size. Each subset has the same probability of being drawn. Hence if order is irrelevant in the problem, it will generally be more convenient to count subsets rather than ordered sequences. On the other hand, in sampling with replacement the size of the subset (the number of distinct items in the sample) is a simple rv. Thus the subsets do not all have the same probability of being drawn and it is more convenient to count ordered sequences.

When we use the term *random sample of size n*, we will mean a sample drawn *without* replacement [i.e., case (b)] unless otherwise specified.

Example 9. Consider a poker hand. Let A be the event that the first two cards dealt are aces and the others are not aces. Let B be the event that the hand has two aces. The event A is relevant for stud poker. Otherwise, B is relevant. In finding $P(A)$ we must consider ordered poker hands, while in finding $P(B)$ order is irrelevant. Thus $P(A) = (4)_2(48)_3/(52)_5$ and $P(B) = \binom{4}{2}\binom{48}{3}\Big/\binom{52}{5}$. Computing with ordered poker hands yields

$$P(B) = (4)_2(48)_3\binom{5}{2}\Big/(52)_5 = \binom{4}{2}\binom{48}{3}(5)_5/(52)_5.$$

Example 10. (Continuous uniform model) Let $\mathcal{S}$ be a subset of $\mathcal{R}^2$ with positive, finite area. For any $A \subset \mathcal{S}$, let $V(A)$ be the area of A. Then $P(A) = V(A)/V(\mathcal{S})$. In particular, if $\mathcal{S}$ is the unit square and A is the interior of the circle of radius $1/2$, centered at $(1/2, 1/2)$, then $P(A) = \pi/4$.

Summary

Sample space: A set of mutually exclusive and exhaustive results of an experiment.
Outcome: Element of a sample space.

Probability model: A sample space, $\mathcal{S}$, and a rule assigning probabilities to subsets of $\mathcal{S}$.

Discrete case: $\mathcal{S}$ is countable.

Uniform case: $P(A) = $ (size of A)/(size of $\mathcal{S}$). If $\mathcal{S}$ has N elements, then each singleton has probability $1/N$.

Sampling with replacement: Each item chosen is replaced before the next is drawn.

Sampling without replacement: Each item chosen is withdrawn before the next is drawn. If n items are so drawn, they constitute a *random sample of size n*.

Events are represented by subsets of the sample space used.

Problems

For Problems 1 to 3, give an adequate sample space $\mathcal{S}$ for which random selection will be a reasonable model. For each event mentioned, identify the event with a subset of $\mathcal{S}$ and compute the probability of the event assuming random selection.

1. A poker hand (five cards) is drawn from an ordinary deck. Let A be the event that all five cards are spades.
2. A class contains 10 students, no two of which have the same height. Three of these students are selected and their heights are measured. Let A be the event that the students are selected in ascending order of heights. Let B be the event that the shortest student in the class is selected. Also consider $A \cup B$ and AB.
3. A coin is tossed five times. Let A be the event that heads and tails alternate. Let B be the event that there are more heads than tails. Let C be the event that all the heads are in succession and all the tails are in succession. Also consider $A \cup B$, AB, $A \cup C$, AC, $B \cup C$, BC, $A \cup B \cup C$, ABC, $A \cup (BC)$, $(AB) \cup C$, A^c, B^c, and C^c.
4. The population under study consists of all U.S. families that have exactly 3 children:

   Case 1: a study of the sex (M or F) of each child (first-born, second-born, third-born) in such families.
   Case 2: a study of the number of male children in such families.
   Case 3: a study of the age differences between consecutive children in such families.

   (a) List the elements of a sample space $\mathcal{S}_1$ suitable for Case 1.
   (b) List the elements of a sample space $\mathcal{S}_2$ suitable for Case 2.
   (c) Describe (as precisely as you can) a suitable sample space $\mathcal{S}_3$ for Case 3.

5. (Refer to Problem 4.) Consider the following events:

A = "Exactly two of the three children are female."
B = "First-born child is female."
C = "All three children are of the same sex."

(a) Which of the events A, B, C are *not* representable as events in $\mathcal{S}_2$? Explain.

For the remaining parts of this problem use $\mathcal{S}_1$, the sample space for Case 1, and list the elements in the followng events:

(b) A
(c) $(A \cup C)^c C$
(d) B^c
(e) $AB^c \cup A^c B$
(f) $A \cup B \cup C$
(g) $AB^c C^c \cup A^c BC^c \cup A^c B^c C^c$

6. For each pair of sample spaces in Example 4, is it possible to use one to obtain the information contained in the other? If so, how?
7. In Example 5, give the event E as a subset of $\mathcal{S}'$.
8. A fast-food chain used the following promotional game. The customer was given a card with nine spaces. Two spaces were marked "zap" and the other seven with prizes. Two of the prize spaces matched. The spaces were covered and the customer uncovered spaces one at a time until either exposing the matched prizes (customer wins) or one of the "zaps" (customer loses). What is the probability that the customer wins? Repeat this problem for a card with three "zaps."
9. A club has 10 members, of which 5 are liars, 5 are lawyers, and 2 are neither lawyers nor liars. Two members are chosen at random and without replacement. Find the probabilities that:
(a) Of the two, at least one is a lawyer.
(b) Of the two, at least one is a lawyer and at least one is a liar.

10. APPENDIX ON COMBINATORICS

We present, without proof, certain counting rules.

Multiplication Rule

Let A and B be sets containing m and n elements, respectively. The number of ordered pairs (a, b) with $a \in A$ and $b \in B$ is mn.

Addition Rule

For each $a \in A$, let B_a be a set with n_a elements and assume that A has m elements. The number of ordered pairs (a, b) with $a \in A$ and $b \in B_a$ is $\Sigma_a\, n_a$.

If each $n_a = n$, we obtain the multiplication rule as a special case.

Example 1.   Eight boxes each contain 20 computer chips. A box is chosen and a chip is chosen from that box. There are $8 \cdot 20 = 160$ ways of choosing a chip.

Example 2.   There are three boxes. The first has 20 computer chips, the second has 40, and the third has 60. A box is chosen and a chip is cosen from that box. There are $20 + 40 + 60 = 120$ ways of choosing a chip.

Generalized Multiplication Rule

For $i = 1, \ldots, r$, let A_i be a set containing m_i elements. The number of ordered r-tuples $(a_1, \ldots, a_r)$ with $a_i \in A_i$ is $\Pi_{i=1}^{r}\, m_i$.

The most frequently encountered applications of the generalized multiplication rule follow.

Sampling with Replacement

Let A be a set with m elements. If we select r elements out of A, replacing each chosen element before choosing the next, the number of ordered selections is m^r.

Permutations

Let A be a set with m elements. The number of ordered selections of $r \leq m$ elements of A without replacement (the number of ordered r-tuples consisting of elements of A and without repetitions) is

$$(m)_r = m(m - 1) \cdots (m - r + 1) = \frac{m!}{(m - r)!}, \tag{1}$$

called the *number of permutations of m things taken r at a time.*

In any expression involving factorials, set $0! = 1$. This leads to the correct value in combinatorial formulas; particularly in (1) when $r = m$ it follows that $(m)_m = m!$.

Combinations

Let A be a set with m elements. The number of subsets of A containing $r \leq m$ elements (the number of unordered selections of r elements of A without replacement) is

$$\binom{m}{r} = \frac{(m)_r}{(r)_r} = \frac{m!}{r!(m-r)!},\tag{2}$$

called the *number of combinations of m things taken r at a time.*

We adopt the convention $\binom{m}{r} = 0$ if $r < 0$ or $r > m$ provided that $m > 0$.

For $i = 1, \ldots, r$, let A_i be a set containing m_i elements. For each i, we select a subset of size k_i from A_i. There are $\Pi_{i=1}^r \binom{m_i}{k_i}$ ways of doing so.

Example 3. A course meets four times a week for 10 weeks. If each week a day is chosen for a quiz, there are 4^{10} different selections of the 10 quiz days.

Example 4. Michigan license plate numbers contain three letters and three digits. There are 26^3 ways of choosing the letters, of which $(26)_3 = 26!/23! = 26 \cdot 25 \cdot 24 = 15{,}600$ contain no repeated letters. There are 10^3 ways of choosing the digits, of which $(10)_3 = 10!/7! = 720$ contain no repeated digits. Thus there are $26^3 \cdot 10^3$ possible Michigan license plate numbers, of which $(26)_3 \cdot (10)_3 = (56{,}600)(720)$ have no repeated letters or digits. The proportion of possible Michigan license plate numbers containing no repeated letters or digits is $(56{,}600)(720)/(26^3 \cdot 10^3) = .639$.

Example 5. The number of poker hands (five-card hands) obtainable from an ordinary 52-card deck is $\binom{52}{5} = 52!/(47!5!) = 2{,}598{,}960$.

(a) To count the number of hands in which all cards are of the same suit (the hand has a flush), consider a two-stage operation. First choose the suit. There are $\binom{4}{1} = 4!/(3!1!) = 4$ ways of doing so. Next choose five cards of that suit. There are $\binom{13}{5} = 13!/(8!\,5!) = 1287$ possible choices. Thus there are $\binom{4}{1}\binom{13}{5} = (4)(1287)$ hands with all cards of the same suit and the proportion of such hands is $\binom{4}{1}\binom{13}{5}\Big/\binom{52}{5} =$ (4)(1287)/2,598,960 = .00198.

(b) To construct a hand with a full house (three of a kind and a pair), consider a three-stage operation. First choose two values in an ordered way. There are $(13)_2$ ways of doing so. There $\binom{4}{3}$ ways of choosing three cards of the first value selected and $\binom{4}{2}$ ways of choosing two cards of the other value. Thus there are $(13)_2\binom{4}{3}\binom{4}{2}$ full-house hands and the proportion of full-house hands is $(13)_2\binom{4}{3}\binom{4}{2}\Big/\binom{52}{5}$.

(c) Suppose that the ace and jack of spades are missing from the deck. Consider the number of hands from this deck which have two pairs (i.e., of the form $xxyyz$). The number of such hands with neither a pair of aces nor a pair of jacks is $\binom{11}{2}\binom{4}{2}^2\binom{42}{1}$; with either (but not both) a pair of aces or a pair of jacks is $\binom{11}{1}\binom{4}{2}\binom{2}{1}\binom{3}{2}\binom{43}{1}$; with both a pair of aces and a pair of jacks is $\binom{2}{2}\binom{3}{2}^2\binom{44}{1}$. The number of two-pair hands is the sum of these three terms which, when divided by $\binom{50}{5}$, yields the proportion of two-pair hands.

Recall that $\binom{m}{r}$ is the number of subsets of size r that can be formed from a set A containing m elements. Thus $\binom{m}{r}$ is the number of ways of partitioning A into a subset containing r elements and a complementary subset containing $m - r$ elements.

Partitioning

Let A be a set containing m elements. The number of ways of partitioning A into k subsets in which the ith subset has r_i elements ($\sum_{i=1}^{k} r_i = m$) is

$$\binom{m}{r_i}\binom{m-r_1}{r_2}\cdots\binom{m-r_1-\cdots-r_{k-1}}{r_k} = \frac{m!}{\prod_{i=1}^{k} r_i!}. \qquad (3)$$

Example 6. For the game of bridge, the 52-card deck is dealt so that each of the four players has 13 cards. This amounts to partitioning the deck into four subsets of 13 cards each. There are

$$\binom{52}{13}\binom{39}{13}\binom{26}{13}\binom{13}{13} = \frac{52!}{(13!)^4}$$

ways of doing so.

Note that (2) is a special case of (3) with $k = 2$, $r_1 = r$, and $r_2 = m - r$. For this reason, the quantity in (3) is often denoted by $\binom{m}{r_1, \ldots, r_{k-1}}$ (or, alternatively, $\binom{m}{r_1, \ldots, r_k}$).

The following theorem may be proved by using the counting techniques developed above.

Theorem 1. (Binomial formula) *For any numbers a and b and any non-negative integer n,*

$$(a + b)^n = \sum_{k=0}^{n} \binom{n}{k} a^k b^{n-k}. \qquad (4)$$

Because of (4), the $\binom{n}{k}$ are also called *binomial coefficients*. Some elementary consequences of (4) are

$$\sum_{k=0}^{n} \binom{n}{k} = 2^n; \qquad (5)$$

$$\sum_{k=0}^{n} (-1)^k \binom{n}{k} = 0; \qquad (6)$$

and

$$\sum_{k=0}^{n} \binom{n}{k} p^k q^{n-k} = 1 \qquad \text{whenever } p + q = 1. \tag{7}$$

We will make extensive use of (7).

Note the useful symmetry property

$$\binom{n}{k} = \binom{n}{n-k}. \tag{8}$$

Another useful relation is the Pascal triangle formula,

$$\binom{n}{k} + \binom{n}{k-1} = \binom{n+1}{k}, \tag{9}$$

which generalizes to

$$\sum_{j=0}^{k} \binom{m}{j} \binom{n}{k-j} = \binom{m+n}{k}. \tag{10}$$

Example 7. A drawer contains n pairs of socks, but the $2n$ individual socks are thoroughly mixed. Suppose that r socks are taken from the drawer. What is the probability that there is at least one pair among the r socks?

Let A be the event that there is at least one pair. If $r = 2, 3$, there is little difficulty in expressing A as a very simple event. For larger r, however, A is more complicated. If $r = 10$, for example, then $A = \cup_{i=1}^{5} B_i$ is the event that there are exactly i pairs.

Observing that A^c is the event that there are no pairs among the r socks, we can easily solve this problem. Let $\mathcal{C}$ denote the partition determined as follows: Each event in $\mathcal{C}$ specifies which r socks are selected out of the $2n$ available. The number of events in $\mathcal{C}$ is the number of subsets of size r obtainable from a set of size $2n$, namely $\binom{2n}{r}$.

Further, we assume that all members of $\mathcal{C}$ have the same probability, namely $\binom{2n}{r}^{-1}$. The probability of any event that is the union of k members of $\mathcal{C}$ has probability $k \Big/ \binom{2n}{r}$.

The event A^c is the union of all members of $\mathcal{C}$ that specify no two socks in the same pair. We can determine how many such members there are in

$\mathcal{C}$ by the following selection argument. We first select r of the n pairs. There are $\binom{n}{r}$ ways of doing this. From each of these r pairs, we must select one of the two socks. There are 2^r ways of doing this. Thus $\binom{n}{r} 2^r$ of the events in $\mathcal{C}$ specify collections of socks with no pair.

Thus $P(A^c) = \binom{n}{r} 2^r \Big/ \binom{2n}{r}$. Then $P(A) = 1 - P(A^c) = 1 - \binom{n}{r} 2^r \Big/ \binom{2n}{r}$.

Problems

Except where otherwise specified, solutions of these problems in terms of combinations, permutations, or powers will suffice.

1. If six people play poker, in how many ways can their hands be dealt? If, for example, we interchange two players' hands, this is to be counted separately.
2. A certain device contains five identical components. A box contains 30 of these components, of which seven are defective. The device is to be made by choosing components at random.
   (a) If the components are in series in the device, find the probability that it functions.
   (b) What would your solution in part (a) be if the components are in parallel in the device?
3. Find the probabilities of the following poker hands:
   (a) Three of a kind (excluding full houses).
   (b) Straight flush (including royal flush).
4. Suppose that your poker hand contains four tens and the king of spades. You have one opponent. What is the probability that your opponent has a better hand (a higher-ranking four of a kind or a straight flush)? *Note*: Consider your opponent's hand as randomly selected from the 47 remaining cards.
5. Suppose that the precinct in which you vote has 1000 registered voters. A sample of 100 is chosen. Solve (a) and (b) for sampling both with and without replacement. In the case of sampling without replacement, reduce the answers to simple fractions.
   (a) What is the probability that you are chosen?
   (b) Suppose that your roommate is also registered to vote in the precinct. What is the probability that at least one of you is chosen? That both of you are chosen?
6. You belong to a club with 20 members. The officers (president, vice-

president, secretary, and treasurer) are selected annually at random from the full membership. Officers may succeed themselves any number of times. What are the probabilities that during a given four-year period:

(a) You serve as president at least once?

(b) You hold office at least once?

7. Prove (2). *Hint*: How many ordered r-tuples (without repetition) are there of the subset of size r?

8. From the integers $\{1, \ldots, m\}$ two are to be selected in order and without replacement. Once the first integer has been chosen, the second cannot be an adjacent integer. In how many ways can the selections be done?

9. Prove (5) by counting subsets.

10. Prove (9) and (10) by sampling arguments.

11. Use a sampling argument to verify that $\sum_{m=r}^{n} \binom{m}{r} = \binom{n+1}{r+1}$.

12. In Example 7, what is the probability that there are at least two pairs among the r socks?

13. Let m and n be positive integers and arrange the integers from 1 to $m + n$ in ascending order. Select m of these integers at random and label them A; the remaining n are labeled B.

(a) How many different orderings of A's and B's are possible?

(b) Assume that $m > 1$ and find the probability that 1 is an A and 2 is a B.

(c) Let $m = 2$, $n = 3$ and let i and j be the A's. For each value of $i + j$, find its probability.

(d) Repeat part (c) for $m = n = 3$.

14. Show that $\binom{r}{k}\binom{N-r}{n-k} \bigg/ \binom{N}{n} = \binom{n}{k}\binom{N-n}{r-k} \bigg/ \binom{N}{r}$. Give a sampling interpretation of this equation.

GENERAL PROBLEMS

1. In a community, suppose that 85% of the families have radios, 70% have television sets, and 10% have neither. What percentage have both? If a disaster occurs that eliminates telephone lines, we may be interested in the proportion of the population who can be reached by either radio or television announcements. Find the desired proportion.

2. Two anticancer drugs, I and II, are to be tested. Three patients are administered drug I and three other patients are administered drug II. We are interested in the event A that more cures are obtained from I than from II.
   (a) Construct a model in which each drug has a 50% chance of effecting a cure on any patient and find the probability of A.
   (b) Let X and Y be the numbers of cures obtained from drugs I and II, respectively. Find $E(X - Y)$.
3. Consider the population consisting of all voters registered in the United States. Take the usual probability model for population proportions. Give a partition of the sample space consisting of at least five events and having some political significance.
4. The population described below consists of 100 worn-out car batteries. The information tabled gives the number of batteries with life length L (in months) for each of three brands X, Y, and Z.

L:	34	35	36	37	38
X	6	9	10	8	5
Y	6	7	14	8	7
Z	3	4	6	4	3

Brands X and Y are expensive ($40); brand Z is inexpensive ($25).
   (a) What probability model would be used to describe life length for batteries in this population?
   (b) What probability model would be used to describe the brand proportions in this population?
   (c) What probability model would be used to describe the life length of expensive batteries in this population?
   (d) Determine the probability that a battery is inexpensive or has a life length of 36 or more months.
   (e) Determine the probability that a battery is expensive and lasts 36 or more months.
5. Consider the population of batteries described in Problem 4. Assume that a random sample of size $n = 5$ is to be chosen from this population (*with replacement*). Define the following events.

   A_1: The sample contains three expensive and two inexpensive batteries.

   A_2: None of the sampled batteries have a life length exceeding 36 months.

   A_3: All brands are represented in the sample.

A_4: The sample contains at least one brand Z battery with life length 36 months.

A_5: The sample contains exactly two inexpensive batteries and their average life length is 37 months.

Determine the probabilities of each of events A_1 to A_5.

6. Repeat Problem 5 for the case of sampling without replacement.

7. What is the probability that k people all have distinct birthdays? Assume *that* each year has 365 days and the birthdays are assigned at random with replacement.

8. In the carnival game chuck-a-luck you pay a $1 to play. You roll three dice. If k sixes result ($k > 0$), you will be paid $\$(k + 1)$. Otherwise, you will be paid 0. Let X be your net gain.
   (a) Let $A_i = [X = i]$. Write X as a linear combination of the I_{A_i}.
   (b) Assuming that the outcomes when the three dice are rolled are equally likely, give the $P(A_i)$ and $E(X)$.
   (c) Suppose that the rules are changed so that you receive a bonus if $k = 3$. What should the bonus be so that $E(X) = 0$?

9. From $\{1, \ldots, N\}$ we choose at random a subset containing n elements. Let X be the smallest member of the subset.
   (a) For each k, find $P[X \geq k]$.
   (b) Determine $E(X)$.
   (c) Let Y be the largest member of the subset and $Z = N + 1 - X$. Use a symmetry argument to show that $E(Y) = E(Z)$ and then evaluate $E(Y)$.

10. A population π contains 10 individuals. Three are drawn at random, with replacement. Let X be the size of the subset of π that is actually drawn.
   (a) Find the probabilities of the events in the partition induced by X.
   (b) Find $E(X)$.

11.† Let A be any event with $0 < P(A) < 1$. The *odds for* and *against A* are $P(A)/P(A^c)$ and $P(A^c)/P(A)$, respectively. When the odds for A are a/b (usually, a and b are integers with no common factors) we say that the *odds for A are a to b* (see Problem 8.1). Consider a game in which you bet d and will win c if A occurs. Let W be your net gain. In each part below find d as a function of odds.
   (a) Find d to make the game fair $[E(W) = 0]$.
   (b) Suppose that you are playing at a gambling house in which the house will win an average of $100p\%$ of all money bet. Solve part (a) in this case.
   (c) Resolve parts (a) and (b) if, instead of being the amount bet, d is the cost of a lottery ticket.

REFERENCES

de Finetti, Bruno (1964). Foresight: its logical laws, its subjective sources, in *Studies in Subjective Probability*, Henry E. Kyburg, Jr., and Howard E. Smokler, eds. New York: John Wiley and Sons, Inc., pp. 93–158. [Translation by Henry E. Kyburg, Jr., of La prévision: ses lois logiques, ses sources subjectives, *Annales de l'Institut Henri Poincaré* **7** (1937), 1–68.] Proof that subjective probabilities derived from rationality assumptions must satisfy the probability axioms.

DeGroot, Morris H. (1970). *Optimal Statistical Decisions*. New York: McGraw-Hill Book Company. Chapters 6 and 7 contain assumptions leading to personal probabilities.

Feller, William (1968). *An Introduction to Probability Theory and Its Applications*, Vol. 1, 3rd ed. New York: John Wiley and Sons, Inc. Chapter I: The development of the model for countable sample spaces. Chapter II: A more complete and more analytic development of combinatorics with many varied and interesting examples and problems.

Hoel, Paul G., Port, Sidney C., and Stone, Charles J. (1971). *Introduction to Probability Theory*. Boston: Houghton Mifflin Company. Chapter 1: The development of the model for general sample spaces. Chapter 2: Combinatorics with some interesting examples and problems (fewer than Feller).

Parzen, Emanuel (1960). *Modern Probability and Its Applications*. New York: John Wiley and Sons, Inc. Detailed analysis of combinatoric problems and their applications. A version of Corollary 5.1 with every a_i equal to ± 1 (pp. 81–83).

2.

Conditioning and Independence

11. SURE AND IMPOSSIBLE EVENTS

The treatment of a given event as sure or impossible may depend on assumptions or the acquisition of new information.

Example 1. In considering a coin toss, it is usually assumed that the event A that the coin stands on end is impossible. We make this assumption because in our experience coins have not stood on end. If, however, the coin is unusually thick, we may assume, instead, that A is, after all, possible.

Example 2. Consider the population consisting of all college students. In selecting a member of this population, the event A that "the student is an undergraduate" is neither sure nor impossible. If the selection is restricted to an undergraduate, then A is sure, whereas if it is restricted to graduate students, then A is impossible.

Example 3. The event A that "the MSU football team will win its opening game next year" is neither sure nor impossible. The day after that football Saturday, this event will become either sure or impossible and must be considered as such in any future assignment of probabilities.

Conditionally Sure, Impossible

If A and B are events, then A is *sure given B* if $A \supset B$ (so that $AB = B$) and A is *impossible given B* if $A \subset B^c$ (so that $AB = \phi$). In the first case, for example, if we know that B has occurred, then A must occur.

Logically Sure, Impossible

Events that remain sure or impossible no matter how assumptions or information may be modified are *logically sure* or *logically impossible*, respectively.

Example 4. If A is any event, then $A \cup A^c$ is logically sure and AA^c is logically impossible.

We will develop the notion of the conditional probability of an event in the next section. If A is sure given B, then whatever $P(A)$ may be, the conditional probability assigned to A given B will be 1. If A is impossible given B, this conditional probability will be 0.

12. CONDITIONAL PROBABILITY AND CONDITIONAL MODELS

This section is about the way that information is used to revise probabilities.

Example 1. A patient undergoes a clinical test, the results of which are subject to error. Suppose that the result indicates that the patient suffers from disease X. How likely is it that the patient actually suffers from X? What information is needed to answer this question?

Example 2. Thunder and Lightning are horses entered in a 10-horse race. We have a side bet. I will win if Thunder wins the race and you will win if Lightning wins the race. The bet is called off if neither wins. What are the proper odds for this bet?

How does the bet (Example 2) compare with a bet on a partisan election outcome in which party nominations have not yet been made? In connection with such problems, it will be useful to define the concept of "the conditional probability of an event A given an event B," which we denote by $P(A \mid B)$.

What properties should $P(A \mid B)$ have? First, since $P(A \mid B)$ is to be a probability, we require that (5.4) be satisfied; that is, for a fixed event, B,

(a) $0 \le P(A \mid B) \le 1$ for any event A;
(b) $P(\Omega \mid B) = 1$ and $P(\phi \mid B) = 0$;   (1)
(c)   If $\{A_1, A_2, \ldots\}$ is a countable partition of A, then $P(A \mid B) = \sum_{n=1}^{\infty} P(A_n \mid B)$.

Two additional requirements are $P(A \mid B) = 1$ for any event A, which is sure given B, and $P(A \mid B) = 0$ for any event A, which is impossible given B. In particular,

$$P(B \mid B) = 1 \quad \text{and} \quad P(B^c \mid B) = 0. \tag{2}$$

Let $\{A_1, A_2\}$ be a partition of B. Using additivity and (2) yields

$$P(A_1 \mid B) + P(A_2 \mid B) = P(B \mid B) = 1. \tag{3}$$

We will also require that for this partition,

$$\frac{P(A_1 \mid B)}{P(A_2 \mid B)} = \frac{P(A_1)}{P(A_2)} \tag{4}$$

whenever $P(A_2) \ne 0$. The following example may help explain the meaning and necessity of (4).

Example 3.   Consider Example 2. Let A_1 be the event "Thunder wins," A_2 be the event "Lightning wins," and $B = A_1 \cup A_2$. Let $P(A_1)$ and $P(A_2)$ be the probabilities of A_1 and A_2, respectively, evaluated at post time. Either of these probabilities may be large or small. The probability that at least one of these horses wins is $P(B) = P(A_1) + P(A_2)$ which may be large or small, depending on the quality of the remaining horses relative to Thunder and Lightning.

Suppose that $P(A_1) = P(A_2)$, meaning that Thunder and Lightning are equally likely to win. If we are informed that one of these two horses did, in fact, win the race, the added information has no effect on the relative magnitudes of the probabilities and hence $P(A_1 \mid B) = P(A_2 \mid B)$.

Similarly, if $P(A_1) = 2P(A_2)$, the knowledge that B has occurred should leave the conditional probability of A_1 twice that of A_2, that is, $P(A_1 \mid B) = 2P(A_2 \mid B)$. Generally, if $P(A_1) = tP(A_2)$, then, necessarily $P(A_1 \mid B) = tP(A_2 \mid B)$, which, with (3), requires that $P(A_1 \mid B) = t/(t + 1)$.

Returning to the general case, solve for $P(A_2 \mid B)$ in (4) and use (3) to obtain

$$P(A_1 \mid B) + \frac{P(A_2)}{P(A_1)} P(A_1 \mid B) = 1$$

or, equivalently,

$$P(A_1 \mid B) = \frac{P(A_1)}{P(A_1) + P(A_2)} = \frac{P(A_1)}{P(B)}.$$

Thus for any $A \subset B$ it follows that

$$P(A \mid B) = \frac{P(A)}{P(B)} \tag{5}$$

provided that $P(B) \neq 0$.

In summary, the conditional probability model given the occurrence of the event B is constructed by rescaling the probability of each subevent of B by the factor $1/P(B)$. From (2), events disjoint from B are assigned conditional probability zero.

Let A be an arbitrary event, not necessarily a subevent of either B or B^c. Since $\{AB, AB^c\}$ is a partition of A, and $AB^c \subset B^c$, it follows by additivity that

$$P(A \mid B) = P(AB \mid B) + P(AB^c \mid B) = P(AB \mid B). \tag{6}$$

Since $AB \subset B$, applying (5) and (6) yields the usual definition:

Conditional Probability

For any event B for which $P(B) > 0$, *the conditional probability of A given B is*

$$P(A \mid B) = \frac{P(AB)}{P(B)}. \tag{7}$$

Observe that $P(A \mid B)$ satisfies (1) and, furthermore, $P(A \mid B) = 1$ if A is sure given B and $P(A \mid B) = 0$ if A is impossible given B. Starting with a probability model using e and P, there results a conditional probability model given B using the same e.

In Problem 10 the reader is asked to relate the definition of conditional probability to the interpretations of probability given in Chapter 1.

Example 4. Two fair dice are rolled. Suppose that the sum of the points on the dice is known to be even. What is the conditional probability that the sum is divisible by 3? By 5? Let A_k be the event that the sum is divisible by k. Then $P(A_3 \mid A_2) = P(A_2 A_3)/P(A_2) = P(A_6)/P(A_2)$ and $P(A_5 \mid A_2) = P(A_2 A_5)/P(A_2) = P(A_{10})/P(A_2)$. There are 18 out of the 36 possible outcomes that result in the occurrence of A_2 so that $P(A_2) = 18/36 = 1/2$. Six of the outcomes result in the occurrence of A_6, so $P(A_6) = 6/36 = 1/6$ and hence $P(A_3 \mid A_2) = (1/6)/(1/2) = 1/3$. Three of the outcomes result in the occurrence of A_{10}, so $P(A_{10}) = 3/36 =$

1/12 and $P(A_5 \mid A_2) = (1/12)/(1/2) = 1/6$. Note that $P(A_3) = 1/3$ and $P(A_5) = 7/36$.

When $P(B) > 0$, (7) yields a relationship between the unconditional model and the conditional model given B. The following example illustrates that the construction of a conditional model given B may be possible even when $P(B) = 0$.

Example 5.† Consider a circular target of radius r and let the origin be the target's center. Suppose that a marksman throws a dart at the target in such a way that the uniform model is appropriate with the target taken as the sample space $\mathcal{S}$. Thus for any $A \subset \mathcal{S}$, we have $P(A) = \text{area}(A)/(\pi r^2)$. In particular, if B is a chord of the target, then $P(B) = 0$ (see Figure 1).

The toss by the marksman produces a pair of coordinates (x, y) as measured from 0. Assume that the value of the x-coordinate becomes known. This amounts to learning that B has occurred. Since $P(B) = 0$, equation (7) cannot be used to construct a conditional model for the y-coordinate.

However, uniformity of P on $\mathcal{S}$ suggests that a reasonable conditional model will be uniform on the set of possible values of the y-coordinate determined by the specified x-coordinate. That is, the conditional model should be uniform on the interval $(-\sqrt{r^2 - x^2}, \sqrt{r^2 - x^2})$. Thus a con-

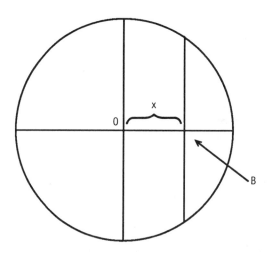

Figure 1

ditional model given B has been obtained even though $P(B) = 0$. The result can be justified by certain limit arguments furnished in a later section.

For any event B, equation (7) can be rewritten to yield the important relation[1]

$$P(AB) = P(A \mid B)P(B) = P(B \mid A)P(A). \tag{8}$$

Equation (8) provides a very useful method of obtaining probabilities of intersections when appropriate conditional probabilities are available. Formula (8) is called the *multiplication* or *chain rule* and extends to intersections of more than two events.

Let A, B, C be events. Then, applying (8) twice,

$$P(ABC) = P(A \mid BC)P(BC) = P(A \mid BC)P(B \mid C)P(C). \tag{9}$$

Alternatively, by reordering, $P(ABC) = P(C \mid AB)P(B \mid A)P(A)$, and so on. The ordering used depends on the conditional information available in the problem.

For any events $A_1, \ldots, A_n$,

$$P\left(\bigcap_{i=1}^{n} A_i\right) = P(A_1)P(A_2 \mid A_1) \cdots P(A_n \mid A_1 \cdots A_{n-1}). \tag{10}$$

Example 6. A population consists of n_i items of type i for $i = 1, 2, 3, 4, 5$. A sequence of seven items is to be chosen without replacement. The probability of observing the sequence of types 2, 3, 2, 5, 1, 1, 2 (i.e., the first item drawn is of type 2, the second of type 3, etc.) is

$$\frac{n_2}{n} \cdot \frac{n_3}{n-1} \cdot \frac{n_2 - 1}{n-2} \cdot \frac{n_5}{n-3} \cdot \frac{n_1}{n-4} \cdot \frac{n_1 - 1}{n-5} \cdot \frac{n_2 - 2}{n-6}.$$

Would this probability change if we change the order (but not the types) in the sequence?

Conditional arguments are particularly useful in multistage experiments, of which the preceding is an example. Here is another.

Example 7. Suppose that a city has M voting precincts and that the ith precinct contains n_i voters of whom k_i intend to vote for candidate X for mayor. A precinct is chosen and, from it, a random sample of size r is

---

[1]We adopt the convention (to be used throughout) that anything undefined, such as perhaps $P(A \mid B)$, multiplied by zero yields zero.

chosen. What is the probability that a majority of this sample favor candidate X?

Example 7 involves a two-stage process. The probability models for the second stage given the result of the first stage are specified in the description of the experiment and can be used to determine the overall probability model.

Since $\{AB, AB^c\}$ is a partition of A, applying additivity and the multiplication rule yields

$$P(A) = P(AB) + P(AB^c) = P(A \mid B)P(B) + P(A \mid B^c)P(B^c). \quad (11)$$

Generalizing (11), let $\{B_1, B_2, \ldots\}$ be a finite or countably infinite partition. Then $\{AB_1, AB_2, \ldots\}$ is a partition of A, so that

$$P(A) = \sum_n P(AB_n) = \sum_n P(A \mid B_n)P(B_n). \quad (12)$$

Equation (12) establishes a correspondence between the unconditional model and conditional models given the members of some countable partition. Formula (12) is called the *total probability law*.

Example 8. A certain community has three supermarkets described by the data in Table 1. With A and the B_i as defined in Table 1,

$$P(A) = (.35)(.2) + (.25)(.3) + (.40)(.1) = .185.$$

That is, 18.5% of the community purchases A-Cola.

Trees are useful tools in problems involving use of the multiplication rule or the total probability law. The tree for Example 8 is illustrated in Figure 2. Each branch of a tree represents some event. Moving from the initial vertex, O, outward, the path followed represents the intersection of the events along that path. Thus the path consisting of the branch B_2

Table 1

Market	Percent of community shopping at market i	Percent of customers purchasing A-Cola
1	$35 = 100P(B_1)$	$20 = 100P(A \mid B_1)$
2	$25 = 100P(B_2)$	$30 = 100P(A \mid B_2)$
3	$40 = 100P(B_3)$	$10 = 100P(A \mid B_3)$

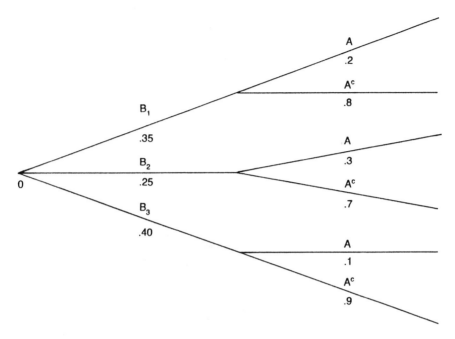

Figure 2

followed by the branch A^c represents B_2A^c. The requirements of a tree diagram are:

(a)   The probability assigned to a branch is the conditional probability of the event represented by the branch given the intersection of every event along the path leading to that branch.
(b)   The events represented by branches from any vertex form a partition (omit branches with zero conditional probability).

The probability of any path is, then, the product of the probabilities of the branches along that path. The probability of any event A is the sum of probabilities of paths leading to A.

Trees may extend to more than two levels but need obey no regularity or symmetry conditions as regards vertices, branches, or lengths of paths. Some possibilities are illustrated in the following two examples. Random sampling without replacement is assumed in both.

Example 9.   A box has one red, two green, and three white balls. Three balls are drawn without replacement. Let R_i be the event that the ith ball drawn is red, and so on. The tree is given in Figure 3.

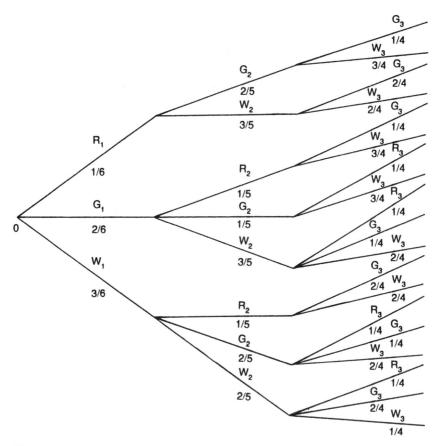

Figure 3

Example 10. You have five keys, one of which will open your door. You try one at a time at random until you find that key. Let A_i be the event that the correct key is the ith tried. The tree is given in Figure 4.

No events other than sure and impossible events have probabilities determined by the rules (5.4). It was shown in Section 11 that even statements that events are sure or impossible may be determined by assumptions or information. In fact, any event is assigned a probability depending on assumptions and/or information. It should be clear, then, that every probability model is, essentially, a conditional model. It is conditional on assumptions and/or information that may involve some degree of uncertainty and subjective judgment.

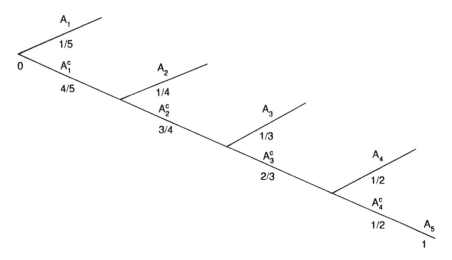

Figure 4

Since, for fixed B, $P(A \mid B)$ applied to events in $\mathcal{C}$ yields a probability model, the results of Sections 5 and 6 are valid for conditional probabilities.

Conditional Expectation

Consider the simple rv $X = \Sigma_{i=1}^{n} a_i I_{A_i}$. The *conditional expectation* of X given the event B is

$$E(X \mid B) = \sum_{i=1}^{n} a_i P(A_i \mid B)$$

provided that $P(A_i \mid B)$ is defined for $i = 1, \ldots, n$. All the properties we obtained in Section 7 for $E(X)$ apply, as well, to $E(X \mid B)$. In particular, by Theorem 7.2, if $X = \Sigma_{j=1}^{m} c_j I_{C_j}$ is another representation of X, then $E(X \mid B) = \Sigma_{j=1}^{m} c_j P(C_j \mid B)$.

A generalization of the total probability law (12) is

$$E(X) = \sum_{n} E(X \mid B_n) P(B_n) \tag{13}$$

for any simple rv X and countable partition $\{B_1, B_2, \ldots\}$.

Example 11. Urn 1 contains seven black and three white balls, urn 2 contains five black and five white balls, and urn 3 contains four black and six white balls. Two balls are to be drawn at random and without replacement from one of the urns. First a fair die is rolled. The balls are drawn from urn 1 if the result is 1, from urn 2 if the result is 2 or 3, and from urn 3 otherwise. Letting U_n be the event the balls are drawn from urn n, then $P(U_1) = 1/6$, $P(U_2) = 2/6$, and $P(U_3) = 3/6$. Let B_i be the event that the ith ball drawn is black ($i = 1, 2$) and $X = I_{B_1} + I_{B_2}$ be the number of black balls drawn. Then

$$P(B_1 \mid U_1) = P(B_2 \mid U_1) = 7/10;$$

$$P(B_1 \mid U_2) = P(B_2 \mid U_2) = 5/10;$$

$$P(B_1 \mid U_3) = P(B_2 \mid U_3) = 4/10,$$

so that $E(X \mid U_1) = 7/5$, $E(X \mid U_2) = 5/5$, and $E(X \mid U_3) = 4/5$. From (13),

$$E(X) = \left(\frac{7}{5}\right)\left(\frac{1}{6}\right) + \left(\frac{5}{5}\right)\left(\frac{2}{6}\right) + \left(\frac{4}{5}\right)\left(\frac{3}{6}\right) = \frac{29}{30}.$$

Compute $E(X)$ by computing the probabilities of the events in the partition induced by X. These events are $C_i = [X = i]$ ($i = 0, 1, 2$). Then $C_0 = B_1^c B_2^c$, $C_1 = B_1 B_2^c \cup B_1^c B_2$, and $C_2 = B_1 B_2$. Their conditional probabilities given the U_n are

$$P(C_0 \mid U_1) = \frac{\binom{3}{2}}{\binom{10}{2}} = \frac{3}{45} \qquad P(C_1 \mid U_1) = \frac{\binom{7}{1}\binom{3}{1}}{\binom{10}{2}} = \frac{21}{45}$$

$$P(C_2 \mid U_1) = \frac{\binom{7}{2}}{\binom{10}{2}} = \frac{21}{45}$$

and similarly, $P(C_0 \mid U_2) = 10/45$, $P(C_1 \mid U_2) = 25/45$, $P(C_2 \mid U_2) = 10/45$, $P(C_0 \mid U_3) = 15/45$, $P(C_1 \mid U_3) = 24/45$, and $P(C_2 \mid U_3) = 6/45$. Applying (12), $P(C_0) = (3/45)(1/6) + (10/45)(2/6) + (15/45)(3/6) = 68/270$, and similarly, $P(C_1) = 143/270$ and $P(C_2) = 59/270$. It follows that $E(X) = 143/270 + 2(59/270) = 29/30$.

Summary

Conditional probability: $P(A \mid B) = P(AB)/P(B)$ if $P(B) > 0$.
Conditional expectation: If $X = \sum_{n=1}^{k} a_n I_{A_n}$, then $E(X \mid B) = \sum_{n=1}^{k} a_n P(A_n \mid B)$ provided that $P(B) > 0$.

In general, $P(A \mid B)$ is a revised value for $P(A)$ assuming the occurrence of B. The formula above gives an explicit way of calculating $P(A \mid B)$ when $P(B) \neq 0$. Similar remarks apply to $E(X \mid B)$.

Multiplication rule:

$$P\left(\bigcap_{i=1}^{n} A_i\right) = P(A_1) \prod_{i=2}^{n} P\left(A_i \mid \bigcap_{j=1}^{i-1} A_j\right).$$

Total probability law: Given a countable partition $\{B_1, B_2, \ldots\}$,

$$P(A) = \sum_{n} P(A \mid B_n)P(B_n),$$

which generalizes to

$$E(X) = \sum_{n} E(X \mid B_n)P(B_n).$$

Problems

1. Two fair dice are rolled. Let A be the event that the sum of the faces is even, B be the event that the sum of the faces is divisible by three, and C be the event that the sum of the faces is divisible by 5.
   (a) Find $P(A \mid B)$ and $P(A \mid C)$ and compare each with $P(A)$.
   (b) Find $P(B \mid C)$ and compare with $P(B)$.
   (c) Find $P(C \mid B)$ and compare with $P(C)$.
2. Let $\mathcal{S}$ be the triangle in the plane with vertices $(0, 0)$, $(0, 1)$, and $(1, 0)$. Let A be the triangle with vertices $(0, 0)$, $(0, 1/2)$, and $(1/2, 0)$. Let B be the subset of $\mathcal{S}$ with $x < 1/4$. For the uniform model on $\mathcal{S}$:
   (a) Find $P(A \mid B)$ and compare with $P(A)$.
   (b) Find $P(B \mid A)$ and compare with $P(B)$.
3.† Let $P(A)$, $P(B) > 0$. We say that B is *favorable* (*unfavorable*) to A if $P(A \mid B) > P(A)$ $[P(A \mid B) < P(A)]$.
   (a) Show that B is (un)favorable to A if, and only if, A is (un)favorable to B.
   (b) Show that if B is favorable to A, then B^c is unfavorable to A.
   (c) Is it true that A (un)favorable to B and B (un)favorable to C implies that A is (un)favorable to C? Prove or give a counterexample.
4.† Show that $P(A \mid BC) = P(AB \mid C)/P(B \mid C)$ if $P(BC) \neq 0$.
5. Suppose that a population consists of 100 students registered in a course. They are classified by sex and grade point average in Table 2. Assume that one student is chosen at random from this population.

Table 2

	Less than 3.0	At least 3.0
Males	45	25
Females	5	15

Suppose that a lottery pays \$1 if the randomly chosen student has grade point average at least 3.0.
  (a) What is the worth of a ticket in this lottery?
  (b) Suppose that it is known that the student is a female. Has the worth of the ticket changed? If so, what is the new worth?

6. Four integers are successively drawn at random and without replacement from the set $\{1, \ldots, 10\}$.
  (a) What is the probability that the integers are drawn in their natural order (smallest to largest)?
  (b) What is the probability that the integers drawn are consecutive?
  (c) What is the probability that the fourth integer drawn is even given that the first one drawn was odd?
  (d) What is the probability that the largest integer drawn is greater than 8 given that the smallest integer is 2?

7. A box has five fair dice. The nth die has faces numbered $1, \ldots, n + 1$.
  (a) A die is chosen at random and rolled. Let X be the face that results. Find $E(X)$.
  (b) Let Y be the sum of the faces that result from two rolls. Compute $E(Y)$ for the following cases:
    (i) Both rolls are from the same die, selected at random.
    (ii) Each roll is from a die selected at random with replacement.
    (iii) As in (ii), but the dice are selected without replacement.

8. Prove (10).

9. Prove (13). Why is (13) a generalization of (12)?

10. Interpret $P(A \mid B)$ in each of the following cases:
  (a) Probabilities are population proportions.
  (b) Probabilities are relative frequencies.
  (c) Probabilities are subjective.

11.† Let A and B be events with probabilities $P(A)$ and $P(B)$. Show that the value of $P(A \mid B)$ determines the probability of every event in the partition generated by $\{A, B\}$.

12.† Suppose that $P(A) > 0$ and $P(A \mid B) = P(A)$. Show that $P(B \mid A) = P(B)$.

13. EXTENSIONS AND APPLICATIONS OF CONDITIONING

Let $\mathcal{S}$ contain N elements, P be uniform on $\mathcal{S}$, and $B \neq \phi$. Then

$$P(A \mid B) = \frac{P(AB)}{P(B)} = \frac{n(AB)/N}{n(B)/N} = \frac{n(AB)}{n(B)},$$

where $n(A)$ is the number of elements in A. Thus the conditional probability model given B must be the equally likely model on B. That is, the conditional probability model given B assigns $1/n(B)$ to each element of B and 0 to each element of B^c.

Example 1. A certain manufactured item is subject to two types of defects: one is external (visible) and the other is internal and can only be detected by destroying the item. Suppose that among 50 such items, 30 are good, 15 have both types of defects, and the remaining 5 have only the internal type. If, from this batch, a sample of size 12 is selected and it is found that exactly 3 of these have visible defects (the event B), what is the conditional probability of A that among the other 9 there is at least one with an internal defect?

Given B, the remaining 9 sampled items must have come from the 35 that are either good (30) or have only internal defects (5). There are $\binom{35}{9}$ subsets of size 9 and $\binom{30}{9}$ of them are without defective items. Hence $P(A \mid B) = 1 - \binom{30}{9} / \binom{35}{9}$. The same result could be obtained from the fact that $P(B) = \binom{35}{9}\binom{15}{3} / \binom{50}{12}$ and $P(A^cB) = \binom{30}{9}\binom{15}{3} / \binom{50}{12}$.

An analogous situation applies to uniform models on sample spaces $\mathcal{S} \subset \mathcal{R}^n$. The conditional model given B with $P(B) > 0$ is uniform on B.

Exercise 1. Verify the statement in the preceding paragraph.

Exercise 2. In the target example, Example 12.5, let $r = 1$. Let B be the event that $\max(|x|, |y|) \leq 1/2$ and A be the event that $|x| + |y| \leq 1/2$. Find $P(A \mid B)$.

The following lemma has useful applications.

Lemma 1. *Let* Π *be a population of size* N *containing elements that are labeled* $1, \ldots, N$. *Assume that* $r \geq k$ *elements of* Π *are selected successively at random and without replacement. Let* $A_{m,k}$ *be the event that the* kth *element drawn has label* m. *Then* $P(A_{m,k}) = 1/N$ *for each* $k = 1, \ldots, r$.

Proof. Let B be the event that the element with label m is not drawn in the first $k - 1$ draws. Then $A_{mk} = BA_{mk}$. Thus

$$P(A_{mk}) = P(B)P(A_{mk} \mid B) = \frac{\binom{N-1}{k-1}}{\binom{N}{k-1}} \cdot \frac{1}{N-k+1} = \frac{1}{N}.$$

Theorem 1. *Let* Π *and the sampling scheme be as in Lemma 1. Let* $B \subset \Pi$ *have* b *elements. Let* C_k *be the event that the* kth *element drawn belongs to* B. *Then* $P(C_k) = b/N$.

Proof. Immediate from Lemma 1 by adding $P(A_{m,k})$ over labels, m, of the elements of B.

Example 2. If a population consists of N items of which D are defective and a random sample of any size is selected, the probability that the kth selected item is defective is D/N.

Bayes' theorem, which follows, is of considerable importance in the philosophy and applications of probability and statistics.

Theorem 2. (Bayes' theorem) *Let* B_1, B_2, $\ldots$ *be a partition. For any* A *with* $P(A) > 0$,

$$P(B_i \mid A) = \frac{P(A \mid B_i)P(B_i)}{\Sigma_n P(A \mid B_n)P(B_n)}. \tag{1}$$

Proof. Recall (12.12), which states that $P(A) = \Sigma_n P(A \mid B_n)P(B_n)$. Then

$$P(B_i \mid A) = \frac{P(AB_i)}{P(A)} = \frac{P(A \mid B_i)P(B_i)}{\Sigma_n P(A \mid B_n)P(B_n)}.$$

for each $i = 1, 2, \ldots$.

Example 3. In a certain population, 2% have lung cancer (C), 1% have tuberculosis (T), and the remainder have healthy lungs (H). A cheap screening test has been devised for which 95% of all patients are correctly diagnosed. Of the lung cancer patients, all incorrect diagnosis will be tuberculosis. The same is true of patients with healthy lungs. Of the tuberculosis patients, 3% will be diagnosed as having lung cancer and 2% as having healthy lungs. If a patient is diagnosed as having lung cancer (the event A), it is desirable to know the conditional probability of each of the three states of his lungs. In probability notation, $P(C) = .02$, $P(T) = .01$, $P(H) = .97$, $P(A \mid C) = .95$, $P(A \mid T) = .03$, and $P(A \mid H) = 0$. Then (1) yields

$$P(C \mid A) = \frac{(.02)(.95)}{(.02)(.95) + (.01)(.03) + (.97)(0)} = \frac{109}{193}$$

and similarly, $P(T \mid A) = 3/193$.

Example 4. There are three cards, one white on both sides (B_1), one black on both sides (B_2), and one black on one side and white on the other (B_3). Out of your sight, the cards are shuffled and one is dealt to you so that you can see one side only. Given that the side you see is white (A), what is the conditional probability that the other side is also white? Note that $P(B_i) = 1/3$ $(i = 1, 2, 3)$, $P(A \mid B_1) = 1$, $P(A \mid B_2) = 0$, and $P(A \mid B_3) = 1/2$, so that the desired conditional probability is $P(B_1 \mid A) = 2/3$. What made you think the answer would be 1/2?

What accounts for the prominence of Bayes' theorem is its use (about which there is considerable debate) in the area of inductive inference, specifically is connection with problems of statistical anaysis. The way in which such an application might arise is as follows: The event A is contained in the sample space $\mathcal{S}$ of a specified experiment whose probability model is not completely known. It is assumed, however, that the correct model belongs to a certain family: each B_i $(i = 1, 2, \ldots)$ represents a possible model for the experiment and $P(B_i)$ is an individual's personal probability (perhaps yours) that model B_i is the correct one. Thus each B_i provides a conditional probability $P(A \mid B_i)$ for any $A \subset \mathcal{S}$. Once the experiment is performed, each $P(B_i)$ is then revised via Bayes' theorem to produce $P(B_i \mid A)$ for the person in question.

In this context, the probability $P(B_i)$ assigned to the model B_i before performing the experiment is called the *prior probability* of B_i. Once the experiment is performed and it is observed that A has occurred, the revised probability, $P(B_i \mid A)$, is called the *posterior probability* of B_i *given A*.

This is a controversial area because of the subjective nature of prior probabilities, which many practitioners find unscientific and irrelevant to an "objective" discipline. Nevertheless, this methodology, referred to as

Bayesian statistics, has a large number of proponents and has had a substantial impact on modern statistical practice.

Example 5. In an effort to estimate the number of fish in a lake, a capture–recapture method is used, of which the following is a particularly simple example. A fish is caught, tagged, and returned to the lake. The next day we catch a fish and observe either A: "It is tagged" or A^c: "It is untagged."

For $m = 25, \ldots, 1000$, let B_m be the model that states that there are m fish in the lake and that each has the same probability of being caught on the second day, so that $P(A \mid B_m) = 1/m$. Assume that $P(B_m) = cm$ ($m = 25, \ldots, 1000$) where $c \sum_{m=25}^{1000} m = 1$.

If, on the second day, the tagged fish is caught, the posterior probability of B_m is

$$P(B_m \mid A) = \frac{(1/m)cm}{\sum_{j=25}^{1000} (1/j)cj} = \frac{1}{976}, \qquad (2)$$

the equally likely model.

If, on the other hand, the fish caught on the second day is untagged, then for $m = 25, \ldots, 1000$,

$$P(B_m \mid A^c) = \frac{(1 - 1/m)cm}{\sum_{i=25}^{1000} (1 - 1/j)cj} = k(m - 1),$$

where $k \sum_{m=25}^{1000} (m - 1) = 1$.

Summary

If P is uniform on $\mathcal{S}$, then for $P(B) > 0$ the conditional model given B is also uniform on B. In random sampling without replacement from a population with N members, the probability model for the kth draw is the same as the probability model for the first draw. If $B_1, B_2, \ldots$ is a partition and $P(A) > 0$, then, according to Bayes' theorem,

$$P(B_i \mid A) = \frac{P(A \mid B_i)P(B_i)}{\sum_j P(A \mid B_j)P(B_j)}.$$

Problems

1. Suppose that your poker hand (five cards) has a pair (two cards of the same value and the other cards of three different values). If you have one opponent, what is the probability that he/she has a pair?
2. In rolling two fair dice, it has been observed that the sum of the two faces is at least five. What is the probability that the sum is seven?

3. In order to compare two anticancer drugs (drug 1 and drug 2), five patients will be treated with drug 1 and five others will be treated with drug 2. The assignment is made by placing 10 balls in an urn, five numbered 1 and five numbered 2. When a patient comes for treatment, a ball is drawn from the urn and the indicated drug is given. Balls are not replaced. For each possible assignment for the first and second patients, what is the conditional probability that the third patient is given drug 1?

4. There are six chests, each with two drawers. Each drawer has one coin. In one chest both coins are gold. In two chests both coins are silver. In the remaining chests one coin is gold and one coin is silver. A chest is chosen at random and from that chest, a drawer is chosen at random and is opened.
   (a) If the observed coin is gold, what is the probability that the other coin in the selected chest is silver?
   (b) If the observed coin is silver, what is the probability that the other coin in the selected chest is gold?
   (c) Compare your answers in parts (a) and (b) with the unconditional probabilities.

5. One of three prisoners, Groucho, Harpo, and Chico, has been sentenced to be executed tomorrow. Groucho does not know who the victim will be but believes that each is equally likely. His guard knows who will be executed but has been ordered not to tell the prisoners until morning. Groucho asks the guard to name one of the other two whose life is to be spared. The guard agrees to do so and names Harpo. Groucho reasons that the probability of his execution is now 1/2. Comment on his reasoning.

6.† A box of five mousetraps has two defective traps. The traps are examined one at a time, without replacement, until both defective traps are found. Find the probability that four traps are examined given that at least three are examined.

7. Table 1 gives the number of students in each class for some course. For each of these classes, the table also gives the number of these students who are majors in some engineering curriculum. For each class find the conditional probability that a student is in that class given that he or she is an engineering major.
   (a) Solve using a conditional probability model.
   (b) Solve using Bayes' theorem.
   (c) Compare your answers with the unconditional probabilities.

8. Modify Example 5 by assuming that two fish are tagged on the first day but that the second day's catch remains one fish.

9. (Polya urn scheme.) A box has N balls, of which s are red and $N -$

Table 1

Class	Students	Engineering majors
Freshman	0	0
Sophomore	2	0
Junior	14	12
Senior	11	8
Masters	8	5
Doctoral	4	1

s are white. Balls are chosen one at a time. Each ball drawn is replaced and c balls of the same color are added. Note that $c = 0$ is sampling with replacement and $c = -1$ is sampling without replacement.

  (a) Show that the probability that the second ball drawn is red in s/N.

  (b) By induction, show that the probability that the kth ball drawn is red is s/N. Note that this generalizes Theorem 1.

10. In a draw poker game, you hold three of a kind and replace the two odd cards by drawing two new ones. What is the probability that your hand improves to either a full house or four of a kind?

14. INDEPENDENT EVENTS

When the conditional probability $P(A \mid B)$ differs considerably from the unconditional probability $P(A)$, knowledge of the occurrence or nonoccurrence of B provides substantial probability information concerning the occurrence or nonoccurrence of A. If no change takes place, that is, if

$$P(A \mid B) = P(A), \tag{1}$$

then B contains no useful information about A and we say that A and B are independent events. Before discussing (1) further, here are some examples.

Example 1. Suppose that $P(A)$ is your probability of A, that MSU will win its next football game. On the eve of the game, you learn of the occurrence of B, that Jones has been injured and will not play. Now, you may ask, "Who is Jones and how is the occurrence of B related to that of A?" If he is MSU's star quarterback, then B is unfavorable to A (see Problem 12.3) and $P(A \mid B) < P(A)$; if he is the opponent's star, then B is favorable to A and $P(A \mid B) > P(A)$; if he is an obscure reserve who rarely plays, his injury could be considered irrelevant to the outcome of the game, in which case $P(A \mid B) = P(A)$, so that A and B are independent events.

Example 2. Two dice (one red and the other blue) are rolled. Let A be the event that the sum of the faces is seven, C be the event that the sum of the faces is nine, and B be the event that the outcome on the blue die is three. It is a simple matter to show that $P(A \mid B) = P(C \mid B) = 1/6$. However, A and B are independent events, whereas C and B are not. In fact, B is favorable to C since $P(C) = 1/9 < 1/6$.

The following definition of independence is mathematically preferable to (1) because it avoids special cases [i.e., $P(B) = 0$] and exhibits the symmetry of the relation. The two definitions are "almost" equivalent (see Remarks 1 and 2).

Independence of Two Events

The events A and B with $P(A)$, $P(B) > 0$ are (*mutually*) *independent* if

$$P(AB) = P(A)P(B). \tag{2}$$

Remark 1. If $P(A)P(B) > 0$, then (1) and (2) are equivalent and either can serve as a definition of independence. However, if either $P(A) = 0$ or $P(B) = 0$, it follows that $P(AB) = 0 = P(A)P(B)$ and from (2), that A and B are independent. Indeed, it follows from (2) that an event of probability 0 (or an event of probability 1) is independent of any other event. In these cases the notion of independence loses its connection with informational content (see, e.g., Example 12.5).

Remark 2. If $P(A) > 0$ and $P(B) > 0$, then (1) implies, by simple calculation (Problem 12.12), that $P(B \mid A) = P(B)$ so that the symmetry of the relation, apparent from (2), is also contained in (1).

Remark 3. If the product relation (2) applies to AB, it also applies to every member of the partition generated by $\{A, B\}$. For example, $P(AB^c) = P(A)P(B^c)$ follows from (2) and $P(AB^c) = P(A) - P(AB)$. Similar equations yield the other product relations (see Problem 12.11). Thus, if A and B are independent, so are A and B^c, A^c and B, A^c and B^c. This is the basis for extending the notion of independence to more than two events [(3), below].

Table 1

		Mother's age	
		Under 30	At least 30
Sex of	Male	$P(A)P(B)$	$P(A^c)P(B)$
baby	Female	$P(A)P(B^c)$	$P(A^c)P(B^c)$

Example 3. The next recorded birth will be either a boy (B) or a girl (B^c). Its mother's age may be less than 30 (A) or at least 30 (A^c). If A and B are independent, which seems theoretically reasonable, the (joint) probabilities would be given in Table 1. Now suppose that in the next N births relative frequencies $f_N(AB)$, $f_N(A)$, and $f_N(B)$ are observed. How close does $f_N(AB)$ have to be to $f_N(A)f_N(B)$ to provide "convincing" evidence that A and B are independent? This is an example of a statistical problem and not a very simple one.

It is tempting to extend the notion of independence to more than two events by requiring pairwise satisfaction of (2) [or even (1)]. That this is unsatisfactory is shown by the following example concerning events A, B, and C.

Example 4. A fair coin is tossed twice. Let A be the event that the first toss results in a head, B be the event that the second toss results in a head, and $C = (AB^c) \cup (A^cB)$, that is, that the results of the two tosses differ.
Then $P(A) = P(B) = P(C) = 1/2$ and $P(AB) = P(AC) = P(BC) = 1/4$, so that each pair consists of independent events. However, $ABC = \phi$, so that $P(ABC) = 0 \neq P(AB)P(C)$. Thus AB and C are not independent, that is, there is some probability information in the pair (A, B) concerning the third event, C. In fact, no member of the partition generated by $\{A, B\}$ is independent of C.
If, in addition to pairwise independence, the events A, B, and C satisfy $P(ABC) = P(A)P(B)P(C)$ (as those in Example 4 do not), then each of the following pairs of events are also independent: (AB, C), (AC, B), and (BC, A). In that case, no pair of events provides probability information concerning the third.

Independence of n Events

Let $\mathcal{D} = \{D_1, \ldots, D_{2^n}\}$ be the partition generated by the events $A_1, \ldots, A_n$. Recall that $D \in \mathcal{D}$ has representation $D = \cap_{i=1}^n G_i$, where

Table 2

$A_1 A_2 A_3$	$p_1 p_2 p_3$	$A_1 A_2^c A_3^c$	$p_1 q_2 q_3$
$A_1 A_2 A_3^c$	$p_1 p_2 q_3$	$A_1^c A_2 A_3^c$	$q_1 p_2 q_3$
$A_1 A_2^c A_3$	$p_1 q_2 p_3$	$A_1^c A_2^c A_3$	$q_1 q_2 p_3$
$A_1^c A_2 A_3$	$q_1 p_2 p_3$	$A_1^c A_2^c A_3^c$	$q_1 q_2 q_3$

for each i, either $G_i = A_i$ or $G_i = A_i^c$. Then $A_1, \ldots, A_n$, are *independent* if

$$P(D) = \prod_{i=1}^{n} P(G_i) \tag{3}$$

for every $D \in \mathcal{D}$.

Example 5. Let A_1, A_2, and A_3 be events with probabilities $P(A_i) = p_i$ and $P(A_i^c) = q_i$. Table 2 displays the members of the partition generated by $\{A_1, A_2, A_3\}$ and their corresponding probabilities assuming independence.

In (3) it appears that 2^n equations must be satisfied. It is, in fact, only necessary to check $2^n - n - 1$ of them (Problems 14 and 15). For example, when $n = 3$ only four equations need to be checked (see Problem 13). It should also be apparent that replacing any A_i by A_i^c preserves the independence relation since the partitions generated are the same. Another important property is that if $A_1, \ldots, A_n$ are independent, then so are any two or more of them (Problem 14).

Summary

Independence: Events A and B are *independent* if $P(AB) = P(A)P(B)$. Events $A_1, \ldots, A_n$ are *independent* if $P(D) = \Pi_{i=1}^{n} P(G_i)$, where $D = \cap_{i=1}^{n} G_i$ and for each i, $G_i = A_i$ or $G_i = A_i^c$.

If $A_1, \ldots, A_n$ are independent events, then:

(i) Any two or more events from $A_1, \ldots, A_n$ are independent.
(ii) If any of the A_i are replaced by A_i^c, the result is still a collection of independent events.

Problems

1. A class of 50 students has 30 men and 10 graduate students. A student is selected at random. What must be the composition of the class in

order that the event that the student is a man and the event that the student is a graduate student are independent?

2. Consider the equally likely model for the 36 outcomes when two dice are rolled. Show that this model is equivalent to assuming that:
   (i) Each die is fair.
   (ii) The results on the two dice are independent.

3. Two fair dice are rolled. Let A_i be the event that the sum of the faces on the dice is divisible by i. For which pairs (i, j) are A_i and A_j independent.

4. (a) Are mutually exclusive events ever independent? If so, under what conditions? In particular, when are A and A^c independent?
   (b) Prove or disprove: $A_1, A_2, \ldots, A_n$ independent and exhaustive (i.e., $\cup_{i=1}^{n} A_i = \Omega$) implies that $P(A_i) = 1$ for some i

5. Let $\mathcal{S}$ be a subset of $\mathcal{R}^2$ with positive, finite area. Let (x, y) be a point in $\mathcal{S}$ chosen uniformly. Let A be the event that $1/3 \le x \le 2/3$ and B be the event that $1/4 \le y \le 1/2$. For each choice of $\mathcal{S}$ below, determine if A and B are independent.
   (a) $\mathcal{S}$ is the square with vertices $(0, 0)$, $(0, 1)$, $(1, 0)$, and $(1, 1)$.
   (b) $\mathcal{S}$ is the triangle with vertices $(0, 0)$, $(0, 1)$, and $(1, 0)$.
   (c) $\mathcal{S}$ is the square with vertices $(1, 1)$, $(1, -1)$, $(-1, 1)$, and $(-1, -1)$.

6. Generalize Problem 5 by letting A be the event that $a_1 \le x \le a_2$ and B be the event that $b_1 \le y \le b_2$. For what $\mathcal{S}$ (not necessarily those in Problem 5) are A and B independent regardless of the choices of a_1, a_2, b_1, and b_2?

7. Repeat Problem 1 for a class of 60 students with 36 men and 12 graduate students. Comment on the possibility of the class having this composition.

8. Suppose that a device consists of n components. Let A_j be the event that the jth component functions, assume that $A_1, \ldots, A_n$ are independent, and set $p_j = P(A_j)$ $(j = 1, \ldots, n)$. Give an expression for the probability that a specified subcollection of components function and all other components fail. Justify your answer. *Hint:* Solve first for $n = 3$ and generalize the method.

9. Let A_1, A_2, A_3, A_4, and B be events related as follows:

$$I_B = 1 - \prod_{i=1}^{4} (1 - I_{A_i}).$$

Assume that the A_i are independent with respective probabilities .5, .4, .3, and .2. Determine $P(B)$.

10. Repeat Problem 9 for the case in which

$$I_B = 1 - (1 - I_{A_1 A_2})(1 - I_{A_1 A_3})(1 - I_{A_2 A_3}).$$

11. Repeat Problem 9 for the case in which

$$I_B = (1 - I_{A_1A_2})(1 - I_{A_1A_3})(1 - I_{A_2A_3})$$
$$- (1 - I_{A_1})(1 - I_{A_2})(1 - I_{A_3}).$$

12. An urn contains four black and eight white balls. Three balls are drawn from the urn without replacement. Consider the following events:

   A = "Exactly one black ball is drawn";
   B = "All balls drawn are the same color"; and
   C = "The third ball drawn is black."

   Are the events in any of the pairs (A, B), (B, C), or (A, C) independent? If so, which ones, and why? If not, why not?

13. Suppose that $P(ABC) = P(A)P(B)P(C)$, $P(ABC^c) = P(A)P(B)P(C^c)$, $P(AB^cC) = P(A)P(B^c)P(C)$, and $P(A^cBC) = P(A^c)P(B)P(C)$. Show that A, B, and C are independent events. Can you exhibit any other systems of four equations which guarantee that A, B, and C are independent?

14. Show that if $n \geq 3$ events are independent, then any $n - 1$ of them are independent and hence any two or more of them are independent events.

15. Given the events $A_1, \ldots, A_n$ ($n \geq 3$), assume that any proper sub-collection of two or more of them consists of independent events and that $P(\cap_{i=1}^n A_i) = \prod_{i=1}^n P(A_i)$. Show that $A_1, \ldots, A_n$ are independent.

16. Suppose that A, B, and C are independent. Show that $A \cup B$ and C are independent.

17. Show each of the following:
   (a) $P(A)P(B) > 0$ implies the equivalence of (1) and (2).
   (b) $P(B) = 1$ implies that (1) and (2) are satisfied for any A.
   (c) Assume that $0 < P(B) < 1$. Then A and B are independent if, and only if, $P(A \mid B) = P(A \mid B^c)$.

15. JOINT EXPERIMENTS AND INDEPENDENT EXPERIMENTS

The multiplication rule (12.10) is a powerful tool for determining probabilities of intersections when the values of all relevant conditional probabilities are known or can be deduced from the context. Often this conditional information comes in the form of assumptions concerning the manner in which an experiment or a sequence of experiments is to be performed, such as whether random sampling is to be made *with* or *without* replacement. Another example follows.

Example 1. Experiments ε and ε^* each begin by tossing a fair coin (ε_1). For ε, the toss is followed by ε_2: one roll of a totally unrelated, ordinary, fair die. For ε^*, the toss is followed by ε_2^*: one roll of D_1 if the outcome of ε_1 is H; one roll of D_2 if the outcome of ε_1 is T. Die D_1 is ordinary except that the 6 is replaced by a 1; Die D_2 is ordinary except that the 1 is replaced by a 6. Note that ε and ε^* have the following in common: the sample space $S = \{(x, y): x = H \text{ or } T, y = 1, 2, 3, 4, 5, \text{ or } 6\}$ and the probability models associated with the first and second components, respectively. For ε, any event concerning the coin toss is independent of any event concerning the outcome on the die, so that $(H, 1)$ has probability $P(H, 1) = P_1(H)P_2(1) = (1/2)(1/6) = 1/12$. In ε^*, however, this probability is $P^*(H, 1) = P^*(H \text{ on coin})P^*(1 \text{ on die}/H \text{ on coin}) = P_1(H)P_2^*(1 \mid H) = (1/2)(2/6) = 1/6$. Note also that $P(H, 6) = 1/12$ while $P^*(H, 6) = 0$. In either case, the probabilities of all outcomes in S are determined from $P(x, y) = P(\text{coin falls } x)P(\text{die falls } y \mid \text{coin falls } x)$.

In general, consider experiments ε_1 and ε_2 with probability models given by S_1, P_1 and S_2, P_2, respectively (to be called *marginal probability models*). The *joint experiment*, denoted by $\varepsilon = \varepsilon_1 \times \varepsilon_2$ consists of performing ε_1 and ε_2, each once, and has sample space $S = S_1 \times S_2$ (see Section 16 on Cartesian products).

Subsets of S are called *product events* if they are of the form $A \times B$. The special product events $A \times S_2$ and $S_1 \times B$ are called *marginal events* and their occurrence is determined solely by ε_1 and ε_2, respectively. Every product event is the intersection of marginal events: $A \times B = (A \times S_2)(S_1 \times B)$.

The probability assignment, P, the subsets of S is determined by the probabilities of product events (see GP 7), and since these are intersections, can be obtained by applying the multiplication rule. To begin with, probabilities of marginal events must satisfy $P(A \times S_2) = P_1(A)$ and $P(S_1 \times B) = P_2(B)$. Then

$$P(A \times B) = P((A \times S_2)(S_1 \times B))$$
$$= P(S_1 \times B \mid A \times S_2)P(A \times S_2) \qquad (1)$$
$$= P(B \mid A)P_1(A),$$

where $P(B \mid A)$ is our abbreviation for "the probability that B occurs in ε_2 given that A occurs in ε_1." Alternatively, we may use $P(A \times B) =$

$P(A \mid B)P_2(B)$. Example 1 illustrates that two experiments may have the same marginals but different joint probability models.

These ideas extend in obvious ways to any number of experiments. In what follows, we examine independence for several events.

Independent Experiments

For $i = 1, \ldots, n$, let $\mathcal{E}_i$ be an experiment with a sample space $\mathcal{S}_i$ and probability assignment P_i on $\mathcal{S}_i$. Let $\mathcal{E} = \times_{i=1}^n \mathcal{E}_1$ be the experiment consisting of one performance of each $\mathcal{E}_i$. Let P be a probability assignment on $\mathcal{S} = \times_{i=1}^n \mathcal{S}_i$, the sample space for $\mathcal{E}$. The experiments are *independent* if $P(\times_{i=1}^n A_i) = \Pi_{i=1}^n P_i(A_i)$ for any choices of events $A_i \subset \mathcal{S}_i$ ($i = 1, \ldots, n$). If all $\mathcal{S}_i = \mathcal{S}_0$ and all $P_i = P_0$, then all $\mathcal{E}_i$ are the same, say $\mathcal{E}_0$; we refer to these independent experiments as *independent trials* of $\mathcal{E}_0$.

Bernoulli Trials

One special case of independent trials is particularly important. Suppose that $\mathcal{E}_0$ consists of observing that some fixed event A does or does not occur. A common choice for $\mathcal{S}_0$ is $\{S, F\}$, where S is called "success" and indicates that A occurs and F is called "failure." Then independent trials of $\mathcal{E}_0$ are called *Bernoulli trials*.

Typical examples of Bernoulli trials are coin tosses and sampling with replacement from dichotomous population (consisting of A's and not A's).

The following examples display situations in which independent experiments or independent trials are appropriate models or approximations.

Example 2. Two anticancer drugs are being compared. Drug 1 is tried on m patients and drug 2 on n patients, all $m + n$ patients being drawn at random from some population. The experiments with drug i are independent trials. The combined $m + n$ experiments are then independent.

Example 3. In an election between Smith and Jones with 1 million registered voters, a sample of 1000 registered voters is selected at random without replacement. Each voter in the sample is asked, "Do you intend to vote for Smith?" Each selection of a voter can be considered as a trial of the same experiment. However, the trials are not independent since sampling is without replacement.

Since 1000 is very small compared with 1 million, the population is not changed much by withdrawing the sample or any part of it. That is, sampling with or without replacement will yield nearly the same result. If sampling were with replacement, the trials would be independent. Thus, in our case, a model of independent trials will very closely approximate the true situation. The approximating model consists, in fact, of Bernoulli trials. An analysis of such approximations is given in Section 39B.

Summary

Joint experiment: $\mathcal{E} = \mathcal{E}_1 \times \mathcal{E}_2$ with sample space $\mathcal{S} = \mathcal{S}_1 \times \mathcal{S}_2$, with a model P, with marginal probabilities $P(A \times \mathcal{S}_2) = P_1(A)$ and $P(\mathcal{S}_1 \times B) = P_2(B)$.

Independent experiments: For $i = 1, \ldots, n$ let $\mathcal{E}_i$ be an experiment with sample space $\mathcal{S}_i$ and probability assignment P_i. The experiment consisting of one performance of each $\mathcal{E}_i$ is denoted by $\mathcal{E} = \times_{i=1}^{n} \mathcal{E}_i$. Then $\mathcal{E}_1, \ldots, \mathcal{E}_n$ are *independent experiments* if $P(\times_{i=1}^{n} A_i) = \Pi_{i=1}^{n} P_i(A_i)$ for any choices of $A_i \subset \mathcal{S}_i$, where P is the probability assignment on $\mathcal{S} = \times_{i=1}^{n} \mathcal{S}_i$, the sample space for $\mathcal{E}$.

Independent trials: If each $\mathcal{E}_i$ is the same experiment, $\mathcal{E}_0$, we refer to independent experiments as *independent trials* of $\mathcal{E}_0$.

Bernoulli trials: When $\mathcal{E}_0$ consists of observing the occurrence or non-occurrence of some event A, independent trials are called *Bernoulli trials*.

Problems

1. Ten students are selected at random from a class and their grade point averages are observed. If sampling is without replacement, are the trials independent? Do you think that "independent trials" is a good approximate model if there are 25 students in the class? What if sampling is with replacement?

2. Consider the equally likely assignment for the 6^n outcomes when n dice are rolled. Show that this is the model for n independent trials. What are the experiment and the model for a single trial?

3. Consider n tosses of a coin. Show that the equally likely model on the 2^n outcomes is equivalent to n Bernoulli trials in which $P(A) = 1/2$.

4. A die is rolled once and a coin is tossed once. Show that the equally likely model on the 12 outcomes is equivalent to independent experiments with equally likely models for each experiment.

5. Suppose that $\mathcal{E}_i$ has m_i outcomes ($i = 1, \ldots, n$). Show that the equally likely model for $\times_{i=1}^{n} \mathcal{E}_i$ on the $\Pi_{i=1}^{n} m_i$ outcomes is equivalent to independent experiments with equally likely models on the m_i outcomes of each $\mathcal{E}_i$.

6. Consider n Bernoulli trials and let B_i be the event that A occurs in the ith trial. Let t be a fixed n-dimensional vector in which k components are 1's and the remaining $n - k$ components are 0's. Find $P[(I_{B_1}, \ldots, I_{B_n}) = t]$.

7. Show that the experiments in any subcollection of n independent experiments are independent.
8. An assembly line produces items that may be either good or defective. Under what conditions does the sequence of items produced, which may be good or defective, constitute Bernoulli trials?
9. Let $\mathscr{E}_1, \ldots, \mathscr{E}_n$ be any independent experiments. Set $\mathscr{F}_1 = \times_{j \in J} \mathscr{E}_j$ and $\mathscr{F}_2 = \times_{j \in K} \mathscr{E}_j$, where J and K are disjoint subsets of $\{1, \ldots, n\}$. Show that $\mathscr{F}_1$ and $\mathscr{F}_2$ are independent experiments.
10. Consider experiments $\mathscr{E}_1$, $\mathscr{E}_2$, and $\mathscr{E}_3$.
   (a) Give the assignment of probabilities to product events.
   (b) Specialize part (a) to the case of $\mathscr{E}_1$ and $\mathscr{E}_2$ independent.
   (c) Give a reasonable definition of "$\mathscr{E}_1$ and $\mathscr{E}_2$ are *conditionally independent* given $\mathscr{E}_3$" and specialize part (a) to this case.

16. APPENDIX ON CARTESIAN PRODUCTS

The *Cartesian product* of the sets A and B is $A \times B = \{(x, y): x \in A, y \in B\}$, that is, the set of ordered pairs (x, y) with $x \in A$, $y \in B$.

Example 1. $\mathscr{R}^2 = \mathscr{R} \times \mathscr{R}$ and Cartesian products of intervals are rectangles. Thus $[a, b] \times [c, d]$ is the rectangle, including boundaries, with corners at (a, c), (a, d), (b, c), and (b, d).

Example 2. The sample space[2] for a coin toss is $\{H, T\}$. The sample space for two tosses is $\{H, T\} \times \{H, T\} = \{(H, H), (H, T), (T, H), (T, T)\}$. This sample space keeps track of the result of each toss.

Example 3. The sample space for a roll of a die is $\{1, 2, 3, 4, 5, 6\}$. The sample space for an experiment consisting of a coin toss and a die roll is

$$\{H, T\} \times \{1, 2, 3, 4, 5, 6\}$$
$$= \{(H, 1), (H, 2), (H, 3), (H, 4), (H, 5), (H, 6),$$
$$(T, 1), (T, 2), (T, 3), (T, 4), (T, 5), (T, 6)\}.$$

In general, the *Cartesian product* of the sets $A_1, \ldots, A_n$ is

$$A_1 \times \cdots \times A_n = \overset{n}{\underset{i=1}{\times}} A_i = \{(x_1, \ldots, x_n): x_i \in A_i \ (i = 1, \ldots, n)\},$$

---

[2] All examples of sample spaces in this section are those usually used for the experiments described.

that is, the set of ordered n-tuples $(x_1, \ldots, x_n)$ with $x_i \in A_i$ for $i = 1, \ldots, n$.

Example 4. $\mathcal{R}^n = \times_{i=1}^{n} \mathcal{R}$ and n-fold Cartesian products of intervals are n-dimensional rectangles.

Let $\mathcal{S}_1, \ldots, \mathcal{S}_n$ be sample spaces and let $A_i \subset \mathcal{S}_i$ for $i = 1, \ldots, n$. Then

$$(A_1 \times \mathcal{S}_2 \times \cdots \times \mathcal{S}_n)(\mathcal{S}_1 \times A_2 \times \mathcal{S}_3 \times \cdots \times \mathcal{S}_n)$$
$$\cdots (\mathcal{S}_1 \times \cdots \times \mathcal{S}_{n-1} \times A_n)$$
$$= \underset{i=1}{\overset{n}{\times}} A_i.$$

Note that the indicator of $\times_{i=1}^{n} A_i$ is a function on $\times_{i=1}^{n} \mathcal{S}_i$ given by $\Pi_{i=1}^{n} I_{A_i}$, where I_{A_i} is a function on $\mathcal{S}_i$.

In particular, for $n = 2$, since $I_{(A_1 \times \mathcal{S}_2)(\mathcal{S}_1 \times A_2)} = I_{A_1 \times \mathcal{S}_2} I_{\mathcal{S}_1 \times A_2} = I_{A_1} I_{\mathcal{S}_2} I_{\mathcal{S}_1} I_{A_2} = I_{A_1} I_{A_2} = I_{A_1 \times A_2}$, it follows that $(A_1 \times \mathcal{S}_2)(\mathcal{S}_1 \times A_2) = A_1 \times A_2$ (see Figure 1).

Consider $\mathcal{R}^n$. For Cartesian products of intervals, that is, for n-dimensional rectangles, the following are useful indicator relations. Let $C = \times_{i=1}^{n} (a_i, b_i]$, $B_i = (-\infty, b_i]$ and $A_i = (-\infty, a_i]$. Since $I_{(a_i, b_i]} = I_{B_i} - I_{A_i}$, it follows that

$$I_C = \prod_{i=1}^{n} (I_{B_i} - I_{A_i}). \tag{1}$$

For $n = 2$, (1) reduces to

$$I_C = (I_{B_1} - I_{A_1})(I_{B_2} - I_{A_2}) = I_{B_1} I_{B_2} - I_{A_1} I_{B_2} - I_{B_1} I_{A_2} + I_{A_1} I_{A_2}$$
$$= I_{B_1 \times B_2} - I_{A_1 \times B_2} - I_{B_1 \times A_2} + I_{A_1 \times A_2}.$$

Let A_i have m_i elements for $i = 1, \ldots, n$. Then $\times_{i=1}^{n} A_i$ has $\Pi_{i=1}^{n} m_i$ elements.

Problems

Hint: For these problems, a figure analogous to Figure 1 may help. Use indicators.

1. Show that $(A_1 \times B_1)(A_2 \times B_2) = (A_1 A_2) \times (B_1 B_2)$.
2. Show that $(A_1 \cup A_2) \times (B_1 \cup B_2) = (A_1 \times B_1) \cup (A_1 \times B_2) \cup (A_2 \times B_1) \cup (A_2 \times B_2)$.
3. Show that $(A \times B)^c = (A^c \times B) \cup (A \times B^c) \cup (A^c \times B^c)$ and that this equation expresses $(A \times B)^c$ as a union of disjoint sets.

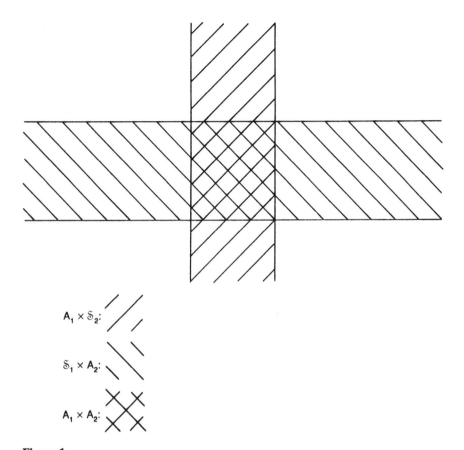

Figure 1

4. Let $\mathcal{S}_1$ and $\mathcal{S}_2$ be sample spaces for $\mathcal{E}_1$ and $\mathcal{E}_2$, respectively, and A and B be events in $\mathcal{E}_1$ and $\mathcal{E}_2$, respectively. Show that $(A \times B)^c = (A^c \times \mathcal{S}_2) \cup (\mathcal{S}_1 \times B^c)$.

GENERAL PROBLEMS

1. Let experiment $\mathcal{E}_1$ consist of choosing an integer from $\{1, 2, 3\}$ at random. If the result of $\mathcal{E}_1$ is n, the experiment $\mathcal{E}_2$ consists of rolling a fair 2^n-sided die.
   (a) Give a model for $\mathcal{E}_1 \times \mathcal{E}_2$ consistent with these assumptions.
   (b) Are $\mathcal{E}_1$ and $\mathcal{E}_2$ independent experiments?
   (c) Give a marginal model for $\mathcal{E}_2$.
2. (a) In Problem 1, assume that the die roll resulted in j. What is the probability that the i-sided die was used?

(b) Repeat part (a) assuming that the die is rolled twice and that the maximum of the two rolls is j.

3. In Problem 1, let X be the number rolled on the die. Find $E(X)$ using conditional expectations.

4. Let $\mathscr{E}_1$ be an experiment with m outcomes and probability assignment p_i to $i = 1, \ldots, m$. Let $\mathscr{E}_2$ be an experiment with n outcomes and probability assignment q_j to $j = 1, \ldots, n$. In $\mathscr{E}_1 \times \mathscr{E}_2$ let r_{ij} be the probability assignment to (i, j).

   (a) Show that $\mathscr{E}_1 \times \mathscr{E}_2$ consists of independent experiments if, and only if, $r_{ij} = p_i q_j$ for each $i = 1, \ldots, m; j = 1, \ldots, n$.

   (b) The condition in part (a) involves mn equations. Show that there is a choice of $(m - 1)(n - 1)$ of these equations which, if satisfied, imply that the remaining equations are satisfied.

5.† Let $S_n = \Sigma_{i=1}^n I_{A_i}$.

   (a) Show that $P[S_n = k] = P(A_n \mid S_{n-1} = k - 1)P[S_{n-1} = k - 1] + P(A_n^c \mid S_{n-1} = k)P[S_{n-1} = k]$.

   (b) Assume that the A_i are independent and that each $P(A_i) = p$. Give $P[S_1 = k]$, $P[S_2 = k]$, $P[S_3 = k]$ and, by induction, give $P[S_n = k]$ for general n. The family of functions $f(k) = P[S_n = k]$ (one for each fixed n and p) is called the *family of binomial probability functions*. This family is discussed in detail in Section 39.

6. Let A and B be independent events. Find $E(I_A I_B)$.

7.† (a) Recall the class $\mathscr{C}$ of events in Section 5 that satisfies:

   (i) If $A, B \in \mathscr{C}$, then $AB \in \mathscr{C}$.

   (ii) If $A \in \mathscr{C}$, there exits a countable partition $\{B_1, B_2, \ldots\}$ of A^c such that each $B_n \in \mathscr{C}$.

   Let $\mathscr{E}_1, \ldots, \mathscr{E}_n$ be experiments with sample spaces $\mathscr{S}_1, \ldots, \mathscr{S}_n$, respectively, and let $\mathscr{S} = \times_{i=1}^n \mathscr{S}_i$. For each $i = 1, \ldots, n$, let $\mathscr{C}_i$ be a class of subsets of $\mathscr{S}_i$ satisfying the conditions of the class $\mathscr{C}$. Let $\mathscr{C}$ be the class of subsets of $\mathscr{S}$ consisting of all subsets of the form $\times_{i=1}^n A_i$ for $A_i \in \mathscr{C}_i$. Show that $\mathscr{C}$ satisfies the conditions of the class $\mathscr{C}$. *Hint*: Use induction.

   (b) Show that the class of all rectangles in $\mathscr{R}^n$ satisfies the conditions of the class $\mathscr{C}$. See Example 5.2.

8. A population of size M contains M_1 individuals labeled "success." A random sample of size n is drawn without replacement. Let A_i be the event that the ith member of the sample is a success and let B_k be the event that the sample contains k successes.

   (a) For $M_1 = 100$, $M = 300$, and $n = 3$, are A_3 and B_1 independent?

   (b) In general, are there any relationships among M_1, M, n, i, and k such that A_i and B_k are independent? If so, identify all such relationships.

A   B

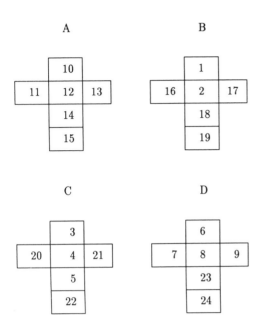

C   D

Figure 1

9. Player 1 tosses a fair coin until he obtains a head. Player 2 does the same. Find the probability that the minimum of the numbers of tosses required by the two players is at least m.

10. (Simpson's paradox)   Show that it is possible for the following three relations to hold simultaneously: AB is favorable to C, A^cB is favorable to C, and B is unfavorable to C. *Note*: More generally, for any n it is possible to find events B and C and a partition of B into n events that are favorable to C, yet have B unfavorable to C.

11. (Efron dice)   Four dice, A, B, C, D, are depicted in Figure 1. I select a die and, with knowledge of my selection, you select one. We then roll our dice (independently) and the one with the larger outcome wins. Show that no matter which die I choose, you can choose one that beats mine with probability $2/3$.

REFERENCES

DeGroot, Morris H. (1986). *Probability and Statistics*, 2nd ed. Reading, Mass.: Addison-Wesley Publishing Co., Inc. Exposition and applications of Bayes' theorem.

Wagner, Clifford H. (1982). Simpson's paradox in real life. *Amer. Statist.* **36**, 46–48. Simpson's paradox (GP 10). Additional references are listed.

Woodroofe, Michael (1975). *Probability with Applications.* New York: McGraw-Hill Book Company. See Chapter 3 for a nice presentation of conditional probability and independence.

3.

Some Elementary Applications

We pause here in the general development of the theory to present a few examples of probability methods brought to bear on questions of practical interest. We have chosen four areas that we hope will suggest the diversity and potential force of probabilistic reasoning. The areas represented are (1) gambling, (2) reliability engineering, (3) sequential choice of experiments, and (4) genetics. The particular tools from the previous sections that are used are partitioning, independence, and conditioning.

17. ROULETTE VS. CRAPS

Suppose that you enter a Las Vegas casino prepared to place one bet and want to choose between betting on a win at craps or betting on red at roulette. Both of these are even money bets; that is, if you bet $1 and win, you win $1. Hence it suffices to compare the probabilities of winning. We described the Nevada routlette wheel in Example 1.2, where we found the probability of red to be $9/19 = .47368$.

In the game of craps you will roll two dice. If you roll 7 or 11, you win immediately. If you roll 2, 3, or 12, you lose immediately. If you roll anything else, say k, then k is your point and you must continue rolling until you either roll k again and win or roll a 7 and lose.

Let B_k be the event that you roll k on the first roll and A be the event that you win. To compute $P(A)$, apply the total probability law (12.12),

$$P(A) = \sum_{k=2}^{12} P(B_k)P(A \mid B_k). \tag{1}$$

Now $P(A \mid B_7) = P(A \mid B_{11}) = 1$ while $P(B_7) = 6/36$ and $P(B_{11}) = 2/36$. Also, $P(A \mid B_2) = P(A \mid B_3) = P(A \mid B_{12}) = 0$. To determine $P(A \mid B_k)$ for other k, first consider a more general problem.

Let C and D be disjoint events and suppose that a sequence of independent trials is performed until $C \cup D$ occurs, at which point we stop. Let T denote the stopping time (i.e., the total number of trials that are performed) and let S_C denote the event that the sequence stops with C. The only information conveyed by $T = k$ is that the sequence stopped at time k, so that $P[S_C \mid T = k] = P(C \mid C \cup D) = P(C)/[P(C) + P(D)]$ and

$$P(S_C) = \frac{P(C)}{P(C) + P(D)} \sum_{n=1}^{\infty} P[T = k] = \frac{P(C)}{P(C) + P(D)}. \tag{2}$$

The last equality comes from $\sum_{k=1}^{\infty} P[T = k] = \sum_{k=1}^{\infty} q^{k-1}p = 1$, where $p = P(C \cup D)$ and $q = 1 - p$.

For craps, let C_k be the event that k is rolled and $D = C_7$. Using (2) gives

$$P(A \mid B_k) = P(S_{C_k}) = \frac{P(C_k)}{P(C_k) + P(D)}. \tag{3}$$

The results of the computations in (3) are given in Table 1.

Apply (1) to obtain

$$P(A) = \frac{3}{36} \cdot \frac{1}{3} + \frac{4}{36} \cdot \frac{2}{5} + \cdots + \frac{2}{36} \cdot 1 = \frac{244}{495} = .49293.$$

Since the probability of red at roulette is .47368, craps is the better bet.

Table 1

k	$P(B_k)$	$P(A \mid B_k)$
4	3/36	3/9
5	4/36	4/10
6	5/36	5/11
8	5/36	5/11
9	4/36	4/10
10	3/36	3/9

If, instead, you enter the Casino de Monte Carlo, the situation is a bit different. The rules for craps are the same; however, the standard European roulette wheel has only one green slot. Hence, in Monte Carlo, the probability of red is $P(A) = 18/37 = .48649$.

However, there is a rule in European roulette that further improves the bettor's chances. If green results, the bettor does not immediately lose. The bet goes "en prison" (in prison) and stays there as long as future trials result in green. If the bet is on red and black is the first nongreen result, the bettor loses. If the first nongreen result is red, the bet is a draw.

To make a valid comparison with the probability of winning at craps, the conditional probability of winning given that the result of the bet is not a draw must be computed. Let A be the event that the bettor wins and B be the event that the result is not a draw. Since $A \subset B$, it follows that $P(A \mid B) = P(A)/P(B)$.

Let C be the event that the first trial is not green. Since $C \subset B$,

$$P(B) = P(C) + P(BC^c) = P(C) + P(C^c)P(B \mid C^c).$$

But $P(C) = 36/37$ and $P(B \mid C^c) = 1/2$, so that

$$P(B) = \frac{36}{37} + \frac{1}{37} \cdot \frac{1}{2} = \frac{73}{74}.$$

Thus

$$P(A \mid B) = \frac{18/37}{73/74} = \frac{36}{73} = .49315.$$

Hence, in Monte Carlo, betting red at roulette is slightly better than betting at craps.

By symmetry, all these conclusions are valid for bets on black, odd, or even (green is neither odd nor even).

Let X be the net amount you win on a \$1 bet so that $X = I_A - I_D$, where A is the event that you win and D is the event that you lose. Note that $D = A^c$ except for European roulette. For craps, $E(X) = .49293 - (1 - .49293) = -.01414$. For Nevada roulette, $E(X) = -.05264$. For European roulette, $E(X \mid B) = -.01370$ and $E(X \mid B^c) = 0$, so that $E(X) = E(X \mid B)P(B) + E(X \mid B^c)P(B^c) = -.01351$.

PROBLEMS

1. In the game chuck-a-luck, three dice are rolled. The player chooses one of the numbers 1, . . . , 6. If he bets \$1, he wins \$$k$ if his number shows

k times ($k = 1, 2, 3$). Otherwise, he loses. Compare chuck-a-luck with the games of this section.

2. In roulette, a winning bet on the first third $(1, \ldots, 12)$ wins \$2 for \$1 bet. Compare this bet with a \$1 bet on red both for Las Vegas and Monte Carlo (with the *en prison* rule).

3. Let A and B be any events in the sample space of an experiment $\mathcal{E}$. What is the probability in a sequence of independent trials of $\mathcal{E}$ that A occurs no later than B?

18. RELIABILITY

The *reliability* of a system or component is, simply, the probability that it works (functions).

Let F be the event that a system functions and let A_i be the event that its ith component, C_i, functions. We have shown in examples and problems how the indicator of F can be expressed in terms of I_{A_i}'s, and now we want to illustrate the calculation of $P(F)$, the system's reliability. The calculation is particularly easy if the components behave independently, that is, if the A_i's are independent, and this is assumed throughout this section.

Both of the following examples concern the use of redundancy of components to increase system reliability. This means replacing a component, say C, with a parallel hook-up of n independent copies of C.

Example 1. A particular device functions if C (e.g., a battery) functions. Install n copies of C in parallel with $P(A_i) = p$. Then $F = \cup_{i=1}^{n} A_i$ so that $I_{F^c} = \Pi_{i=1}^{n} I_{A_i^c}$ and

$$P(F^c) = \prod_{i=1}^{n} P(A_i^c) = (1 - p)^n. \tag{1}$$

How should n be chosen if it is required that $P(F) \geq \gamma$, where γ is a prescribed reliability for the device? By complementation, the requirement becomes $P(F^c) = (1 - p)^n \leq 1 - \gamma$ and hence

$$n \geq \frac{\log(1 - \gamma)}{\log(1 - p)}. \tag{2}$$

For economy, let n be the smallest integer satisfying (2).

If $p = .5$ and $\gamma = .99$, then $n \geq \log(.01)/\log(.5) = 6.64$, so tht $n = 7$ suffices.

Example 2. A device has two components, C_A and C_B, in series. Put m copies of C_A in parallel and do the same with n copies of C_B. These subsystems, S_A and S_B, are put in series to build a device with redundant components. Then $F = (\cup_{i=1}^{m} A_i)(\cup_{j=1}^{n} B_j) = F_A F_B$, where A_i is the event "ith copy of C_A functions" and B_j is similarly defined.

Set $P(A_i) = \alpha$ and $P(B_j) = \beta$. Using (1) to determine $P(F_A^c)$ and $P(F_B^c)$ yields

$$P(F^c) = P(F_A^c) + P(F_B^c) - P(F_A^c)P(F_B^c)$$

$$= (1 - \alpha)^m + (1 - \beta)^n - (1 - \alpha)^m(1 - \beta)^n.$$

If it is required that $P(F) \geq \gamma$, then any pair m, n satisfying

$$(1 - \alpha)^m + (1 - \beta)^n - (1 - \alpha)^m(1 - \beta)^n \leq 1 - \gamma \qquad (3)$$

will suffice. Pairs m, n satisfying (3) can be found by fixing values of m and solving for n. Thus

$$(1 - \beta)^n \leq \frac{1 - \gamma - (1 - \beta)^m}{1 - (1 - \alpha)^m}$$

or

$$n \geq \frac{\log[1 - \gamma - (1 - \alpha)^m] - \log[1 - (1 - \alpha)^m]}{\log(1 - \beta)}.$$

Suppose that $\alpha = .5$, $\beta = .6$, and $\gamma = .99$. Then

$$n \geq \frac{\log[.01 - (.5)^m] - \log[1 - (.5)^m]}{\log(.4)}. \qquad (4)$$

For n take the smallest integer solution of (4), a nonincreasing function of m. But (3) implies that $(1 - \alpha)^m < 1 - \gamma$, so that $m > \log(.01)/\log(.5)$, which implies that $m \geq 7$. Similarly, $n > \log(.01)/\log(.4)$, which implies that $n \geq 6$. Thus start with $m = 7$, increase m in steps of 1, and stop when $n = 6$.

If C_A and C_B have equal cost, then $m + n$ is to be minimized subject to (4). If C_B is twice as costly as C_A, then $m + 2n$ is to be minimized subject to (4). Both cases are considered in Table 1. In computing Table 1, m was fixed and the smallest integer, n, satisfying (3) was found.

Table 1

m	n	$m + n$	$m + 2n$
7	7	14	21
8	6	14	20

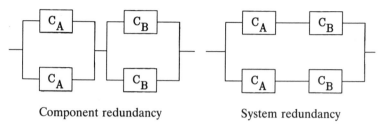

Component redundancy   System redundancy

Figure 1

Hence in the equal-cost case, there are thus two solutions, $m = n = 7$ and $m = 8, n = 6$. If C_B is twice as costly as C_A, there is a unique solution, $m = 8, n = 6$.

Problems

1. NASA is about to attempt a mission and has available two first-stage rockets, C_1 and C_2, with reliabilities p_1 and p_2, respectively, and two second-stage rockets, D_1 and D_2, with reliabilities q_1 and q_2, respectively. Either C_1 will be paired with D_1 and C_2 will be paired with D_2 or C_1 will be paired with D_2 and C_2 will be paired with D_1. Assume that the four components behave independently and that $p_1 > p_2$ and $q_1 > q_2$. The mission will be a success if at least one pair used functions. Which pairings should NASA use?
2. Suppose that C_A has reliability .5 and costs \$15 and C_B has reliability .7 and costs \$20. A device is to be built as in Example 2. What is the most economical choice of m and n if we wish the device to have reliability .99?
3. A device has two components, C_A and C_B, with reliabilities α and β, respectively. Both must function for the device to function. To increase reliability suppose that two copies of each are put in the device and wired in one of the two ways shown in Figure 1. Which is better?
4. Generalize the solution of Problem 3 to the case of m copies of each of n components. *Hint*: Compare the events "system fails for component redundancy" and "system fails for system redundancy." If the reliabilities of the copies of the same component are not all equal, would your conclusion change?

19. SEQUENTIAL CHOICES

A rat is placed in a T-maze (see Figure 1) and must choose the left or the right arm. If the left arm is chosen, the rat is rewarded with probability

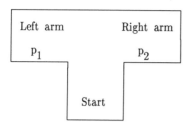

Figure 1

p_1, and otherwise, receives an electric shock. For the right arm, the situation is the same, except that p_2 is the probability of a reward.

Suppose that the rat is placed in the maze three times and adopts the following strategy?

(i)   Choose an arm at random for the first time (equal probabilities).
(ii) If the rat is rewarded on the nth trial, it returns to the same arm on the $(n + 1)$st trial.
(iii) If the rat is shocked on the nth trial, it goes to the other arm on the $(n + 1)$st trial.

We call this the *play the winner* strategy.

The probability of the event A, that the rat is rewarded on a majority of the trials (at least twice), is of interest. Let L be the event that the rat chooses the left arm on the first trial. By (12.11)

$$P(A) = P(L)P(A \mid L) + P(L^c)P(A \mid L^c)$$
$$= \frac{1}{2} P(A \mid L) + \frac{1}{2} P(A \mid L^c). \tag{1}$$

Let R_i be the event that the rat is rewarded on the ith trial. Then

$$P(A \mid L) = P(R_1 R_2 \mid L) + P(R_1 R_2^c R_3 \mid L) + P(R_1^c R_2 R_3 \mid L)$$
$$= p_1^2 + p_1(1 - p_1)p_2 + (1 - p_1)p_2^2 \tag{2}$$
$$= p_1^2 + p_2(1 - p_1)(p_1 + p_2).$$

Interchanging p_1 and p_2 in (2) gives

$$P(A \mid L^c) = p_2^2 + p_1(1 - p_2)(p_1 + p_2). \tag{3}$$

Substituting (2) and (3) in (1) yields, after some simplification,

$$P(A) = p_1^2 + p_2^2 + p_1 p_2(1 - p_1 - p_2).$$

In the special case $p_2 = 1 - p_1$, formula (3) simplifies to $P(A \mid L) = p_1^2 + (1 - p_1)^2 = P(A \mid L^c)$. This equality holds for five trials, but not generally.

Let X be the number of times the rat is rewarded. Then

$$X = I_{R_1 R_2^c R_3^c} + I_{R_1^c R_2 R_3^c} + I_{R_1^c R_2^c R_3} + 2I_{R_1 R_2 R_3^c} + 2I_{R_1 R_2^c R_3}$$
$$+ 2I_{R_1^c R_2 R_3} + 3I_{R_1 R_2 R_3}$$

and

$$E(X \mid L) = p_1(1 - p_1)(1 - p_2) + (1 - p_1)p_2(1 - p_2)$$
$$+ (1 - p_1)(1 - p_2)p_1 + 2p_1^2(1 - p_1)$$
$$+ 2p_1(1 - p_1)p_2 + 2(1 - p_1)p_2^2 + 3p_1^3$$
$$= (1 - p_1)(1 - p_2)(2p_1 + p_2)$$
$$+ 2(1 - p_1)(p_1^2 + p_1 p_2 + p_2^2) + 3p_1^3.$$

By symmetry,

$$E(X \mid L^c) = (1 - p_2)(1 - p_1)(2p_2 + p_1)$$
$$+ 2(1 - p_2)(p_2^2 + p_1 p_2 + p_1^2) + 3p_2^3.$$

In the case $p_2 = 1 - p_1$, after simplification,

$$E(X \mid L) = 2[1 - 2p_1(1 - p_1)] + p_1$$

and by symmetry,

$$E(X \mid L^c) = 2[1 - 2p_1(1 - p_1)] + 1 - p_1.$$

Thus, in this special case,

$$E(X) = \frac{1}{2} E(X \mid L) + \frac{1}{2} E(X \mid L^c) = \frac{1}{2}[5 - 8p_1(1 - p_1)]$$

and $3/2 \le E(X) \le 5/2$, the minimum being achieved if, and only if, $p_1 = 1/2$.

The play the winner strategy might only seem reasonable for a rat with a short memory. However, it has been applied as a strategy in clinical trials for comparing two drugs where L represents drug A, L^c represents drug B, and R_n represents a cure for the nth patient.

Problems

1. Assume that the rat has five trials and $p_2 = 1 - p_1$. Compute $P(A)$ and $E(X)$ in this case.

2. Assume that the rat has three trials and chooses an arm at random for each trial. Evaluate $E(X)$ in this case. Compare the two strategies.
3. Suppose that the rat makes an infinite sequence of trials. Let P_n be the probability that the rat chooses the left arm on the nth trial and assume that P_1 is arbitrary
   (a) Give a recursion relation for P_n.
   (b) Show that $p_2 = 1 - p_1$ implies that $P_n = p_1$ for $n = 2, 3, \ldots$.
   (c) Show that $p_1 = p_2 = 0$ implies that $P_{2n} = 1 - P_1$ and $P_{2n+1} = P_1$ for $n = 1, 2, \ldots$.
   (d) Show that $p_1 = p_2 = 1$ implies that $P_n = P_1$ for $n = 2, 3, \ldots$.
   (e) Obtain a recursion relation for $P_{n+1} - P_n$.
   (f) Assume that $p_1, p_2 \neq 0, 1$. Show that $\lim_{n \to \infty} P_n$ exists and evaluate the limit.
   (g) Recall that R_n is the event that the rat is rewarded on the nth trial. Evaluate $P(R_n)$ under the conditions in parts (b), (c), and (d). Evaluate $\lim_{n \to \infty} P(R_n)$ under the conditions in part (f).

20. SELECTIVE BREEDING

Consider the simplest possible genetic model. A certain characteristic, such as flower color, is determined by a pair of genes. Each gene in the pair is one of two varieties, say red (R) or white (W). Thus an individual can be any of three varieties (genotypes): RR, RW, or WW. An RR individual produces red flowers and a WW individual produces white flowers. Assume that R is dominant over W so that an RW individual produces red flowers. Assume that red flowers are very desirable and white flowers are to be eliminated. In each generation the breeder will eliminate individuals with white flowers and only permit those producing red flowers to propagate. Assume that, among the latter, mating is random.

Let $p_n = 1 - q_n$ be the conditional probability that an individual in the nth generation is RR given that it produces red flowers. We develop a recursion formula for p_n. Consider the nth generation before the elimination of WW individuals. Let A_1 be the event that an individual is the offspring of two RR parents, A_2 be the event that it is the offspring of an RR and an RW parent, and A_3 be the event that it is the offspring of two RW parents. Then

$$P(A_1) = p_{n-1}^2;$$
$$P(A_2) = 2p_{n-1}q_{n-1};$$
$$P(A_3) = q_{n-1}^2.$$

Let B be the event that the individual is RR and C be the event that the individual is RW. Then

$$P(B \mid A_1) = 1$$

$$P(B \mid A_2) = \frac{1}{2} \qquad P(C \mid A_2) = \frac{1}{2}$$

$$P(B \mid A_3) = \frac{1}{4} \qquad P(C \mid A_3) = \frac{1}{2}.$$

Thus

$$P(B) = p_{n-1}^2 + p_{n-1}q_{n-1} + \frac{q_{n-1}^2}{4} = \frac{(1 + p_{n-1})^2}{4}$$

and

$$P(C) = p_{n-1}q_{n-1} + \frac{q_{n-1}^2}{2} = \frac{1 - p_{n-1}^2}{2},$$

so that

$$P(B \cup C) = \frac{(1 + p_{n-1})(3 - p_{n-1})}{4}.$$

Hence

$$p_n = P(B \mid B \cup C) = \frac{1 + p_{n-1}}{3 - p_{n-1}}. \tag{1}$$

The expression in (1) shows that $p_n > p_{n-1}$ unless $p_{n-1} = 1$, in which case $p_n = 1$.

The p_n are increasing and bounded above by 1. Hence $\lim_{n \to \infty} p_n = p_0$ exists. Passing to the limit in (1), we obtain

$$p_0 = \frac{1 + p_0}{3 - p_0}, \tag{2}$$

from which $p_0 = 1$.

Roughly, p_n will be the proportion of RR individuals among those with red flowers. This breeding procedure guarantees that the proportion of RR individuals will grow and, after a sufficient number of generations, will be arbitrarily close to 1.

If the initial generation has proportions $1/4$, $1/2$, and $1/4$ of RR, RW, and WW, respectively, then $p_1 = 1/3$. From (1) we obtain $p_2 = 1/2$, $p_3 = 3/5$, $p_4 = 2/3$, $p_5 = 5/7$, and $p_6 = 3/4$, so that after six generations about $3/4$ of the individuals allowed to reproduce will be RR.

It can be shown that

$$q_n = \frac{2q_1}{n + 1 - (n - 1)p_1} \tag{3}$$

and hence that

$$p_n = \frac{n - 1 - (n - 3)p_1}{n + 1 - (n - 1)p_1}.$$

Problems

1. Suppose that an R gene passed to an offspring has probability p of mutating to a W gene. No mutations of W genes occur. Otherwise, assume the model and notation of this section.
   (a) Show that

   $$p_n = \frac{(1 - p)(1 + p_{n-1})}{3 + p - (1 - p)p_{n-1}}.$$

   (b) Show that $(p - p_{n-1})(p - p_n) \geq 0$ and $(p - p_{n-1})(p_n - p_{n-1}) \leq 0$.
   (c) Show that $p_0 = \lim_{n \to \infty} p_n$ exists and $p_0 = (1 - \sqrt{p})/(1 + \sqrt{p})$.
   (d) Show that $(1 - p_0)^2/(1 + p_0)^2 = p$.
2. Under the conditions of Problem 1, suppose that an individual produces white flowers. In terms of p_{n-1} and p, give the conditional probability that at least one of the individual's genes was mutated.
3. Let p_n, q_n, and r_n be the nth-generation probabilities of genotypes RR, RW, and WW, respectively, when no selection is made during breeding (i.e., WW individuals are not eliminated).
   (a) Obtain recursion relations for p_n, q_n, and r_n and show that they can be expressed in terms of $p_{n-1} + q_{n-1}/2$ and $r_{n-1} + q_{n-1}/2$.
   (b) Show that $p_n + q_n/2 = p_{n-1} + q_{n-1}/2$.
   (c) What does the observation in part (b) imply for p_n, q_n, and r_n?
   (d) Suppose that initially 20% of the population are RR, 70% are RW, and 10% are WW. What would the proportions be in the twentieth generation?
   (e) Interpret $p_n + q_n/2$ in words.
4. Use induction to prove (3).

GENERAL PROBLEMS

1.† Certain diseases are diagnosed by testing a sample of blood. Assume that a proportion p of individuals in the population have the disease.

Suppose that n individuals are being tested. A pooled blood sample will be formed by mixing blood from each individual. The test on the pooled sample will show the disease if at least one individual has the disease. In that case, each individual's blood will be tested separately. Let X be the number of tests performed.
(a) Find $E(X)$.
(b) For what values of p is it more economical (in the sense of smaller expectation) to use this method rather than testing each individual's blood separately at the beginning?

2. In Problem 1 suppose that $n = 2m$. The procedure begins as in Problem 1, but if the first test shows the disease, take half the individuals and put their blood in one pool and the other half in a second pool. If the first of these shows the disease, test each individual in the group and proceed in the same way to the second. Otherwise, test each individual in the second group. Let Y be the number of tests performed.
(a) Find $E(Y)$.
(b) For what values of p is it more economical to use this method rather than the one in Problem 1?

3. A trainee will have n training sessions to learn a particular task. If he has learned the task by any given session, he will perform it correctly from then on. If he has not learned the task by the ith session, he performs it correctly with probability p and he learns it in that session with probability r.
(a) What is the probability that he learns the task by the ith session?
(b) What is the probability that he performs the task correctly on the ith session?
(c) Let X be the number of times he performs the task correctly. Find $E(X)$.

4. In a single-elimination tournament between $n = 2^m$ players, the players are randomly paired for the first round. The $n/2$ first-round winners are randomly paired for the second round. This continues until, on the last (the mth) round, there is one game whose winner wins the tournament. Let the players be numbered from 1 to n and assume that if players i and j ($i < j$) play, player i will win. Thus player 1 will win the tournament.
(a) Find the probability that players 1 and 2 play the last match (so that player 2 is the runner-up).
(b) Let X be the number of matches won by player 2. Find $E(X)$.

5. Consider the situation in Section 20, but assume that all plants can reproduce. A plant is grown from a seed taken from a known plant. It is known that the pollen came from one of two plants, one with red and the other with white flowers. Let p and q be the proportions of R and W genes, respectively.

(a) Under each of the assumptions below, for what values of p do you think the red-flowered plant is more likely?
(i) Both the source of the seed and the offspring produce red flowers.
(ii) Both plants in (i) produce white flowers.
(iii) The source of the seed and the offspring produce red and white flowers, respectively.
(iv) The reverse of (iii).
(b) How might the results in part (a) provide evidence in a paternity case?

REFERENCES

Feller, William (1968). *An Introduction to Probability Theory and Its Applications*, Vol. 1, 3rd ed. New York: John Wiley and Sons, Inc. Contains a wide variety of applications.

Hillier, Fredrick S., and Lieberman, Gerald F. (1990). *Introduction to Operations Research*, 5th ed. New York: McGraw-Hill Book Company. Chapter 21 is concerned with reliability.

Ross, Sheldon M. (1985). *Introduction to Probability Models*, 3rd ed. Orlando, Fla.: Academic Press, Inc. Chapter 9 is concerned with reliability.

Zelen, M. (1969). Play the winner rule and the controlled clinical trial. *J. Amer. Statist. Assoc.* **64**, 131–146. Gives properties of the play the winner rule and modifications. Uses some mathematical tools not covered here.

4.

Probability Models on $\mathscr{R}^n$

21. RANDOM VARIABLES AND RANDOM VECTORS

Chapters 1 and 2 dealt largely with abstractions: arbitrary collections of events, abstract sample spaces. In the special cases to which we now turn, our concern will be with the uncertainty regarding the value or values of one or more quantities. A single quantity will be referred to as a *random variable* (rv); an ordered collection of (n) random variables will be referred to as a (*n-dimensional*) *random vector*. Simple rv's, which were defined in Section 7, only take one of a finite set of possible values. In the general case there is no restriction on the set of possible values for a rv. All events defined in terms of a rv are subsets of $\mathscr{R}$ (the real line), and events defined in terms of a random vector are subsets of $\mathscr{R}^n$ (n-dimensional Euclidean space).

Events defined by quantities become the objects of our attention in various ways: (a) an experiment whose outcome is a number (e.g., physical measurements such as height, weight, distance, time, etc.); (b) a number whose value is determined by the outcome of a nonnumerical experiment (e.g., the number of successes in a sequence of Bernoulli trials); and (c) any quantity whose specific value may be uncertain (e.g., the fiftieth digit in the decimal expansion for π, the number of fatal accidents on Route 80 in 1987, or the exact number of plays written by William Shakespeare).

In addition to these examples, following are some others to illustrate the type of phenomena we have in mind.

Example 1. A drug is administered to a cancerous rat, and after a period of time, the volume of the tumor is measured. If the experiment is repeated on another rat, the outcome may be a different number. Additional uncertainty in the measured volume is caused by measurement errors.

The minimum dosage required to eliminate a tumor is another example of a rv. Its value for a given patient may be difficult to determine exactly. Once the tumor has been eliminated, all that is known is an upper bound for the minimal required dose.

Example 2. A population consists of a certain voting precinct. From a list of the precinct's registered voters, 50 will be selected at random without replacement and interviewed to obtain certain numerical data. The sample space for this experiment can be chosen to be the class of all subsets of size 50 from the list of registered voters. The focus of our interest, however, is the collection of numbers the interviews yield (e.g., income, number of children, length of time at this address, etc.). The number of voters in the sample favoring proposal A, for example, is a simple rv.

Example 3. An election is about to be held for mayor. A week before the election the proportion, p, of voters who will vote for candidate A is unknown. One way of expressing uncertainty about p is through a probability model on $[0, 1]$. This might be a subjective model. We may also be concerned with the total turnout, k, for the election. The pair (p, k) would require a probability model on a two-dimensional set.

The common thread in all these examples is that we consider one or more quantities whose values determine and are determined by the occurrences of events.

The simple rv $X = \sum_{i=1}^{n} a_i I_{A_i}$ is constant on each member of the partition generated by $\{A_1, \ldots, A_n\}$. This may be generalized as follows.

Random Variable, Random Vector

A *random variable* (rv) is a real-valued function on some partition. A *random vector* is a vector whose components are rv's.

The one-element subsets of a sample space $\mathcal{S}$ for an experiment (Section 9) constitute a partition of $\mathcal{S}$ that is a refinement (see Problem 4.14) of any other partition of $\mathcal{S}$. Hence a rv related to an experiment can be defined to be a function on $\mathcal{S}$.

There is nothing random or variable about a function. There is uncertainty about the specific value a rv takes, and in repeated trials of an experiment its value may vary.

Let $\mathbb{S}(X)$ denote a subset of $\mathbb{R}$ that contains all values of the rv X. As in the case of sample spaces, our specific choice of $\mathbb{S}(X)$ is a matter of convenience, as long as it is a sure event. If $\mathbf{X}$ is a random vector, then $\mathbb{S}(\mathbf{X}) \subset \mathbb{R}^n$. If, for example, X is the proportion of voters favoring a certain candidate, then $\mathbb{S}(X) = [0, 1]$ would always suffice. For a set of 10 voters, $\{0, 1/10, 2/10, \ldots, 1\}$ would also suffice.

Model (Distribution) for a Random Vector

A *probability model for an n-dimensional random vector* $\mathbf{X}$ (rv if $n = 1$) is a probability model for an experiment with sample space $\mathbb{S}(\mathbf{X})$ (possibly $\mathbb{R}^n$). It is also called the *distribution* of the random vector.

In what follows, we consider the special case for which X is a function on a sample space $\mathbb{S}$ for an experiment, $\mathbb{E}$. The sample space for $\mathbb{E}$ may, or may not, be numerical. In this case, $\mathbb{S}(X) = \{X(s): s \in \mathbb{S}\}$ is the smallest subset of $\mathbb{R}$ that would suffice. If the only concern is with X, then this choice of $\mathbb{S}(X)$ is an adequate, but possibly inconvenient sample space. See Example 4.

The probabilities on $\mathbb{S}$ given by, say, P_1 induce probabilities P_2 on subsets of $\mathbb{S}(X)$. The only assignment P_2 of probabilities on $\mathbb{S}(X)$ that is consistent with the assignment P_1 on $\mathbb{S}$ is obtained as follows. Let $A \subset \mathbb{S}(X)$ and consider the event $A^* = \{s: X(s) \in A\} \subset \mathbb{S}$. To every event $A \subset \mathbb{S}(X)$ there corresponds, by this construction, a unique event $A^* \subset \mathbb{S}$. Thus A and A^* are alternative expressions for the same event and therefore must be assigned the same probability. That is, set $P_2(A) = P_1(A^*)$.

Example 4. A committee consists of two Federalists (Washington and Adams) and three Whigs (Jefferson, Madison, and Monroe). A subcommittee of size 2 is to be formed at random. A sample space $\mathbb{S}$ consists of all possible subcommittees of size 2, of which there are $\binom{5}{2} = 10$. If X is the number of Federalists on the subcommittee, X is a function on $\mathbb{S}$ with $\mathbb{S}(X) = \{0, 1, 2\}$. If the subcommittee is any of {Jefferson, Madison}, {Jefferson, Monroe} or {Madison, Monroe}, then $X = 0$. Thus $P_2(\{0\}) = 3/10$. Similarly, $P_2(\{1\}) = 6/10$ and $P_2(\{2\}) = 1/10$. Incidentally, $\mathbb{S}$ is the most convenient sample space because the probabilities on $\mathbb{S}(X)$ are easily determined from the equally likely model on $\mathbb{S}$.

The probability assignment P_2 to subsets of $\mathcal{S}(X)$ can be extended to subsets of $\mathcal{R}$. For $B \subset \mathcal{S}(X)^c$, set $P_2(B) = 0$, so that $A \subset \mathcal{R}$ implies that $P_2(A) = P_2(A\mathcal{S}(X))$. Similarly, an n-dimensional random vector defined on a sample space $\mathcal{S}$ results in a probability assignment to subsets of $\mathcal{R}^n$.

At first glance, it may appear that this assignment can be made to every $A \subset \mathcal{R}$. Unfortunately, A^*, the counterpart of A, may be an event (i.e., a subset of $\mathcal{S}$) to which P_1 does not assign probability. We avoid this difficulty by requiring that $P_1(A^*)$ is defined whenever A is an interval.

The class $\mathcal{C}$ of intervals in $\mathcal{R}$, including ϕ and $\mathcal{R}$, satisfies the conditions of Section 5 (see Example 5.3). Hence a probability assignment on $\mathcal{C}$ can be extended uniquely to subsets of $\mathcal{R}$ obtainable from $\mathcal{C}$ by countable set operations. These are the only events we need to consider. In $\mathcal{R}^n$, an analogous statement applies to the class of n-dimensional rectangles (see GP 2.7).

For $A \subset \mathcal{R}^n$, the notation of Section 7 will be used in a more general fashion. Thus $[\mathbf{X} \in A]$ is the event that the random vector $\mathbf{X}$ (rv if $n = 1$) takes a value in A. For $n = 1$ and $A = [a, b]$, for example, $[X \in A] = [a \le X \le b]$. For $n = 2$ and $A = [a, b] \times [c, d]$, write $[\mathbf{X} \in A] = [a \le X_1 \le b][c \le X_2 \le d] = [a \le X_1 \le b, c \le X_2 \le d]$ (the comma is read as "and").

We will no longer write $P_2(A)$, but $P[X \in A]$, and so on. Thus in Example 5, $P[X = 0] = 3/10$, $P[X = 1] = 6/10$, and $P[X = 2] = 1/10$.

We have dealt with probabilities induced on $\mathcal{R}$ by a rv defined on a sample space. Generally, if X is a rv for which $P[X \in A]$ is given or obtainable for any interval A, a probability model on $\mathcal{R}$ is determined.

Example 5. A channel is transmitting a teletype message. The channel is noisy in the sense that any character may be transmitted incorrectly. Consider a message consisting of a large number of characters and suppose that the appropriate probability model for this phenomenon is that of a sequence of Bernoulli trials with probability .99 of success (correct character transmitted). Let A_i be the event that the ith character is correctly transmitted ($i = 1, 2, \ldots$) and let X be the index of the first erroneous character. For any $k = 1, 2, \ldots$ we can evaluate $P[X = k]$. Note that $[X = k] = (\cap_{i=1}^{k-1} A_i)A_k^c$, and since the A_i are independent, $P[X = k] = (.99)^{k-1}(.01)$. Here $\mathcal{S}(X) = \{1, 2, \ldots\}$ and $P[X \in A] = \Sigma_{k \in A} P[X = k]$ for any interval A. For example, $P[3 \le X \le 10] = \Sigma_{k=3}^{10} (.99)^{k-1}(.01) = (.99)^2 - (.99)^{10}$ and $P[X > 5] = (.99)^5$.

If X is a rv and h is a real-valued function on $\mathcal{R}$, then $h(X)$ is the rv that takes the value $h(r)$ when $X = r$.

The probability model on $\mathcal{R}$ for $h(X)$ can be obtained in exactly the same manner used earlier in this section to obtain the model for a rv defined as a function on an abstract sample space. The particular sample space is $\mathcal{R}$ itself or, possibly, a subset of $\mathcal{R}$ so that

$$P[h(X) \in B] = P[X \in A], \tag{1}$$

where $A = \{a: h(a) \in B\}$. When B is an interval, the probability in the right hand side of (1) must be defined.

Capital letters from the latter part of the alphabet (such as X, Y, Z) will be used to denote rv's and random vectors. Subscripts will be used as needed to denote components of random vectors. Thus X_i is the ith component of the random vector $\mathbf{X} = (X_1, X_2, \ldots, X_n)$.

Representation

The notions of rv's and random vectors involve models (or distributions) on $\mathcal{R}^n$. A *representation* of a model on $\mathcal{R}^n$ is any function that can be used to calculate probabilities on n-dimensional rectangles (intervals when $n = 1$). Some representations are discussed in the next four sections.

We shall distinguish between two main cases: (1) $\mathcal{S}(X)$ is finite or countably infinite (X is called a *discrete* rv), and (2) $\mathcal{S}(X)$ is uncountable. Within the latter class only special subclasses will be considered.

Continuous Random Variables

One special subclass we will consider consists of *continuous* rv's. The model is also called continuous. These are characterized by $P[X = r] = 0$ for all $r \in \mathcal{R}$. Only events containing nondegenerate intervals may have positive probability. More elaborate discussions appear in Sections 23 and 24.

Example 6. Let X denote the number of rolls required to complete a game of craps (see Section 17), let Y be the number of innings required to complete a baseball game, and let Z be the number of traffic fatalities on Labor Day weekend. Each of these is an example of a discrete rv.

Example 7. Let X be the "true" weight of a nominal ton of bricks, and let Y denote the boiling point of water at the top of Mt. Everest. Quantities such as weight and temperature are generally treated as continuous variables, so that neither X nor Y is discrete. Their measured values, which are dependent on the devices used to obtain them, could be considered discrete (in fact, simple) rv's. Typically, however, when dealing with reasonably precise measurements, continuous models are adequately realistic and easier to use.

Theorem 1. *A rv X is discrete if, and only if, there is a countable partition $\{A_1, A_2, \ldots\}$ and a sequence $\{a_1, a_2, \ldots\}$ such that*

$$X = \sum_{i=1}^{\infty} a_i I_{A_i}.$$

Proof. Any X of this form is clearly discrete since there are at most countably many a_i's and the A_i's are disjoint. Conversely, if X is discrete, then $\mathcal{S}(X) = \{a_1, a_2, \ldots\}$, a countable set, and the events $A_i = [X = a_i]$, $i = 1, 2, \ldots$ must form a partition. Since $X = \Sigma_i\, a_i I_{A_i}$, the conclusion follows.

Partition Induced by X

The *partition induced by* a rv X consists of all the events of the form $[X = r]$ for $r \in \mathcal{R}$.

As shown in Section 7, the same rv may be represented in many ways. There we were dealing with simple rv's. In the following example we have a nonsimple rv represented as a linear combination of indicators of events that do not form a partition.

Example 8. Let X be the number of customers who arrive at a store during a day, a day being the time interval $(0, 1]$. Clearly, X is a discrete rv and $X = \Sigma_{k=0}^{\infty} k I_{A_k}$, where $A_k = [X = k]$. Another representation of X is $X = \Sigma_{j=0}^{\infty} j I_{B_j} + \Sigma_{m=0}^{\infty} m I_{C_m} = Y + Z$, where B_j is the event that j customers arrive during $(0, 1/2]$ and C_m is the event that m customers arrive during $(1/2, 1]$. The events B_j and C_m for $j, m = 0, 1, \ldots$ are not disjoint. The rv's Y and Z have obvious meanings.

The theorem below, a generalization of Theorem 7.1, gives a convenient representation of a nonnegative, integer-valued rv in terms of tail events.

Theorem 2. *Let X be a nonnegative, integer-valued rv. Then $X = \Sigma_{k=0}^{\infty} I_{B_k}$, where $B_k = [X > k]$.*

Proof. Left as a problem.

If every value in an interval is a possible value of X, then X is not discrete. Even such a rv can be written as a countable linear combination of indicators, but the events will not form a partition.

Example 9.† Any $c \in [0, 1]$ has binary representation $c = \sum_{n=1}^{\infty} a_n(1/2)^n$, where each $a_n = 0$ or 1. Consider an infinite sequence of Bernoulli trials (see Section 15) and let A_n be the event "success on the nth trial." Then $X = \sum_{n=1}^{\infty} (1/2)^n I_{A_n}$ is a rv that can take any value in $[0, 1]$. However, $\{A_1, A_2, \ldots\}$ is not a partition. The construction in this example can easily be extended to more general rv's and leads to the conclusion that *every rv is a countable linear combination of indicators.*

The following examples demonstrate the calculation of probabilities of intervals for continuous rv's.

Example 10. Consider the triangle with vertices $(0, 0)$, $(2, 0)$, and $(2, 1)$ which has area 1. Pick a point at random in the interior of this triangle. Denote this random vector by (X, Y). Then $P[X \leq a] = (1/2)a(a/2) = a^2/4$ for $0 < a < 2$, so that $P[a < X \leq b] = b^2/4 - a^2/4$ for $0 < a < b < 2$.

Example 11.† A point is chosen at random from the interior of a circle of radius 1. Let X be the distance of the point from the center. Then

$$P[a \leq X \leq b] = b^2 - a^2 \quad \text{if } 0 \leq a \leq b \leq 1;$$
$$= b^2 \quad \text{if } a < 0 \leq b < 1;$$
$$= 1 - a^2 \quad \text{if } 0 < a < 1 \leq b;$$
$$= 1 \quad \text{if } a \leq 0, b \geq 1;$$
$$= 0 \quad \text{if } a \geq 1 \text{ or } b \leq 0.$$

In Examples 10 and 11 the models on $\mathcal{R}$ were determined by the probabilities in sample spaces for experiments. If X is the income of an individual in some population, the model on $\mathcal{R}$ is determined by the proportions of various incomes in the population. For other rv's the models may be determined by personal assessment of probabilities.

Summary

Random variable (rv): A real-valued function on a partition.
Random vector: A vector whose components are rv's.
Distribution: A model on $\mathcal{R}^n$.
Discrete rv: Any rv that can take values in a finite or countably infinite
   set $\{r_1, r_2, \ldots\}$.
Discrete random vector: Each component is a discrete rv.

Continuous rv X: $P[X = r] = 0$ for all $r \in \mathscr{R}$.

Partition induced by a rv X: The class of events of the form $[X = r]$.

Representation of a model on $\mathscr{R}^n$:   Any function on $\mathscr{R}^n$ that can be used to calculate probabilities of n-dimensional rectangles.

Any discrete rv can be written in the form $X = \Sigma_i r_i I_{A_i}$, where $\{A_1, A_2, \ldots\}$ is a partition. If X is a nonnegative, integer-valued rv, then $X = \Sigma_{k=0}^{\infty} I_{B_k}$, where $B_k = [X > k]$.

Problems

For each experiment in Problems 1 to 8, give a rv or random vector that may be of interest.

1. A poker hand (five cards) is dealt.
2. Five cancer patients are administered drug I and five are administered drug II.
3. The Dow Jones Industrial Average is observed for a week.
4. A coin is tossed five times.
5. A point is selected in the subset of the plane bounded by the lines $x = 0$, $y = 0$, and $x + y = 1$.
6. A point is selected in the interior of the sphere in $\mathscr{R}^3$ of radius 1 centered at the origin.
7. Three cars are selected from today's production at Power Motor Corp.
8. A coin is tossed until a head is obtained.
9. In an infinite sequence of Bernoulli trials, let X be the number of trials until the first success. Let A_i be the event success on the ith trial.
    (a) Write the events in the partition induced by X in terms of the A_i.
    (b) Write the tail events for X in terms of the A_i.
    (c) Give two expressions for X as a linear combinations of indicators using parts (a) and (b).
10. Let Y be as defined in Example 10. Find $P[a < Y \le b]$ for $0 < a < b < 1$.
11. (a) Let X be a nonnegative rv, $U = [X]$, and $V = X - U$, so that U and V are the integer and fractional parts of X, respectively. Use Example 9 to express X as a countable linear combination of indicators.
    (b) Extend part (a) to the case in which X need not be nonnegative.
12. Prove Theorem 2.

In Problems 13 to 15, prove the stated extensions of Theorem 2.

13.† If X is an integer-valued rv, then $X = \Sigma_{k=0}^{\infty} I_{C_k} - \Sigma_{k=1}^{\infty} I_{D_k}$ where $C_k = [X > k]$ and $D_k = [X \le -k]$.

14. If X is a nonnegative, discrete rv with values $0 = a_0 < a_1 < a_2 < \cdots$, then $X = \Sigma_{k=0}^{\infty} (a_{k+1} - a_k)I_{C_k}$, where $C_k = [X > a_k]$.

15.† If X is a discrete rv with values $0 = a_0 < a_1 < a_2 < \cdots$ and $0 = b_0 > b_1 > b_2 > \cdots$, then $X = \Sigma_{k=0}^{\infty} (a_{k+1} - a_k)I_{C_k} - \Sigma_{j=0}^{\infty} (b_j - b_{j+1})I_{D_{j+1}}$, where $C_k = [X > a_k]$ and $D_j = [X < b_j]$.

22. DISCRETE MODELS ON $\mathscr{R}^n$

A. Probability Functions

Probability models for discrete random variables or vectors are most often given by specifying a countable set $\mathcal{S}(\mathbf{X})$ and the probabilities of all points in $\mathcal{S}(\mathbf{X})$.

Probability Function

The function on $\mathscr{R}$ with values

$$f(r) = P[X = r] \tag{1}$$

is called the *probability function of the rv* X. The definition applies as well to random vectors $\mathbf{X} = (X_1, \ldots, X_n)$, but then the argument is $\mathbf{r} = (r_1, \ldots, r_n) \in \mathscr{R}^n$. In the case of random vectors (when $n \ge 2$) f is also called the *joint probability function* of the rv's $X_1, \ldots, X_n$. The same terminology is used if each X_i is a random vector.

The events $[\mathbf{X} = \mathbf{r}]$ for $\mathbf{r} \in \mathcal{S}(\mathbf{X})$ (regardless of dimension) constitute the partition induced by $\mathbf{X}$, so that f satisfies

(i) $f(\mathbf{r}) \ge 0$ and equals 0 if $\mathbf{r} \notin \mathcal{S}(\mathbf{X})$.

(ii) $\displaystyle\sum_{\mathbf{r} \in \mathcal{S}(\mathbf{X})} f(\mathbf{r}) = 1.$ $\tag{2}$

Any function satisfying (2) for a countable set $\mathcal{S}(\mathbf{X}) = \mathcal{S} \subset \mathscr{R}^n$ is a probability function and represents a probability model on $\mathscr{R}^n$. The probability of any subset $A \subset \mathscr{R}^n$ is

$$P(A) = \sum_{\mathbf{r} \in A\mathcal{S}} f(\mathbf{r}) = P[\mathbf{X} \in A], \tag{3}$$

where $\mathbf{X}$ is a random vector with probability function f.

When a formula for a probability function is given it will be understood that $f(\mathbf{r}) = 0$ for any value of $\mathbf{r} \in \mathcal{R}^n$ omitted from the expression. However, it is essential to specify the set of values for which the formula is to be applied. Thus

$$f(x) = \frac{x + 1}{3} \qquad \text{if } x = 0, 1$$

is a probability function and so is

$$f(x) = \frac{x + 1}{3} \qquad \text{if } x = -\frac{1}{2}, \frac{1}{3}, \frac{1}{6}.$$

The set of all possible x-values *must* always be given. The function $f(x) = (x + 1)/3$ is not a probability function until the applicable x's are given.

At times the probability function of the random vector $\mathbf{X} = (X_1, \ldots, X_n)$ will be denoted by $f_{\mathbf{X}}$ or by $f_{X_1,\ldots,X_n}$.

Example 1.†   Example 21.5 concerned a committee consisting of two Federalists and three Whigs. Let X be the number of Federalists on a subcommittee of size 2 chosen at random. The probability function of the simple rv X is

$$f(r) = \frac{\binom{2}{r}\binom{3}{2 - r}}{\binom{5}{2}} \qquad \text{if } r = 0, 1, 2 \tag{4}$$

$$= 0 \qquad\qquad \text{otherwise.}$$

This is the probability function of a *hypergeometric model* and X is a *hypergeometric rv*.

In the following, X is discrete but not simple.

Example 2.   Your first roll in a game of craps is a 4. Let X be the number of additional rolls to complete the game. Then the probability function of X is $f(r) = (1/4)(3/4)^{r-1}$ for $r = 1, 2, \ldots$. This follows from the fact that $X = r$ if, and only if, the first $r - 1$ rolls failed to produce either a 4 or a 7 and the rth roll did result in a 4 or a 7. The function f belongs to the family of *geometric* probability functions.

From f, probabilities of various events concerning X can be computed. For example,

$$P[X \text{ even}] = \sum_{k=1}^{\infty} f(2k) = \left(\frac{1}{4}\right) \sum_{k=1}^{\infty} \left(\frac{3}{4}\right)^{2k-1} = \frac{3}{7}$$

and

$$P[3 \le X \le 7] = \left(\frac{1}{4}\right) \sum_{k=3}^{7} \left(\frac{3}{4}\right)^{k-1} = \left(\frac{3}{4}\right)^2 - \left(\frac{3}{4}\right)^7.$$

Probability functions on $\mathcal{R}^n$ for $n \ge 2$ give rise to probability functions on lower-dimensional spaces. This is illustrated for $n = 2$ in the following example.

Example 3.†   Let X_1 and X_2 be the minimum and maximum, respectively, of the results when two fair dice are rolled and consider the two-dimensional random vector $\mathbf{X} = (X_1, X_2)$. Then, for any given pair of integers $1 \le r_1 \le r_2 \le 6$,

$$
\begin{aligned}
f_{\mathbf{X}}(r_1, r_2) &= P[X_1 = r_1, X_2 = r_2] \\
&= 2/36 \quad \text{if } r_1 < r_2 \\
&= 1/36 \quad \text{if } r_1 = r_2.
\end{aligned}
$$

The factor 2 in the case $r_1 < r_2$ results from the fact that the outcomes can come in either of two orders.

Explicitly, these probabilities are given in Table 1.

Observe that the functions $g(r_1) = \Sigma_{r_2} f_{\mathbf{X}}(\mathbf{r})$ and $h(r_2) = \Sigma_{r_1} f_{\mathbf{X}}(\mathbf{r})$ are probability functions. In fact, g and h are the probability functions of X_1 and X_2, respectively.

Table 1

r_1	r_2 1	2	3	4	5	6	$\Sigma_{r_2} f_{\mathbf{X}}(\mathbf{r})$
1	1/36	2/36	2/36	2/36	2/36	2/36	11/36
2	0	1/36	2/36	2/36	2/36	2/36	9/36
3	0	0	1/36	2/36	2/36	2/36	7/36
4	0	0	0	1/36	2/36	2/36	5/36
5	0	0	0	0	1/36	2/36	3/36
6	0	0	0	0	0	1/36	1/36
$\Sigma_{r_1} f_{\mathbf{X}}(\mathbf{r})$	1/36	3/36	5/36	7/36	9/36	11/36	1

To see that g is the probability function of X_1, note that the events $[X_1 = r_1, X_2 = 1], \ldots, [X_1 = r_1, X_2 = 6]$ form a partition of the event $[X_1 = r_1]$. A similar argument shows that h is the probability function of X_2.

All probability functions on lower-dimensional spaces can be obtained in a similar manner.

Marginal Probability Function

Let $\mathbf{X}$ and $\mathbf{Y}$ be random vectors with joint probability function $f_{\mathbf{X},\mathbf{Y}}$. The probability function $f_{\mathbf{X}}(\mathbf{r}) = \Sigma_{\mathbf{s}} f_{\mathbf{X},\mathbf{Y}}(\mathbf{r}, \mathbf{s})$ is called the *marginal probability function* of $\mathbf{X}$, a term suggested by the display of these probability functions in margins as in Table 1.

Following is an example with $n = 3$ whose only purpose is to illustrate techniques of manipulation for higher-dimensional discrete random vectors.

Example 4. Let $\mathbf{W} = (X, Y, Z)$ be a random vector with probability function

$$f(x, y, z) = (x - 1) \left(\frac{1}{2}\right)^z \qquad \text{if } 2 \leq x < y < z < \infty,$$

where x, y, and z are integers. Then

$$f_{X,Y}(x, y) = \sum_{z=y+1}^{\infty} (x - 1) \left(\frac{1}{2}\right)^z = (x - 1) \left(\frac{1}{2}\right)^y \sum_{z-y=1}^{\infty} \left(\frac{1}{2}\right)^{z-y}$$

$$= (x - 1) \left(\frac{1}{2}\right)^y \qquad \text{if } y > x \geq 2.$$

Also,

$$f_X(x) = \sum_{y=x+1}^{\infty} (x - 1) \left(\frac{1}{2}\right)^y = (x - 1) \left(\frac{1}{2}\right)^x \qquad \text{if } x = 2, 3, \ldots.$$

The proof that f_X is a probability function is the subject of Problem 7.

At times it will be convenient to express a probability function in the form $f(r) = cg(r)$, where g is a nonnegative function with finite sum and c is a constant whose value can be determined from the relation $\Sigma_r g(r) = 1/c$.

Example 5. If $g(r) = 1/(r + 1)$ for $r = 1, 2, \ldots$, then g is nonnegative. Since $\sum_{r=1}^{\infty} g(r) = \infty$, no multiple of g is a probability function. However, $\sum_{r=1}^{\infty} g(r)g(r + 1) < \infty$, so that for suitable c, $f(r) = cg(r)g(r + 1)$ is a probability function.

Example 6. Let

$$f_{\mathbf{X}}(s, t, u) = f_{X_1, X_2, X_3}(s, t, u)$$

$$= c(s + 2t) \quad \text{if } s, t, u = 0, 1, 2 \quad \text{and} \quad s \le u \le t.$$

The marginal probability function of (X_1, X_2) is $f_{X_1, X_2}(s, t) = c(t - s + 1)(s + 2t)$ for $s, t = 0, 1, 2$ and $s \le t$. The marginal probability function of X_1 is $f_{X_1}(s) = c(16 + 2s/3 - 9s^2/2 + 5s^3/6)$ for $s = 0, 1, 2$. Simple calculations show that $c = 1/35$.

The next two examples illustrate computations of probabilities from probability functions.

Example 7. Let T_1 be the waiting time (number of tosses) for coin C_1 to produce a head, and let T_2 be the corresponding waiting time for coin C_2. If the respective probabilities of heads are p_1 and p_2 and if all tosses are independent, the joint probability function of T_1 and T_2 is

$$f_{T_1, T_2}(r, s) = p_1 q_1^{r-1} p_2 q_2^{s-1} \quad \text{if } r, s = 1, 2, \ldots,$$

where $q_i = 1 - p_i$ $(i = 1, 2)$. The probability that C_2's waiting time is longer than C_1's is

$$P[T_1 < T_2] = p_1 p_2 \sum_{r=1}^{\infty} \left(q_1^{r-1} \sum_{s=r+1}^{\infty} q_2^{s-1} \right)$$

$$= p_1 p_2 \sum_{r=1}^{\infty} q_1^{r-1} \frac{q_2^r}{1 - q_2} = \frac{p_1 q_2}{1 - q_1 q_2}.$$

Furthermore, by symmetry,

$$P[T_1 > T_2] = \frac{q_1 p_2}{1 - q_1 q_2}$$

and hence

$$P[T_1 = T_2] = 1 - \frac{p_1 q_2}{1 - q_1 q_2} - \frac{q_1 p_2}{1 - q_1 q_2} = \frac{p_1 p_2}{1 - q_1 q_2}.$$

Example 8. Referring to Example 2, what is the probability that having rolled a 4, you will win the game? This problem was solved previously in Section 17 and we give an alternative solution here.

Let T_k be the number of rolls required to produce the result k. The problem is to find $P[T_4 < T_7]$. For $r < s$, the event $[T_4 = r$ and $T_7 = s]$ occurs if, and only if, the first $r - 1$ results are neither 4 nor 7, the rth result is 4, the next $s - r - 1$ results are not 7, and the sth result is 7. Thus

$$P[T_4 = r, T_7 = s] = \left(\frac{27}{36}\right)^{r-1}\left(\frac{3}{36}\right)\left(\frac{30}{36}\right)^{s-r-1}\left(\frac{6}{36}\right).$$

This would be enough to solve the problem, but for completeness, observe that for $r > s$,

$$P[T_4 = r, T_7 = s] = \left(\frac{27}{36}\right)^{s-1}\left(\frac{6}{36}\right)\left(\frac{33}{36}\right)^{r-s-1}\left(\frac{3}{36}\right).$$

Thus the joint probability function of T_4 and T_7 is

$$g(r, s) = \frac{1}{72}\left(\frac{3}{4}\right)^{r-1}\left(\frac{5}{6}\right)^{s-r-1} \qquad \text{if } 1 \le r < s < \infty;$$

$$= \frac{1}{72}\left(\frac{3}{4}\right)^{s-1}\left(\frac{11}{12}\right)^{r-s-1} \qquad \text{if } 1 \le s < r < \infty.$$

Then

$$P[T_4 < T_7] = \frac{1}{72}\sum_{r=1}^{\infty}\sum_{s=r+1}^{\infty}\left(\frac{3}{4}\right)^{r-1}\left(\frac{5}{6}\right)^{s-r-1}$$

$$= \frac{1}{72}\left[\sum_{j=0}^{\infty}\left(\frac{3}{4}\right)^{j}\right]\left[\sum_{k=0}^{\infty}\left(\frac{5}{6}\right)^{k}\right] = \frac{1}{3}.$$

B. Applications to Sampling

This subsection contains a theorem and two corollaries that will be useful later. Section 10 contains the combinatorial notation used below.

The remainder of this subsection broadens the analysis of sampling models begun in Theorem 13.1.

Theorem 1. *Suppose that a random sample of size n is taken from $J = \{1, \ldots, M\}$. Let U_i be the ith number drawn, $U = (U_1, \ldots, U_n)$, and $J^n = \times_{i=1}^{n} J$. For sampling without replacement,*

$$f_U(t) = \frac{1}{(M)_n} \qquad \text{if } t \in A_n, \tag{5}$$

and for sampling with replacement,

$$f_{\mathbf{U}}(\mathbf{t}) = \frac{1}{M^n} \text{ if } \mathbf{t} \in J^n, \tag{6}$$

where $A_n \subset J^n$ consists of those elements with distinct components.

Proof. We will prove (5) and leave the proof of (6) as a problem. When sampling is without replacement (so that $n \le M$), each element of A_n has the same probability of being chosen and $P(A_n) = 1$. There are $(M)_n$ elements $\mathbf{t} \in A_n$.

Corollary 1. *Under the conditions of Theorem 1, every r-dimensional sub-vector, $\mathbf{V}$, of $\mathbf{U}$ has the same marginal probability function,*

$$f_{\mathbf{V}}(\mathbf{s}) = \frac{1}{(M)_r} \qquad \text{if } \mathbf{s} \in A_r \tag{7}$$

for sampling without replacement, and

$$f_{\mathbf{V}}(\mathbf{s}) = \frac{1}{M^r} \qquad \text{if } \mathbf{s} \in \underset{i=1}{\overset{r}{\times}} J \tag{8}$$

for sampling with replacement.

Proof. To obtain $f_{\mathbf{V}}(\mathbf{s})$, the marginal probability function of $\mathbf{V}$, sum $f_{\mathbf{U}}(\mathbf{t})$ over those $\mathbf{t}$ such that the r components of $\mathbf{t}$ corresponding to $\mathbf{s}$ are all fixed. The number of terms in the sum depends only on r and on whether sampling is with or without replacement. Expressions (7) and (8) are obtained by setting $\mathbf{s} = (t_1, \ldots, t_r)$ and applying (5) and (6), respectively.

Corollary 2. *Let $w_1, \ldots, w_M$ be arbitrary numbers, not necessarily all distinct. Let $X_i = \sum_{j=1}^{M} w_j I_{[U_i=j]}$ and let $\mathbf{Y} = (X_{i_1}, \ldots, X_{i_r})$. Then, under the conditions of Theorem 1, the marginal probability function of $\mathbf{Y}$ is the same for any choice of distinct indices $i_1, \ldots, i_r$.*

Proof. Immediate from Corollary 1.

A less abstract way of describing the situation in Corollary 2 is to consider a population of size M in which the jth member is assigned the value w_j (e.g., income of the jth person in East Lansing or milk production of the jth cow in Farmer Jones' barn). A random sample of size n is taken and X_i is the observed value of the ith item in the sample. Corollary 2 states that each X_i has the same distribution, each pair (X_i, X_j) $(i \ne j)$ has the same distribution, and so on.

C. Commonly Encountered Probability Functions

We close this section with a brief list of some standard families of probability functions on $\mathcal{R}$. A more complete discussion is given in Chapter 7. (Section numbers are given in parentheses next to the name.)

1. *Binomial* (39): $f(x) = \binom{n}{x} p^x q^{n-x}$, where $x = 0, 1, \ldots, n; n = 1,$ $2, \ldots; 0 \leq p \leq 1$; and $q = 1 - p$. The binomial is the distribution of the number of successes in n Bernoulli trials.

2. *Hypergeometric* (39): $f(x) = \binom{M_1}{x}\binom{M_2}{n-x} \Big/ \binom{M}{n}$, where $x = 0,$ $1, \ldots, n; M_1, M_2 = 1, 2, \ldots; M = M_1 + M_2$. The hypergeometric is the distribution of the number of successes in a random sample drawn without replacement from a population of size M containing M_1 successes.

3. *Poisson* (39): $f(x) = (\mu^x/x!)e^{-\mu}$, where $x = 0, 1, \ldots; \mu > 0$. The Poisson approximates the binomial when n is large and p is small. Under certain assumptions the Poisson is the distribution of the number of occurrences of an event in a fixed period of time.

4. *Geometric* (40): $f(x) = pq^{x-1}$, where $x = 1, 2, \ldots; 0 < p \leq 1$; $q = 1 - p$. The geometric is the distribution of the waiting time for one success in a sequence of Bernoulli trials.

5. *Negative binomial* (40): $f(x) = \binom{x-1}{r-1} p^r q^{x-r}$, where $x = r, r +$ $1, \ldots; r = 1, 2, \ldots; 0 < p \leq 1; q = 1 - p$. The negative binomial is the distribution of the waiting time for the rth success in a sequence of Bernoulli trials.

6. *Uniform on the integers* $1, \ldots, N$: $f(x) = 1/N$, where $x = 1,$ $\ldots, N$.

Summary

Probability function of a discrete random vector $\mathbf{X}$: $f_{\mathbf{X}}$ defined by $f_{\mathbf{X}}(\mathbf{r}) = P[\mathbf{X} = \mathbf{r}]$ for $\mathbf{r} \in \mathcal{S}(\mathbf{X})$.
Marginal probability function of $\mathbf{Y}$: $f_{\mathbf{Y}}(\mathbf{u}) = \Sigma_\mathbf{v} \, f_{\mathbf{X}}(\mathbf{u}, \mathbf{v})$, where $\mathbf{X} = (\mathbf{Y}, \mathbf{Z})$.

A function f on $\mathcal{R}^n$ is a probability function if, and only if, (1) $f \geq 0$, (2) $\{\mathbf{r}: f(\mathbf{r}) > 0\}$ is countable, and (3) $\Sigma_\mathbf{r} \, f(\mathbf{r}) = 1$.
If f is the probability function of the rv or random vector $\mathbf{X}$ and $A \subset \mathcal{R}^n$, then $P[\mathbf{X} \in A] = \Sigma_{\mathbf{r} \in A\mathcal{S}} \, f(\mathbf{r})$, where $\mathcal{S} = \mathcal{S}(\mathbf{X})$.

Let $X_1, \ldots, X_n$ be a sample from a numerical population. Corollary 2 states that each X_i has the same distribution, each pair (X_i, X_j) $(i \neq j)$ has the same distribution, and so on.

Problems

1. For each of the functions f given below, answer the following questions:
    (i) Is f a probability function?
    (ii) Does there exist a number c such that $cf(x)$ is a probability function? If so, find it.
    (a) $f(x) = 1/2^x$ if $x = 0, 2, \ldots$
    (b) $f(x) = 1/2^x$ if x is an integer in $(-\infty, \infty)$
    (c) $f(x) = x$ if $|x| = 1, 2, 3, 4, 5$
    (d) $f(x) = |x|$ if $|x| = 1, 2, 3, 4, 5$
2. Let $f(r) = c(1/3)^{|r|}$ if r is an integer.
    (a) Find c so that f is a probability function.
    (b) Let X be a rv with probability function f and compute $P[|X| > 10]$ and $P[X \text{ is even}]$.
3. Let $f(n) = (n - 1)(1/2)^n$ for $n = 2, 3, \ldots$ (a special case of the negative binomial).
    (a) Verify that f is a probability function.
    (b) For a rv X with probability function f, find $P[X \geq k]$.
4. Two players play the following game. Player 1 tosses a fair coin until a head results. Player 2 then does the same. Let m and n be the required numbers of tosses. Player 1 then pays $\$n$ to player 2 if $n > m$ and receives $\$n$ from player 2 if $n < m$. If $n = m$, the payoff is 0. Let X be the net gain to player 1.
    (a) Find the probability function of X.
    (b) Find $P[X > 0]$.
5. Consider a sequence of Bernoulli trials in which the probability of success on a single trial is p $(0 < p < 1)$. Let X be the number of trials required to obtain the first success.
    (a) Show that the probability function of X is geometric.
    (b) Find $P[X \text{ even}]$ and verify $P[X \text{ even}] < 1/2$.
    (c) Can you conclude that $P[X \text{ even}] < 1/2$ without computing $P[X \text{ even}]$? *Hint*: Evaluate $P[X = 2n]/P[X = 2n - 1]$.
6. Verify the statement about $g(r)g(r + 1)$ in Example 5. Evaluate c. *Hint*: Express $g(r)g(r + 1)$ as a difference.
7. Find the remaining two- and one-dimensional probability functions in Example 4.

8. A fair die is rolled and a fair coin is tossed twice. Let X_1 be the resulting face on the die and X_2 be the number of heads resulting from the coin tosses.
   (a) Give the equally likely probability model for this experiment.
   (b) Give the resulting probability function of the two-dimensional random vector X.
   (c) Find the marginal probability functions of X_1 and of X_2.

9. A small town has a population of 50 voters of whom 20 favor proposition Q, 25 are opposed and 5 are undecided. A random sample of 5 voters is taken. Let X_1 be the number of voters in the sample who favor Q and X_2 be the number opposed to Q.
   (a) Find the joint probability function of X_1 and X_2.
   (b) Find $P[X_1 > X_2]$.

10.† Consider a sequence of rolls of a fair die. Let X be the number of rolls until a 1 or 2 results and Y be the number of rolls until a 1, 2, or 3 results.
   (a) Find the joint probability function of X and Y.
   (b) Find the two marginal probability functions.

11. A certain city has square blocks. Consider the Cartesian coordinate system with the origin at one of the intersections of streets and axes along these streets. Let the unit of measurement be the block length. You start walking at the origin. At each intersection, you choose one of the four directions at random. Let X and Y be the coordinates of your location after you have walked five blocks.
   (a) Find the joint probability function of X and Y.
   (b) Find the marginal probability function of X.
   (c) Do the two marginal probability functions differ? Explain.

12. An urn has N balls numbered $1, \ldots, N$. Balls are selected at random with replacement until ball 1 is drawn. Let X be the number of draws required and Y be the largest number drawn.
   (a) Find the joint probability function of X and Y. *Hint*: $P[X = k, Y = l] = P[X = k]P(Y = l \mid X = k)$.
   (b) Find the marginal probability function of Y.

13. In Example 8, find:
   (a) The marginal probability functions of T_4 and of T_7.
   (b) $P(T_4 < T_7 \mid T_7 = s)$.

14. Use the joint probability function of T_4 and T_7 in Example 8 to find the probability function of X, the rv defined in Example 2 as the waiting time for termination given that the first roll was 4.

15. Consider a population of M items of which M_1 are defective. We sample at random, without replacement, until all M_1 defectives have

been drawn. Let T_i be the number of draws required to produce the ith defective. Let $\Delta_1 = T_1$ and for $i = 2, \ldots, M_1$, let $\Delta_i = T_i - T_{i-1}$. Let $\Delta_{M_1+1} = M + 1 - T_{M_1}$. Show that any subvector of $\Delta = (\Delta_1, \ldots, \Delta_{M_1+1})$ has a uniform distribution, depending only on its dimension and M.

16. Let A_1, A_2, A_3 be any events.
    (a) Give a table of the joint probability function of $I_{A_1}, I_{A_2}, I_{A_3}$.
    (b) Find the two-dimensional marginal probability functions.
    (c) Find the one-dimensinal marginal probability functions.
17. Prove (6).
18. Complete the proof of Theorem 2.

23. CONTINUOUS MODELS ON $\mathcal{R}^n$

A. Densities

The basic procedure for calculating probabilities of events for a discrete model is summation. Thus if X is a discrete rv, each point in $\mathcal{S}(X)$ is assigned a probability and the probability of any other event is obtained by adding the probabilities of the points contained in it.

Another basic procedure for producing probabilities of subsets of $\mathcal{R}$ or, more generally, $\mathcal{R}^n$, is integration.

Density

Let f be a real-valued function on $\mathcal{R}$ satisfying

$$f(t) \geq 0 \quad \text{for all } t \in \mathcal{R} \quad \text{and} \quad \int_{-\infty}^{\infty} f(t)\, dt = 1. \tag{1}$$

Any function satisfying (1) is called a *density* on $\mathcal{R}$.

Let A be an interval with endpoints a and b, $a \leq b$. We allow the possibility that $a = -\infty$ or $b = \infty$. Furthermore, A can include either or both endpoints when they are finite. To A we assign probability

$$\int_a^b f(t)\, dt. \tag{2}$$

Every set consisting of a single element (and hence any countable set) has probability 0. Probabilities assigned to intervals by continuous models do not depend on whether or not endpoints are included. Furthermore, changing f on a countable set does not affect the probability model.

The assignment given by (2) is obviously between 0 and 1 and the mathematical properties of integration on $\mathcal{R}$ guarantee that the rules (5.4) of probability will be satisfied. Hence a function satisfying (1) represents a probability model on $\mathcal{R}$.

Let X be a continuous rv for which $P[X \in A]$ is given by (2) when A is an interval with endpoints $a \le b$. Then f is called the *density of X*. As explained in Section 25, not every continuous rv has a density.

From (2) we see that if $\varepsilon > 0$ is small and f is continuous at a, then

$$P\left[a - \frac{\varepsilon}{2} < X < a + \frac{\varepsilon}{2}\right] \approx \varepsilon f(a). \tag{3}$$

In the continuous case $P(A) = 0$ should not be interpreted to mean that A is impossible. Continuous models are used when individual values of a rv are so exceedingly unlikely that no substantial error is made by assuming that they have probability 0 and assigning positive probabilities only to intervals.

Example 1. Consider a sequence of n random digits preceded by a decimal point and independently chosen from $\{0, 1, \ldots, 9\}$. Every n-digit number in $[0, 1)$ has probability $1/10^n$ of being selected. For n large, an approximate model is given by the density $f(t) = 1$ for $0 \le t < 1$ and $f(t) = 0$ otherwise. Letting $f(1) = 1$ instead of 0 would not change the model. This is the *uniform density* on $[0, 1)$. Uniform models are discussed more fully in Section 43.

Many of the densities we consider are positive only on a proper subset $\mathcal{S}$ of $\mathcal{R}$. As in our treatment of probability functions in Section 22, when we give an expression for $f(t)$ for $t \in \mathcal{S}$, it will be understood that $f(t) = 0$ for $t \notin \mathcal{S}$.

Example 2.† Let $f(t) = 1 - |t|$ for $|t| < 1$. The function f (see Figure 1) is nonnegative on $\mathcal{S} = [-1, 1]$ and

$$\int_{-1}^{1} f(t)\, dt = \int_{-1}^{1} (1 - |t|)\, dt = 2 \int_{0}^{1} (1 - t)\, dt = 1.$$

Hence f is a density if it is assumed that $f(t) = 0$ for $|t| \ge 1$.

Let X be a rv with density f. Then

$$P\left[-\frac{1}{2} < X < \frac{3}{4}\right] = \int_{-1/2}^{3/4} (1 - |t|)\, dt$$

$$= \int_{-1/2}^{0} (1 + t)\, dt + \int_{0}^{3/4} (1 - t)\, dt = \frac{27}{32}.$$

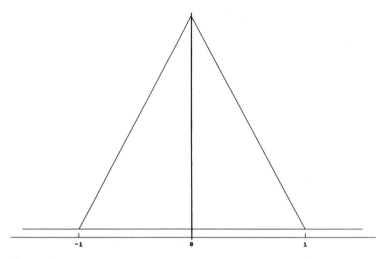

Figure 1

Now consider continuous models in n-dimensional space. Let f be a real-valued function on $\mathcal{R}^n$ satisfying

$$f(\mathbf{t}) \geq 0 \quad \text{for all } \mathbf{t} \in \mathcal{R}^n \quad \text{and} \quad \int_{\mathcal{R}^n} f(\mathbf{t}) \, d\mathbf{t} = 1. \tag{4}$$

Any function satisfying (4) is called a *density on* $\mathcal{R}^n$. The integral in (4) is an abbreviation for the n-fold integral

$$\int_{-\infty}^{\infty} \cdots \int_{-\infty}^{\infty} f(t_1, \ldots, t_n) \, dt_1 \cdots dt_n.$$

We will often use the briefer notation in (4) and its analog for integrals over $A \subset \mathcal{R}^n$. For example, let $n = 2$ and let A be the rectangle $[a_1, b_1] \times [a_2, b_2]$. Then

$$\int_A f(\mathbf{t}) \, d\mathbf{t} = \int_{a_1}^{b_1} \int_{a_2}^{b_2} f(x, y) \, dy \, dx = \int_{a_2}^{b_2} \int_{a_1}^{b_1} f(x, y) \, dx \, dy.$$

To any n-dimensional rectangle A assign

$$P(A) = \int_A f(\mathbf{t}) \, d\mathbf{t}. \tag{5}$$

The assignment given by (5) is between 0 and 1 and the class $\mathcal{C}$ of n-dimensional rectangles satisfies the conditions of Section 5. Again, the properties of integration on $\mathcal{R}^n$ guarantee that the rules (5.4) of probability will be satisfied. Hence a density on $\mathcal{R}^n$ represents a probability model.

Joint Density

Let $\mathbf{X} = (X_1, \ldots, X_n)$ be an n-dimensional random vector for which $P[\mathbf{X} \in A]$ is given by (5) when A is an n-dimensional rectangle. Then f is called the *density* of $\mathbf{X}$ or the *joint density* of $X_1, \ldots, X_n$. The same terminology is used if each X_i is a random vector. If $\mathbf{X}$ has a density on $\mathcal{R}^n$ and A has dimension less than n, then $P[\mathbf{X} \in A] = 0$. For example, if $n = 2$, then $P[a < X_1 < b, X_2 = c] = 0$. In general, for any A for which the integral is defined,

$$P[\mathbf{X} \in A] = \int_A f(\mathbf{t}) \, d\mathbf{t}. \tag{6}$$

From (6), it follows that if A is an n-dimensional cube each side of which has small length $\varepsilon > 0$, $t \in A$, and f is continuous at t, then

$$P[\mathbf{X} \in A] \approx \varepsilon^n f(t). \tag{7}$$

Random vectors with densities have marginal densities. The following illustrates the method of obtaining marginal densities when $n = 2$. Let f be the density of $\mathbf{X} = (X_1, X_2)$. First compute $P[a < X_1 < b] = P[\mathbf{X} \in A]$, where A is the shaded region in Figure 2. Using (6) gives

$$P[a < X_1 < b] = P[a < X_1 < b, -\infty < X_2 < \infty]$$

$$= \int_a^b \left[\int_{-\infty}^{\infty} f(t_1, t_2) \, dt_2 \right] dt_1 = \int_a^b g(t_1) \, dt_1,$$

where

$$g(t_1) = \int_{-\infty}^{\infty} f(t_1, t_2) \, dt_2. \tag{8}$$

Thus g, defined in (8), is the density of X_1.

Marginal Density

Let $\mathbf{X}$ and $\mathbf{Y}$ be random vectors with joint density $f_{\mathbf{X},\mathbf{Y}}$. Assume that $\mathbf{Y}$ is k-dimensional. The density $f_{\mathbf{X}}(\mathbf{u}) = \int_{\mathcal{R}^k} f_{\mathbf{X},\mathbf{Y}}(\mathbf{u}, \mathbf{v}) \, d\mathbf{v}$ is called the *marginal density of* $\mathbf{X}$.

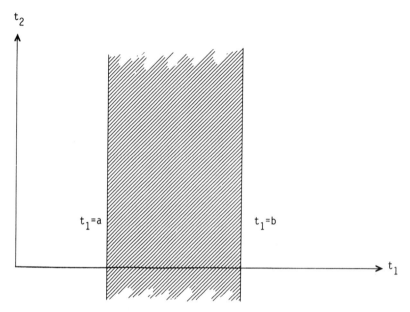

Figure 2

Example 3.† Let $\mathbf{X} = (X_1, X_2)$ have density

$$f(\mathbf{t}) = t_1^2 e^{-t_1(1+t_2)} \qquad \text{if } t_1, t_2 > 0, \quad \mathbf{t} \in \mathcal{R}^2. \tag{9}$$

The marginal density of X_1 is

$$g(t_1) = t_1^2 e^{-t_1} \int_0^\infty e^{-t_1 t_2} \, dt_2 = t_1 e^{-t_1} \qquad \text{if } t_1 > 0.$$

The marginal density of X_2 is

$$h(t_2) = \int_0^\infty t_1^2 e^{-t_1(1+t_2)} \, dt_1 = \frac{2}{(1 + t_2)^3} \qquad \text{if } t_2 > 0.$$

That g and h (shown in Figure 3) are both densities follows from the fact that f is a density. Conversely, the fact that f is a density can be confirmed by showing that the integral of either g or h is 1 (nonnegativity of f, g, and h being obvious).

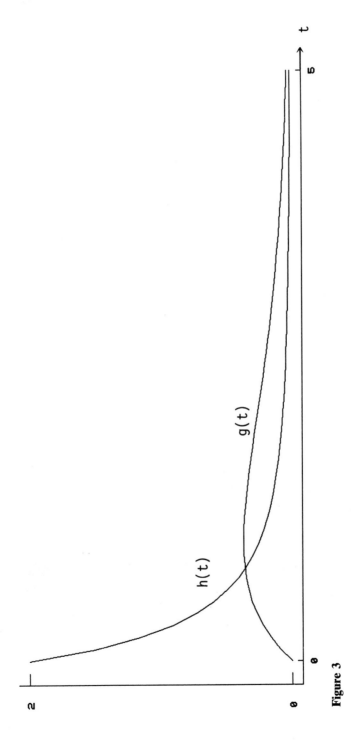

Figure 3

Now consider using (9) to compute $P[X_1 X_2 \leq 1]$. For ease of integration compute $P[X_1 X_2 \geq 1]$. Thus

$$P[X_1 X_2 \geq 1] = P\left[X_2 \geq \frac{1}{X_1}\right] = \int_0^\infty t_1^2 e^{-t_1} \left[\int_{1/t_1}^\infty e^{-t_1 t_2}\, dt_2\right] dt_1$$

$$e^{-1} \int_0^\infty t_1 e^{-t_1}\, dt_1 = e^{-1} = .368.$$

The density was integrated over the shaded region in Figure 4. Hence $P[X_1 X_2 \leq 1] = .632$.

We use the notation $f_{\mathbf{X}} = f_{X_1,\ldots,X_n}$ for the density of the random vector $\mathbf{X} = (X_1, \ldots, X_n)$ and similar notation for marginal densities.

Example 4.   Let

$$f_{\mathbf{X}}(s, t, u) = f_{X_1,X_2,X_3}(s, t, u) = c(s + 2t) \qquad \text{if } 0 \leq s \leq u \leq t \leq 2,$$

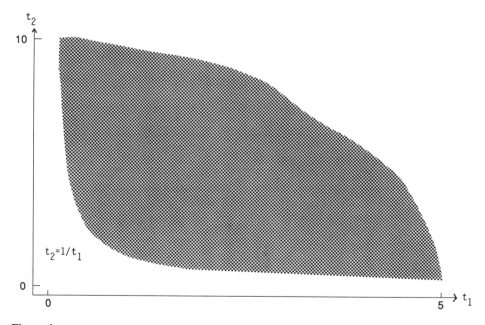

Figure 4

where c is the constant that makes $f_\mathbf{X}$ a density. The marginal density of (X_1, X_2) is

$$f_{X_1,X_2}(s, t) = c(t - s)(s + 2t) \qquad \text{if } 0 \le s \le t \le 2.$$

The marginal density of X_1 is

$$f_{X_1}(s) = c\left(\frac{16}{3} - 2s - 2s^2 + \frac{5s^3}{6}\right) \qquad \text{if } 0 \le s \le 2.$$

Integration of f_{X_1} would yield $c = 3/14$.

B. Histograms; Empirical Densities

A histogram is a special type of density with a variety of descriptive and analytic uses. Its most familiar form is that of a graph consisting of non-overlapping rectangles over a set of contiguous intervals (a bar chart) as in Figure 5. The scales are chosen so that the sum of the areas of the rectangles is 1.

Often the purpose of such a graph is to convey the visual impression of a density when the model being described is, in fact, discrete. This is done by "spreading" the positive mass of a value (or of several values) over a suitable interval.

The analytic form of a histogram is as follows.

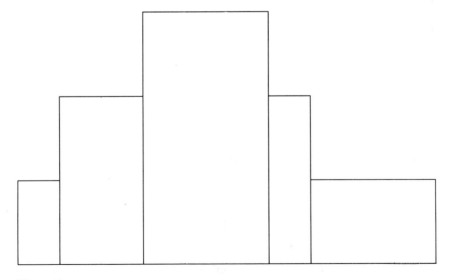

Figure 5

Histogram

Let $a_0 < a_1 < \cdots < a_r$ be numbers, let $A_i = (a_{i-1}, a_i]$ for $i = 1, \ldots,$ r and let $\{p_1, \ldots, p_r\}$ be probabilities adding up to 1. The length of the interval A_i is $a_i - a_{i-1}$, which we denote by Δ_i. (Most often the Δ_i are chosen to be equal.) The density, h (or its graph), whose values are

$$h(x) = \sum_{i=1}^{r} \frac{p_i}{\Delta_i} I_{A_i}(x), \tag{10}$$

is called a *histogram*.

Generally speaking, histograms act as a simple analytic bridge between discrete and continuous probability models. The connection is described in the following remarks and examples.

Remark 1. To a given histogram h there corresponds a model on a finite set whose probability function η is given by

$$\eta(m_i) = p_i \qquad \text{for } i = 1, \ldots, r, \tag{11}$$

where $m_i = (a_{i-1} + a_i)/2$ is the midpoint of the interval A_i.

Remark 2. A histogram can be (and often is) constructed from a given finite or discrete model through appropriate choices of the A_i's and the p_i's. This arises in two settings, an empirical one and a theoretical one. In the empirical case the object is to produce a graphical (or tabular) summary of a set of numbers. The resulting histogram may be used to suggest (or confirm) that the numbers are a sample from a certain continuous probability model (see Example 5). In this case a histogram becomes an empirical density.

In the theoretical setting to which we referred, a histogram may be used to examine the fit of an approximating density when the correct discrete model is analytically difficult to manage. This will be illustrated in Figure 42.1.

Remark 3. Let X be a rv with density f and let $\varepsilon > 0$ be fixed but arbitrary. By an appropriate choice of $a_0 < a_1 < \cdots < a_n$ we can construct a histogram, h (with $A_i = (a_{i-1}, a_i]$) satisfying

(i) $p_1 = \int_{-\infty}^{a_1} f(x)\, dx$ and $p_n = \int_{a_{n-1}}^{\infty} f(x)\, dx.$
(ii) $p_i = P[X \in A_i] = \int_{a_{i-1}}^{a_i} f(x)\, dx$ if $i = 2, \ldots, n - 1.$
(iii) $\int_{-\infty}^{\infty} |h(x) - f(x)|\, dx < \varepsilon.$

In other words, all probabilities determined from f by (6) can be approximated within specified limits by those determined from a suitably chosen histogram (Example 6). Property (iii) is closely related to the definition of integral as a limit of sums.

In view of these properties and Remark 1, the model for X is seen to be closely approximated by the model for a simple rv Y, whose probability function η is determined in (i) and (ii). It is to be expected, therefore, that all properties of X that depend only on its probability model will be quite similar to those of Y.

Example 5. The melting points of 50 filaments were obtained with results summarized in Table 1, where n_i is the number of these observation in the interval $A_i = (a_{i-1}, a_i]$. Figure 6 displays the histogram with $p_i = n_i/50$, the relative frequency of values in A_i. The density

$$f(x) = \frac{x - 300}{400} \quad \text{if } 300 < x \le 320$$

$$= \frac{340 - x}{400} \quad \text{if } 320 < x < 340$$

is also displayed. This is a moderately good fit for such a small sample.

Example 6. Let X have the exponential density $f(x) = e^{-x}$ for $x > 0$. Then for any $\Delta > 0$ and any $a \ge 0$,

$$P[a < X \le a + \Delta] = \int_a^{a+\Delta} e^{-x}\, dx = e^{-a}(1 - e^{-\Delta})$$

and

$$P[X > a] = e^{-a}.$$

Table 1

i	A_i	n_i	$n_i/50$
1	(302.5, 307.5]	1	.02
2	(307.5, 312.5]	6	.12
3	(312.5, 317.5]	10	.20
4	(317.5, 322.5]	14	.28
5	(322.5, 327.5]	11	.22
6	(327.5, 332.5]	7	.14
7	(332.5, 337.5]	1	.02

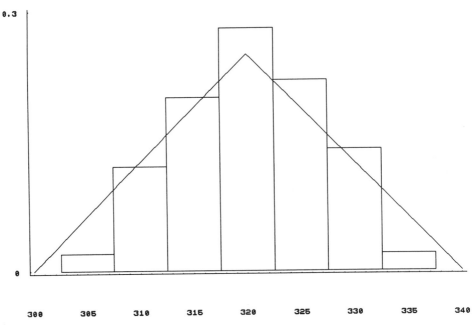

Figure 6

Now let $a_i = i\Delta$ for $i = 0, 1, \ldots, r + 1$, where r will be chosen so that $e^{-r\Delta}$ is suitably small. Then $A_i = ((i - 1)\Delta, i\Delta]$ and we define a histogram h by

$$h(x) = \frac{e^{-i\Delta}(1 - e^{-\Delta})}{\Delta} \qquad \text{if } x \in A_i, \quad i = 0, 1, \ldots, r$$

$$= \frac{e^{-(r+1)\Delta}}{\Delta} \qquad \text{if } x \in A_{r+1}.$$

The density f and the superimposed histogram h are shown in Figure 7 for $\Delta = 1$ and $r = 5$. The "distance" between f and h for any r and Δ is defined by

$$\int_0^\infty |f(x) - h(x)| \, dx = \sum_{i=0}^{r-1} \int_{i\Delta}^{(i+1)\Delta} \left| e^{-x} - \frac{e^{-i\Delta}(1 - e^{-\Delta})}{\Delta} \right| dx$$

$$+ \int_{r\Delta}^{(r+1)\Delta} \left| e^{-x} - \frac{e^{-r\Delta}}{\Delta} \right| dx + \int_{(r+1)\Delta}^\infty e^{-x} \, dx$$

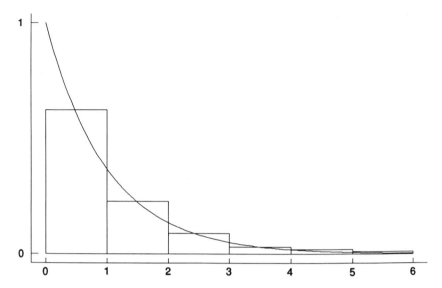

Figure 7

$$\int_0^\infty |f(x) - h(x)| \, dx = \sum_{i=0}^{r-1} e^{-i\Delta} \int_0^\Delta \left| e^{-x} - \frac{1 - e^{-\Delta}}{\Delta} \right| dx$$

$$+ e^{-r\Delta} \int_0^\Delta \left| e^{-x} - \frac{1}{\Delta} \right| dx + e^{-(r+1)\Delta}.$$

Each term on the right can be made arbitrarily small. The first term can be made small uniformly in r by choosing Δ sufficiently small and the other two can be made small for specified Δ by choosing r sufficiently large. In two special cases the results are: (1) for $\Delta = 1$ and $r = 5$ the distance between f and h is .25 and (2) for $\Delta = .5$ and $r = 10$ the distance is .13.

Finally, let Y be a rv whose probability function is

$$\eta\left(\Delta\left(i + \frac{1}{2}\right)\right) = e^{-i\Delta}(1 - e^{-\Delta}) \qquad \text{if } i = 0, 1, \ldots, r - 1$$

$$= e^{-r\Delta} \qquad \text{if } i = r.$$

Then X and Y can be expected to have very similar properties if Δ is small and r is large.

C. Commonly Encountered Densities

We close this section with a brief list of some standard densities on $\mathcal{R}$. A more complete discussion is provided in Chapter 7. (Section numbers are given in parentheses next to the name.)

1. *Exponential* (40): $f(x) = (1/\beta)e^{-x/\beta}$, where $x, \beta > 0$. Under certain assumptions, the exponential is the distribution of the waiting time for one occurrence of an event. The exponential is also used as a simplified lifetime model.

2. *Gamma* (40): $f(x) = [1/\beta^\alpha\Gamma(\alpha)]x^{\alpha-1}e^{-x/\beta}$, where $x\ \alpha, \beta > 0$; $\Gamma(\alpha) = \int_0^\infty t^{\alpha-1}e^{-t}\,dt$. Note that $\Gamma(\alpha) = (\alpha - 1)!$ for α an integer. Under certain assumptions the gamma is the distribution of the waiting time for the αth occurrence of an event.

3. *Normal* (42): $f(x) = (1/\sqrt{2\pi}\sigma)\exp[-(x - \mu)^2/(2\sigma^2)]$, where $\sigma > 0$. The normal plays a unique role in probability and statistics for approximations and as a frequently assumed model.

4. *Uniform* (43): $f(x) = 1/(b - a)$, where $a < x < b$. The uniform with $a = 0, b = 1$ is used in simulations.

5. *Beta* (44): $f(x) = [\Gamma(\alpha + \beta)/(\Gamma(\alpha)\Gamma(\beta))]x^{\alpha-1}(1 - x)^{\beta-1}$, where $0 < x < 1$; $\alpha, \beta > 0$. The beta has a variety of uses that will be discussed later.

Summary

Density: A nonnegative function f on $\mathcal{R}^n$ such that $\int_{\mathcal{R}^n} f(\mathbf{t})\,d\mathbf{t} = 1$.
Marginal density: $g(\mathbf{u}) = \int f(\mathbf{u}, \mathbf{v})\,d\mathbf{v}$, where f is the density of $\mathbf{X} = (\mathbf{Y}, \mathbf{Z})$.
Histogram: A density that is constant on intervals.

If f is the density of the n-dimensional random vector $\mathbf{X}$, then $P[\mathbf{X} \in A] = \int_A f(\mathbf{t})\,d\mathbf{t}$ for any $A \subset \mathcal{R}^n$ for which the integral is defined.

Problems

1. For each of the functions f given below, answer the following questions.
   (i) Is f a density?
   (ii) Does there exist a number c such that $cf(x)$ is a density?
   (a) $f(x) = x$ if $0 < x < 1$
       $\qquad\quad = 2 - x$ if $1 \le x < 2$
   (b) $f(x) = x$ if $0 < x < 2$
       $\qquad\quad = 2$ if $2 \le x < 3$

    (c) $f(x) = 1/x^2$ if $x > 1$
    (d) $f(x) = 1/x$ if $x > 1$

2. Let $f(t) = (1/2)e^{-|t|}$.
    (a) Verify that f is a density.
    (b) Let X be a rv with density f. Evaluate $P[|X| > 2]$, $P[-1 < X < 3]$, and $P[X > 5]$.

3. Let X be a rv with density

$$f(t) = t \qquad \text{if } 0 < t < 1$$

$$= \frac{1}{2} \qquad \text{if } 1 \leq t < 2.$$

    Evaluate $P[1/2 < X < 3/2]$, $P[X < 7/4]$, and $P[1 < X < 5]$.

4. Let X be a rv with density $f(t) = 2/(1 + t)^3$ for $t \geq 0$.
    (a) Let $A = (1, 2]$ and $B = (4, 5]$. Find $P[X \in (A \cup B)^c]$.
    (b) Construct a histogram h approximating f so that the distance between f and h is at most .1. *Hint*: See Example 6.

5. The proportions of defective items produced by a certain factory were observed for 25 days with the (simulated) results shown in Table 2.
    (a) Give a histogram for these data with intervals of length .02.
    (b) Find the distance between the histogram and the density $f(x) = 20(1 - 10x)$ for $0 < x < .1$.

6. Verify the calculations in Example 2 without integrating.

7. For $\mathbf{t} \in \mathcal{R}^2$, let

$$f(\mathbf{t}) = e^{-(t_1 + t_2)} \qquad \text{if } t_1, t_2 > 0.$$

    (a) Verify that f is a density.
    (b) Find the marginal densities.
    (c) Let $\mathbf{X} = (X_1, X_2)$ be a random vector with density f. Find $P[\min(X_1, X_2) < 3]$.

8. For $\mathbf{t} \in \mathcal{R}^2$, let $f(\mathbf{t}) = c$ if $0 < t_1 < t_2 < 1$.
    (a) Find c so that f is a density.
    (b) Find the marginal densities.

Table 2

.008	.016	.442	.086	.016
.068	.026	.046	.056	.008
.012	.024	.018	.022	.056
.040	.026	.024	.018	.072
.046	.080	.022	.026	.050

(c) Let $\mathbf{X} = (X_1, X_2)$ be a random vector with density f. Find $P[X_1 > X_2]$.

9.† Let f be a density on $\mathcal{R}$. Let $\mathcal{S}$ be the subset of $\mathcal{R}^2$ bounded below by the horizontal axis and above by the graph of f. A random vector $\mathbf{X} = (X_1, X_2)$ is chosen according to the uniform model on $\mathcal{S}$. Verify that f is the density of X_1.

10. For $\mathbf{t} \in \mathcal{R}^3$, let

$$f(\mathbf{t}) = ce^{-(t_1 + 2t_2 + 3t_3)} \qquad \text{if } t_1, t_2, t_3 > 0.$$

(a) Find the two- and one-dimensional marginal densities.
(b) Let $\mathbf{X} = (X_1, X_2, X_3)$ be a random vector with density f. Evaluate $P[X_1 < X_2 < X_3]$.

11. Let $f(\mathbf{t}) = 1$ for $\mathbf{t} \in \mathcal{R}^2$ with $0 < t_1, t_2 < 1$.
(a) Verify that f is a density.
(b) Let $\mathbf{X} = (X_1, X_2)$ be a random vector with density f. Find $P[(X_1 - 1/2)^2 + (X_2 - 1/2)^2 = 1/4]$ and $P[(X_1 - 1/2)^2 + (X_2 - 1/2)^2 > 1/4]$.

24. DISTRIBUTION FUNCTIONS ON $\mathcal{R}^n$

A. Definitions and Properties

Every random variable or vector (discrete, continuous, or neither) has a model that can be represented by its distribution function, to be defined below. For rv's, this function and some of its properties were the subject of Example 6.4 and Problem 6.13. In this section we resume the analysis of this representation and extend it to $\mathcal{R}^n$.

Distribution Function

The function on $\mathcal{R}$ with values

$$F(r) = P[X \le r] \qquad (1)$$

is called the *distribution function* of the rv X. The definition applies as well to random vectors where $[\mathbf{X} \le \mathbf{r}] = [X_1 \le r_1, \ldots, X_n \le r_n]$. The function F is also called the *joint distribution function of* $X_1, \ldots, X_n$.

At times the distribution function of $\mathbf{X} = (X_1, \ldots, X_n)$ will be denoted by $F_{\mathbf{X}}$ or by $F_{X_1, \ldots, X_n}$.

Representation of models on $\mathcal{R}^n$ ($n \ge 2$) through distribution functions is often so unwieldy that we generally prefer other representations, if available. However, they are frequently used implicitly in derivations concerning the other representations. They are most useful when $n = 1$ and for the time being we focus on this case.

The following are necessary and sufficient conditions for F to be a distribution function on $\mathcal{R}$.

$$F(-\infty) = \lim_{r \to -\infty} F(r) = 0; \tag{2}$$

$$F(\infty) = \lim_{r \to \infty} F(r) = 1; \tag{3}$$

$$F(r) \geq F(s) \quad \text{for } r \geq s; \quad \text{and} \tag{4}$$

$$\lim_{t \downarrow r} F(t) = F(r) \quad \text{for all } r. \tag{5}$$

Property (5) is called *continuity from the right*. By "$\lim_{t \downarrow r}$," we mean that the limit is taken with t decreasing to r (i.e., from the right).

Let $F(r+) = \lim_{t \downarrow r} F(t)$ and $F(r-) = \lim_{t \uparrow r} F(t)$, that is, the limit as t increases to r (i.e., from the left). Then (5) is: $F(r+) = F(r)$. The difference $F(r) - F(r-)$ is the *jump* at r.

Properties (2) to (5) are the subject of Problem 6.13. Property (4) follows from $[X \leq s] \subset [X \leq r]$. The remaining relations all involve monotone sequences of events and follow from Theorem 6.3.

To show sufficiency of (2) to (5) it must be shown that we can construct a probability model on $\mathcal{R}$ for which $P[X \leq x] = F(x)$. To do this, observe that for $a < b$,

$$P[a < X \leq b] = F(b) - F(a) \tag{6}$$

and probabilities of other intervals are obtained (Problem 4) by adjustments for endpoints using

$$P[X = r] = F(r) - F(r-) \quad \text{for all } r \in \mathcal{R}, \tag{7}$$

that is, the jump of F at r. Verification of (7) uses the result in Example 6.4. Notice that a model on $\mathcal{R}$ is continuous ($P[X = r] = 0$ for all $r \in \mathcal{R}$) if, and only if, its distribution function is continuous. While F assigns numbers between 0 and 1 to every interval, more advanced mathematics is required to show that this assignment is countably additive (5.4c).

The point x is a *point of increase* of the distribution function F if $F(x + \varepsilon) - F(x - \varepsilon) > 0$ for all $\varepsilon > 0$. If F has a jump at x, then x is a point of increase of F. In the discrete case, every point of increase is a jump. A distribution function is determined by its values at its points of increase. (It is also determined by its values at points of continuity.) When thinking of a discrete distribution function, the picture to keep in mind is that of a step function that has jumps at the points of increase and is constant in between. If a discrete distribution function has only a single jump at c, it is called *degenerate* at c and represents the model of a rv that is identically equal to c (i.e., a constant).

When we give an expression for F it will be in either of two forms: (1) $F(t)$ is specified wherever $F(t) > 0$, or (2) F is specified only at its points of increase, provided that no confusion is possible.

In the discrete case we can compute probability functions on $\mathfrak{R}$ from distribution functions and conversely. Let X be a discrete rv. If f_X is given, then

$$F_X(t) = \sum_{r \le t} f_X(r). \tag{8}$$

On the other hand, if F_X is given, applying (7) yields

$$f_X(t) = F_X(t) - F_X(t-). \tag{9}$$

When X is integer valued, (9) becomes

$$P[X = r] = f_X(r) = F_X(r) - F_X(r - 1). \tag{10}$$

Similar relationships exist in the continuous case when f_X is a density. Then

$$F_X(t) = P[X \le t] = \int_{-\infty}^{t} f_X(u)\, du \tag{11}$$

[apply (23.2) with $a = -\infty, b = t$]. From elementary properties of integrals, it can be seen that F_X is continuous and we may choose

$$f_X(t) = F_X'(t) \tag{12}$$

for all t such that $F_X'(t)$ exists.

Conversely, let F be a distribution function and $f(t) = F'(t)$, where $F'(t)$ exists. Elsewhere, let $f(t) \ge 0$ be arbitrary. If (11) is satisfied (omitting subscripts, X), then f is a density corresponding to F. This case is usually referred to as the *absolutely continuous case*.

Before turning to multivariate models, we give examples.

Example 1. In Example 22.1 we considered a rv X with probability function

$$f(r) = \frac{3}{10} \qquad \text{if } r = 0$$

$$= \frac{6}{10} \qquad \text{if } r = 1$$

$$= \frac{1}{10} \qquad \text{if } r = 2.$$

The distribution function of X obtained from (10) is

$$F(r) = \frac{3}{10} \quad \text{if } 0 \le r < 1$$

$$= \frac{9}{10} \quad \text{if } 1 \le r < 2$$

$$= 1 \quad \text{if } r \ge 2.$$

See Figure 1, where F is displayed. Note that f can be recovered by examining the heights of the jumps.

Example 2. Consider a random sample of size 3 drawn with replacement from the integers $1, \ldots, N$. Let X be the maximum number drawn. The points of increase of F_X are $1, \ldots, N$, and at these points, $F_X(k) = (k/N)^3$. By (9),

$$f_X(k) = \left(\frac{k}{N}\right)^3 - \left(\frac{k-1}{N}\right)^3 = \frac{3k^2 - 3k + 1}{N^3} \quad \text{if } k = 1, \ldots, N.$$

Alternatively, a direct sampling argument can be used to find f_X. Of course, F_X can then be recovered by summing.

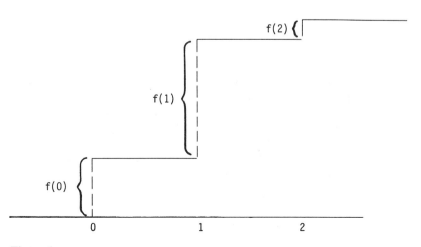

Figure 1

Example 3.   Consider the density $f(t) = te^{-t}$ if $t > 0$. The corresponding distribution function is

$$F(t) = \int_{-\infty}^{t} f(x)\, dx = \int_{0}^{t} xe^{-x}\, dx = 1 - (1 + t)e^{-t} \quad \text{if } t > 0.$$

Note that $f(t) = F'(t)$.

Example 4.   The density is $f(t) = 1 - |t|$ (see Figure 23.1). By elementary geometry (areas of triangles), the corresponding distribution function is

$$F(t) = \frac{1}{2}(1 + t)^2 \qquad \text{if } -1 < t \le 0$$

$$= 1 - \frac{1}{2}(1 - t)^2 \qquad \text{if } 0 < t < 1$$

$$= 1 \qquad \text{if } t \ge 1.$$

Briefly, here are the essential properties of distribution functions on $\mathcal{R}^n$. In the discrete case,

$$F_{\mathbf{X}}(\mathbf{t}) = P[\mathbf{X} \le \mathbf{t}] = \sum_{r_1 \le t_1} \cdots \sum_{r_n \le t_n} f_{\mathbf{X}}(\mathbf{r}) \tag{13}$$

and if X has a density, then

$$F_{\mathbf{X}}(\mathbf{t}) = \int_{-\infty}^{t_1} \cdots \int_{-\infty}^{t_n} f_{\mathbf{X}}(\mathbf{r})\, d\mathbf{r}. \tag{14}$$

Let $\mathbf{r} = (r_1, \ldots, r_n)$. The analogs of (2), (3), and (5), proved similarly, are

$$\lim_{r_i \to -\infty} F(\mathbf{r}) = 0 \qquad \text{for any } i = 1, \ldots, n. \tag{15}$$

If $r_i \to \infty$ for each i, then $F(\mathbf{r}) \to 1$. $\tag{16}$

If $t_i \downarrow r_i$ for each i, then $F(\mathbf{t}) \to F(\mathbf{r})$. $\tag{17}$

Since every n-dimensional rectangle must have nonnegative probability, expansion of the product in (16.1) (in the section on Cartesian products) imposes an inequality on F that is the analog of monotonicity when $n = 1$ [as in (4)]. For $n = 2$ we can rewrite (16.2) in the form

$$I_{(a,b] \times (c,d]} = I_{(-\infty,b] \times (-\infty,d]} - I_{(-\infty,a] \times (-\infty,d]} - I_{(-\infty,b] \times (-\infty,c]} + I_{(-\infty,a] \times (-\infty,c]}.$$

Hence the distribution function F must satisfy

$$F(b, d) - F(a, d) - F(b, c) + F(a, c) \ge 0 \tag{18}$$

for any rectangle $(a, b] \times (c, d]$. For larger n, the condition becomes more complicated (Problem 6).

From (14)

$$f(\mathbf{t}) = \frac{\partial^n}{\partial t_1 \cdots \partial t_n} F(\mathbf{t}) \tag{19}$$

whenever this derivative exists. Again, we can take any value for f wherever the derivative does not exist.

A differentiable function F on $\mathfrak{R}^n$ is a distribution function if, and only if, f computed in (19) is a density. In the discrete case, the probability function is obtained by shrinking an n-dimensional rectangle to a point.

Marginal models are easy to identify from the joint distribution function. Let $\mathbf{X}$ and $\mathbf{Y}$ be random vectors where $\mathbf{Y}$ is $\mathbf{k}$-dimensional. Let $F_{\mathbf{X},\mathbf{Y}}$ be their joint distribution function.

Marginal Distribution Function

The *marginal distribution function* of $\mathbf{X}$ is $F_{\mathbf{X}}(x) = \lim_{\mathbf{y} \to \infty} F_{\mathbf{X},\mathbf{Y}}(\mathbf{x}, \mathbf{y})$, where "$\mathbf{y} \to \infty$" means that $y_i \to \infty$ for each $i = 1, \ldots, k$.

Example 5. For $\mathbf{t} \in \mathfrak{R}^2$, let

$$
\begin{aligned}
F(\mathbf{t}) &= 2t_1 t_2 && \text{if } t_1, t_2 > 0, t_1 + t_2 < 1; \\
&= 2(t_1 + t_2) - (t_1^2 + t_2^2) - 1 && \text{if } 0 < t_1, t_2 < 1, t_1 + t_2 \geq 1; \\
&= t_1(2 - t_1) && \text{if } 0 < t_1 < 1 \leq t_2; \\
&= t_2(2 - t_2) && \text{if } 0 < t_2 < 1 \leq t_1; \\
&= 1 && \text{if } t_1, t_2 \geq 1.
\end{aligned}
$$

Let

$$
\begin{aligned}
f(\mathbf{t}) &= \frac{\partial^2}{\partial t_1 \, \partial t_2} F(\mathbf{t}) \\
&= 2 && \text{if } t_1, t_2 > 0, t_1 + t_2 < 1 \\
&= 0 && \text{otherwise.}
\end{aligned}
$$

This is the density of the uniform distribution on the triangle $\mathcal{S}$ bounded by $t_1 = 0$, $t_2 = 0$, $t_1 + t_2 = 1$. That is, the probability assigned by the model to $A \subset \mathcal{S}$ is proportional to its area (twice the area). Since f is a density, F is a distribution function. The family of uniform models is discussed in detail in Section 43.

The marginal distribution functions are

$$G(t_1) = \lim_{t_2 \to \infty} F(t) = t_1(2 - t_1) \qquad \text{if } 0 < t_1 < 1$$

and

$$H(t_2) = \lim_{t_1 \to \infty} F(t) = t_2(2 - t_2) \qquad \text{if } 0 < t_2 < 1.$$

Hence the marginal densities are

$$g(u) = G'(u) = 2(1 - u) \qquad \text{if } 0 < u < 1$$

and by symmetry, $h(u) = H'(u) = g(u)$.

B. Application to Failure Rates

Let f and F be the density and distribution function, respectively, of a nonnegative rv X, and set $\overline{F}(t) = 1 - F(t)$. Then $r(t) = f(t)/\overline{F}(t)$ is the *failure rate* of the model (or of any of its representations).

Let X be the lifetime of some object (an animal, a computer chip, etc.). Then $\overline{F}(t)$ is the probability that the object survives to time t and

$$G(x \mid t) = \frac{\overline{F}(t) - \overline{F}(t + x)}{\overline{F}(t)}$$

is the conditional probability that the object survives an additional time x given that it has survived to time t. Then $r(t) = \dfrac{d}{dx} G(x \mid t)\big|_{x=0}$.

Since $r(t) = -\overline{F}'(t)/\overline{F}(t)$, it follows that

$$\overline{F}(t) = e^{-\int_0^t r(x)\,dx}. \tag{20}$$

Thus the failure rate determines the distribution function.

A model f has *increasing failure rate* (IFR) if r is a nondecreasing function and *decreasing failure rate* (DFR) if r is a nonincreasing function. If f has both IFR and DFR, then f has *constant failure rate* (CFR).

Theorem 1. *A density f has CFR if, and only if, f is an exponential density $f(t) = \lambda e^{-\lambda t}$ for $x > 0$ for some $\lambda > 0$.*

Proof. Let $f(t) = \lambda e^{-\lambda t}$ if $t > 0$, so that $\overline{F}(t) = e^{-\lambda t}$ for $t > 0$ and $r(t) = \lambda$. By (20), if f has CFR, then $\overline{F}(t) = e^{-\lambda t}$, which is the right-tail probability of an exponential distribution.

Example 6. Set $\overline{F}_\alpha(t) = e^{-(\lambda t)^\alpha}$ if $t > 0$, where $\lambda, \alpha > 0$. Then the corresponding density is $f_\alpha(t) = \alpha\lambda(\lambda t)^{\alpha-1} e^{-(\lambda t)^\alpha}$ if $t > 0$ and $r(t) = \alpha\lambda(\lambda t)^{\alpha-1}$.

For $\alpha = 1$, the density is exponential and CFR. For $\alpha > 1$, the density is IFR, and for $\alpha < 1$, it is DFR.

The family in Example 6 is called the *Weibull family*.

Many physical objects have IFR, deteriorating as they age. Some metals have DFR, hardening as they age. Section 40 will give assumptions leading to exponential densities and hence to CFR. Electronic components are often in this category. Living organisms generally have U-shaped FR, with high infant mortality, and DFR to some age followed by IFR.

Example 7. A highly contagious disease has entered a certain population. Let $F(t)$ be the proportion of the population that has had the disease at time t. The initial proportion ($t = 0$) is 0 and eventually ($t \to \infty$) every member of the population will have had the disease. As the proportion having had the disease increases, the exposure of healthy persons increases. On the other hand, as the proportion having had the disease grows, the proportion yet to be infected declines. Represent these observations by the assumption that

$$F'(t) = cF(t)[1 - F(t)], \tag{21}$$

where $c > 0$. Now (21) is equivalent to

$$\frac{F'(t)}{F(t)} + \frac{F'(t)}{1 - F(t)} = c,$$

which has the general solution

$$\log \frac{F(t)}{1 - F(t)} = a + ct$$

or

$$F(t) = \frac{de^{ct}}{1 + de^{ct}} \quad \text{if } t > 0, \tag{22}$$

where $d = e^a$. Set $f(t) = F'(t)$. From (21) and (22),

$$f(t) = \frac{c\,de^{ct}}{(1 + de^{ct})^2} \quad \text{if } t > 0. \tag{23}$$

Clearly, $f \geq 0$, $F(-\infty) = 0$, and $F(\infty) = 1$, so that f is a density.

The family of densities given in (23) is called the *logistic family*.

The failure rate for a member of the logistic family is

$$r(t) = \frac{de^{ct}}{1 + de^{ct}},$$

which is increasing.

Summary

Distribution function: $F(\mathbf{r}) = P[X_1 \le r_1, \ldots, X_n \le r_n]$.
Point of increase, x: For a one-dimensional model, $F(x + \varepsilon) - F(x - \varepsilon) > 0$ for all $\varepsilon > 0$.
Marginal distribution function of $(X_1, \ldots, X_{n-1})$: $F(r_1, \ldots, r_{n-1}, \infty)$.
Other marginal distribution functions are defined similarly.
Failure rate: For f and F the density and distribution function, respectively, of a nonnegative rv, $r(t) = f(t)/\overline{F}(t)$, where $\overline{F}(t) = 1 - F(t)$.

If $\mathbf{X}$ is an n-dimensional random vector with distribution function F, then

(i)   If $\mathbf{X}$ is discrete with probability function f, then

$$F(\mathbf{t}) = \sum_{r_1 \le t_1} \cdots \sum_{r_n \le t_n} f(\mathbf{r}).$$

If $n = 1$, then $f(t) = F(t) - F(t-)$, the jump at t. For the case of X integer valued, $f(t) = F(t) - F(t - 1)$ for integer t.

(ii)   If $\mathbf{X}$ is continuous with density f, then

$$F(\mathbf{t}) = \int_{-\infty}^{t_1} \cdots \int_{-\infty}^{t_n} f(\mathbf{u}) \, du_n \cdots du_1$$

and

$$f(\mathbf{t}) = \frac{\partial^n}{\partial t_1 \cdots \partial t_n} F(\mathbf{t})$$

wherever this derivative exists.

For $n = 1$, the function F is a distribution function if, and only if, (1) $F(-\infty) = 0$, (2) $F(\infty) = 1$, (3) F is nondecreasing, and (4) $F(r+) = F(r)$ for all r.

Problems

1. Find the distribution function corresponding to each density.
    (a) $f(t) = 12t(1 - t)^2$ if $0 < t < 1$
    (b) $f(t) = 2/t^3$ if $t > 1$

(c) $f(t) = t/2$ if $0 < t < 1$

      $= \dfrac{1}{2}$ if $1 \le t < 2$

      $= (3 - t)/2$ if $2 \le t < 3$

(d) For $\mathbf{t} \in \mathcal{R}^2$, $f(\mathbf{t}) = 2e^{-(t_1 + t_2)}$ if $t_1 > t_2 > 0$.

(e) For $\mathbf{t} \in \mathcal{R}^2$, $f(\mathbf{t}) = t_1 + t_2$ if $0 < t_1, t_2 < 1$.

2. For the distribution functions given below:

  (i) Is the distribution function discrete or continuous?

  (ii) Give the probability function or density.

  (a) $F(t) = t/2$ if $0 \le t < 1$

        $= 1/2$ if $1 \le t < 2$

        $= 1 - \dfrac{1}{2} e^{-(t-2)}$ if $t \ge 2$

  (b) $F(t) = (k + 1)/20$ if $k \le t < k + 1$, $k = 0, 1, \ldots, 9$

          $= 1 - (1/2)^{k-8}$ if $k \le t < k + 1$, $k = 10, \ldots, 14$

          $= 1$ if $t \ge 15$

  (c) $F(t) = \sum_{k=0}^{\infty} [k/(k + 1)] I_{A_k}(t)$, where $A_k = [k, k + 1)$.

  (d) $F(t) = (1 - \cos t)/8$ if $0 \le t < \pi$

          $= (5 + 3 \cos t)/8$ if $\pi \le t < 2\pi$

          $= 1$ if $t \ge 2\pi$

  (e) For $\mathbf{t} \in \mathcal{R}^2$,

$$F(\mathbf{t}) = \frac{t_2(1 - 1/t_1^2)}{2} \qquad \text{if } 0 < t_2 < 1 \le t_1;$$

$$= 1 - \frac{1/t_2 + t_2/t_1^2}{2} \qquad \text{if } 1 \le t_2 < t_1;$$

$$= 1 - \frac{1}{t_1} \qquad\qquad \text{if } 1 \le t_1 \le t_2.$$

     In addition, find the marginal distribution functions and, from them, the marginal probability functions or densities.

3. A sample of size s is drawn at random from the set of integers $\{1, \ldots, N\}$. Let X be the maximum of the s integers drawn. Determine the distribution function and probability function of X.

  (a) When the draws are made with replacement.

  (b) When the draws are made without replacement.

  (c) What happens in these cases as s gets large?

4. Find expressions analogous to (6) for all other possible intervals.

5. Give the analog of (18) for $n = 3$.

6. Let $F(x, y) = 1$ for $x + y \ge 0$; $x, y \ge -1$. Show that

  (a) F satisfies (15), (16), and (17);

  (b) $\lim_{x \to \infty} F(x, y)$ and $\lim_{y \to \infty} F(x, y)$ are distribution functions;

  (c) F is not a distribution function.

7. A point is chosen at random from the interior of a circle and X is the distance of the point from the center (see Example 21.11). Find F_X and the density f_X.

8. For $\mathbf{x} \in \mathcal{R}^2$, let

$$F(\mathbf{x}) = \frac{(1 - e^{-x_1})(1 - e^{-x_2})}{2} + \frac{x_1 x_2}{2(1 + x_1)(1 + x_2)} \qquad \text{if } x_1, x_2 > 0.$$

   (a) Verify that F is a continuous distribution function.
   (b) Find the marginal distribution functions.
   (c) Find the corresponding density and marginal densities.

9. For $\mathbf{x} \in \mathcal{R}^n$, let $F(\mathbf{x}) = \Pi_{i=1}^n (1 - \exp(-x_i))$ for $x_1, \ldots, x_n > 0$.
   (a) Show that F is a continuous distribution function.
   (b) Find the marginal distribution function of the first k components and the corresponding marginal density.

10. Find and classify the failure rates for the following densities.
   (a) The density of Problem 23.7.
   (b) $f(t) = \lambda^2 t e^{-\lambda t}$ if $t > 0$ where $\lambda > 0$.
   (c) $f(t) = (t - 1)^2 e^{-1/3} e^{-(t-1)^3/3}$ if $t > 0$.

11. Let the failure rate be $r(t) = \alpha + \beta t$. Find the corresponding distribution function and density. What restrictions, if any, must be put on α and β?

25. OTHER TYPES OF MODELS ON $\mathcal{R}^n$

Within the class of probability models on $\mathcal{R}$ (or $\mathcal{R}^n$), some differ in a very substantial way from those presented so far. Some of these are quite applicable (in a practical sense), whereas others are primarily of mathematical interest.

Every distribution function on $\mathcal{R}$ is a weighted average (*mixture*) of three distribution functions: one discrete, one with a density, and one of a third type called "singular." A *singular* distribution function, F, is continuous but has no density [i.e., no f satisfying (24.13) and (24.14)]. Such distribution functions can be shown to exist[1] but have virtually no practical applications. However, distribution functions that are averages of the other two types may be quite useful, as in the following example.

Example 1.† Suppose that a device is turned on at the beginning of a day with the intention of allowing it to operate for a full working day (eight

---

[1] The most widely used example involves the Cantor set.

hours). Turning on the device may cause it to fail, say with probability p. If the device survives this initial shock, the time T, in hours, to failure, has density

$$f(y) = \frac{1}{10} e^{-y/10} \qquad \text{if } y > 0,$$

so that its distribution function is

$$F(y) = 1 - e^{-y/10} \qquad \text{if } y > 0.$$

Let X be the total time in hours that the device operates on a given day. Then $0 \le X \le 8$ and

$$X = I_A(TI_{[T \le 8]} + 8I_{[T > 8]}),$$

where A is the event that the device survives the shock $[P(A) = 1 - p]$. Since

$$P[X \le y] = P(X \le y \mid A^c)P(A^c) + P(X \le y \mid A)P(A),$$

the distribution function of X is

$$G(y) = p + (1 - p)(1 - e^{-y/10}) \qquad \text{if } 0 \le y < 8$$
$$= 1 \qquad \text{if } y \ge 8.$$

The distribution function G has jumps of height p at 0 and $(1 - p)e^{-4/5}$ at 8. The jump at 0 comes from the possibility of immediate failure. The jump at 8 is due to the fact that if the device survives that long, it is turned off.

Then $G(y) = \alpha H_1(y) + (1 - \alpha)H_2(y)$, where

$$H_1(y) = \frac{p}{1 - (1 - p)(1 - e^{-4/5})} \qquad \text{if } 0 \le y < 8;$$
$$= 1 \qquad \text{if } y \ge 8,$$
$$H_2(y) = \frac{1 - e^{-y/10}}{1 - e^{-4/5}} \qquad \text{if } 0 \le y \le 8;$$
$$= 1 \qquad \text{if } y \ge 8,$$

and

$$\alpha = 1 - (1 - p)(1 - e^{-4/5}).$$

The distribution functions H_1 and H_2 are, respectively, discrete and continuous.

Another departure from the simple cases of the earlier sections arises in connection with higher-dimensional ($n \ge 2$) models. Consider a random vector in which the components are of different types, specifically the case in which some marginal models are discrete and others are continuous. Such models arise frequently in connection with subjective probability.

Example 2.† For $\mathbf{t} \in \mathcal{R}^2$, let

$$
\begin{aligned}
F(\mathbf{t}) &= \frac{1}{2} [1 - (1 - t_1)^2] && \text{if } 0 \le t_1, t_2 < 1 \\
&= t_1 && \text{if } 0 \le t_1 < 1, t_2 \ge 1 \\
&= \frac{1}{2} && \text{if } t_1 \ge 1, 0 \le t_2 < 1 \\
&= 1 && \text{if } t_1, t_2 \ge 1.
\end{aligned}
$$

The marginal distribution functions are

$$
\begin{aligned}
G(t_1) &= \lim_{t_2 \to \infty} F(t) \\
&= t_1 && \text{if } 0 \le t_1 < 1 \\
&= 1 && \text{if } t_1 \ge 1
\end{aligned}
$$

and

$$
\begin{aligned}
H(t_2) &= \lim_{t_1 \to \infty} F(t) \\
&= \frac{1}{2} && \text{if } 0 \le t_2 < 1 \\
&= 1 && \text{if } t_2 \ge 1.
\end{aligned}
$$

Then G is continuous with density $g(t_1) = 1$ for $0 < t_1 < 1$ and H is discrete with probability function $h(t_2) = 1/2$ for $t_2 = 0, 1$.

Another representation for a model in which some marginals are discrete and others are continuous is a function f that combines the attributes of both probability functions and densities. Let $n = 2$ and, as in Example 2, suppose that X_1 is continuous and X_2 is discrete. The representation we now have is a nonnegative function f on $\mathcal{R}^2$ such that:

(i) $g(u_1) = \sum_{u_2} f(\mathbf{u})$ is the marginal density of X_1.
(ii) $h(u_2) = \int_{-\infty}^{\infty} f(\mathbf{u}) \, du_1$ is the marginal probability function of X_2.

Note that for each $\mathbf{u} = (u_1, u_2) \in \mathcal{R}^2$ there can only be a countable set of values of u_2 for which $f(\mathbf{u}) \neq 0$.

The corresponding distribution function is

$$
F(\mathbf{t}) = \sum_{u_2 \le t_2} \int_{-\infty}^{t_1} f(\mathbf{u}) \, du_1.
$$

Example 3.† In Example 2, let

$$f(\mathbf{u}) = 1 - u_1 \quad \text{if } 0 < u_1 < 1; \quad u_2 = 0$$
$$= u_1 \quad \text{if } 0 < u_1 < 1; \quad u_2 = 1.$$

This satisfies (i) and (ii). Then $g(u_1) = 1$ for $0 < u_1 < 1$ is the marginal density corresponding to the marginal distribution function G, and $h(u_2) = 1/2$ for $u_2 = 0, 1$ is the marginal probability function corresponding to the marginal distribution function H.

The model in Examples 2 and 3 applies to the following experiment:

1. Select a number between 0 and 1 according to the distribution function G.
2. If the result in 1 is t_1, toss a coin once where the coin has probability t_1 of producing a head.

Alternatively, this is the model for tossing a coin with unknown probability p of a head and G represents a personal probability model for p.
The proof of these statements entails methods described in Section 28.

Summary

There are three basic types of distribution functions. Any distribution function on $\mathcal{R}$ is a mixture of three distribution functions, one of each type. For distribution functions on $\mathcal{R}^n$ $(n \geq 2)$, it is possible to have marginal distribution functions of different types.

Problems

1. Complete Example 1 by obtaining the probability function of H_1 and the density of H_2.
2.† Consider the same setup as Example 1, but imagine that shocks to the system (in addition to the initial one at time 0) also occur at times 2 and 6. Each shock causes the system to fail with probability p, and the outcomes of different shocks are independent. Until turn-off (at time 8) the time to failure (undisturbed) has the density specified.
   (a) Determine the distribution function F of X, the total time the device functions on a given day.
   (b) Obtain the analogs of H_1 and H_2 as in Example 1 and repeat Problem 1 for this case.
3. Consider the situation in Examples 2 and 3, but suppose that the coin is tossed twice. Determine the marginal distribution function and probability function of the number of heads tossed.

GENERAL PROBLEMS

1. The three components of a device have lifetimes T_1, T_2, and T_3, respectively, with joint density.

$$f(\mathbf{u}) = ce^{-(u_1 + 2u_2 + u_3)} \qquad \text{if } u_1, u_2, u_3 > 0.$$

   (a) Evaluate c.
   (b) Evaluate the one-dimensional marginal densities and distribution functions.
   (c) Let Y_1 be the index of the smallest T_i, Y_3 be the index of the largest T_i and Y_2 be the index of the remaining T_i. Find the joint probability function of Y_1, Y_2, and Y_3.
   (d) Use part (c) to determine the one-dimensional marginal probability functions of Y_1, of Y_2, and of Y_3.
   (e) If the system is hooked up in series, the lifetime T_0 of the system is given by $T_0 = \min(T_1, T_2, T_3)$. Determine the distribution function and density of T_0. *Hint*: Begin by obtaining $P[T_0 > v]$.
2. Let the density of X be the exponential

$$f(r) = e^{-r} \qquad \text{if } r > 0.$$

   Set $Y = [X]$. Verify that Y is a discrete rv and find the probability function and distribution function of Y. *Note*: $[a]$ is the largest integer less than or equal to a. Thus $[1] = [1.5] = 1$.
3. Suppose that F and G are continuous, strictly increasing distribution functions on $\Re$ and set $H(x, y) = \min(F(x), G(y))$.
   (a) Define $y(x)$ by $G(y(x)) = F(x)$ and show that $y(x)$ is unique and increases in x.
   (b) Show that H is a continuous distribution function that assigns probability 1 to the graph of $y(x)$.

REFERENCES

Hoel, Paul G., Port, Sidney C., and Stone, Charles J. (1971). *Introduction to Probability Theory*. Boston: Houghton Mifflin Company.

Hogg, R. V., and Ledolter, J. (1987). *Engineering Statistics*. New York: Macmillan Publishing Company. The data in Example 23.5 were taken from problem 1.2-3 of this reference.

Woodroofe, Michael (1975). *Probability with Applications*. New York: McGraw-Hill Book Company.

5.

Conditional Models and Independent Random Variables

26. REPRESENTATIONS OF CONDITIONAL MODELS ON $\mathcal{R}^n$

The concepts of conditional probability and conditional model were introduced in Chapter 2 and were shown to be quite useful in a variety of ways: (a) to express the change in uncertainty concerning an event when new information becomes available, (b) to determine the probability model for joint experiments (or intersections of events) when conditional models (or probabilities) are specified, and (c) to evaluate the "total" probability of an event [as in (12.12)] by first conditioning on each member of some appropriate partition. These concepts and related developments will now be applied within the framework of random variables and vectors. In particular, this section and the next are devoted to rv (or vector) versions of (a) and (b) above. Section 29 explores corresponding modifications of (c). The examples are all two-dimensional, although the theoretical development is for arbitrary random vectors.

To begin with, let $\mathbf{X}$ and $\mathbf{Y}$ (with values in $\mathcal{R}^m$ and $\mathcal{R}^n$, respectively) be random vectors and let $B \subset \mathcal{R}^n$ be such that $P[\mathbf{Y} \in B] > 0$. By direct application of the definition (12.7), every rectangle $A \subset \mathcal{R}^m$ can be assigned the conditional probability

$$P(\mathbf{X} \in A \mid \mathbf{Y} \in B) = \frac{P[\mathbf{X} \in A, \mathbf{Y} \in B]}{P[\mathbf{Y} \in B]}. \tag{1}$$

The numerator in (1) is determined by the joint distribution of X and Y; the denominator depends only on the marginal model for Y. The model on $\mathcal{R}^m$ determined by (1) is the conditional distribution of X given the event $[Y \in B]$. Thus, as long as $P[Y \in B] > 0$, (1) can be used to evaluate various representations of the conditional model. These will be called the conditional distribution function, the conditional probability function (if the conditional model is discrete), or the conditional density (if the conditional distribution function has one) of X given the occurrence of $[Y \in B]$.

The most interesting and useful cases arise when the conditioning event is of the form $[Y = v]$, a member of the partition induced by Y. The resulting conditional model describes the change, if any, in the uncertainty concerning X provided by the information $[Y = v]$. If Y is discrete, (1) can be applied to obtain a conditional model for X given $[Y = v]$ for all relevant v [i.e., for all v such that $f_Y(v) = P[Y = v] > 0$]. If, however, Y is continuous, then $P[Y = v] = 0$ for every v and (1) becomes useless. Nevertheless, even in this case it will be possible under conditions to be described below to establish the relation between the joint probability model and the conditional model. This was done in an ad hoc manner in Example 12.5.

The definitions that follow give conditional probability functions and conditional densities in terms of the corresponding joint and marginal representations.

Conditional Probability Functions, Conditional Densities

Let X and Y be random vectors. For any fixed v such that $f_Y(v) > 0$, the function g whose values are

$$g(u \mid v) = \frac{f_{X,Y}(u, v)}{f_Y(v)} \qquad \text{if } u \in \mathcal{S}(X) \qquad (2)$$

is called (a) the *conditional probability function of* X *given* $[Y = v]$ if $f_{X,Y}$ is the joint probability function of the discrete random vector (X, Y) or (b) the *conditional density of* X *given* $[Y = v]$ if $f_{X,Y}$ is the joint density of the continuous random vector (X, Y). In either case, the value of $g(u \mid v)$ is undefined and, in fact, irrelevant whenever $f_Y(v) = 0$. The function g defined in (2) will usually be denoted by $f_X(u \mid Y = v)^1$ or, when unambiguous, by $f_X(u \mid v)$. Relation (2) also defines, with obvious

---

[1]In some cases we will abuse notation by failing to distinguish between a function and its values. This is done only when more precise notation might lead to confusion.

modifications, conditional probability functions and densities when either **X** is discrete and **Y** is continuous, or vice versa.

When **Y** is discrete, (2) results from a direct application of (1) whether **X** is discrete or continuous. When **Y** is continuous, however, relation (2) is not immediately obvious. The definition is based, in part, on the following reasoning. Let (X, Y) be a two-dimensional random vector. For small ε and δ,

$$P\left[u - \frac{\varepsilon}{2} < X < u + \frac{\varepsilon}{2}, v - \frac{\delta}{2} < Y < v + \frac{\delta}{2}\right] \approx \varepsilon\delta f_{X,Y}(u, v) \quad (3)$$

provided that $f_{X,Y}$ is continuous on the specified rectangle. Also,

$$P\left[v - \frac{\delta}{2} < Y < v + \frac{\delta}{2}\right] \approx \delta f_Y(v). \quad (4)$$

From (3), (4), and the definition of conditional probability,

$$P\left(u - \frac{\varepsilon}{2} < X < u + \frac{\varepsilon}{2} \,\middle|\, v - \frac{\delta}{2} < Y < v + \frac{\delta}{2}\right) \approx \frac{\varepsilon f_{X,Y}(u, v)}{f_Y(v)}.$$

This crude argument can be extended to higher dimensions and its conclusion suggests the definition (2). A rigorous justification of the definition requires advanced mathematical concepts.

Example 1. If two fair dice are rolled, the joint probability function of the minimum X and the maximum Y (see Example 22.3) is

$$f(x, y) = \frac{1}{36} \quad \text{if } 1 \le x = y \le 6$$

$$= \frac{2}{36} \quad \text{if } 1 \le x < y \le 6.$$

The marginal probability function of Y is

$$f_Y(y) = \frac{2y - 1}{36} \quad \text{if } 1 \le y \le 6.$$

Hence

$$f_X(x \mid Y = y) = \frac{2}{2y - 1} \quad \text{if } x = 1, \ldots, y - 1$$

$$= \frac{1}{2y - 1} \quad \text{if } x = y$$

and

$$F_X(x \mid Y = y) = \frac{2x}{2y - 1} \qquad \text{if } x = 1, \ldots, y - 1$$

$$= 1 \qquad \text{if } x = y.$$

Example 2.   Let the density of $\mathbf{X} = (X_1, X_2)$ be

$$f(\mathbf{t}) = t_1^2 e^{-t_1(1+t_2)} \qquad \text{if } t_1, t_2 > 0.$$

This is the density considered in Example 23.3. The marginal density of X_2 was found to be $h(t_2) = 2/(1 + t_2)^3$ for $t_2 > 0$ and the marginal density of X_1 was found to be $g(t_1) = t_1 e^{-t_1}$ for $t_1 > 0$. The conditional density of X_1 given $X_2 = t_2$ is, therefore,

$$f_{X_1}(t_1 \mid t_2) = \frac{1}{2} t_1^2 (1 + t_2)^3 e^{-t_1(1+t_2)} \qquad \text{if } t_1 > 0,$$

which, as can be seen, depends on t_2 and consequently, differs from the marginal density of X_1.

Remark 1.   Let $\mathbf{X}$ be a random vector and let A be an arbitrary event. The conditional distribution function of $\mathbf{X}$ given A (i.e., given $I_A = 1$) will be denoted by $F_{\mathbf{X}}(\mathbf{u} \mid A)$ and the conditional probability function or density given A will be denoted by $f_{\mathbf{X}}(\mathbf{u} \mid A)$.

Remark 2.   In the determination of conditional probability functions or densities, it is often unnecessary to exhibit the marginal of the conditioning variable explicitly. If the joint model is represented by $f(\mathbf{u}, \mathbf{v})$, the conditional model for $\mathbf{X}$ given $\mathbf{Y} = \mathbf{y}_0$ will be represented by $f_{\mathbf{X}}(\mathbf{u} \mid \mathbf{v}_0) = \gamma(\mathbf{y}_0) f(\mathbf{u}, \mathbf{v}_0)$, where $\gamma(\mathbf{v}_0)$ is uniquely determined by the fact that $f_{\mathbf{X}}(\mathbf{u} \mid \mathbf{v}_0)$ represents a probability model.

Example 3.   Let

$$f_{X,Y}(u, v) = 3u \qquad \text{if } 0 < u \le v < 1;$$

$$= 4uv \qquad \text{if } 0 < v < u < 1.$$

Then

$$f_X(u \mid v) = 3\gamma(v)u \qquad \text{if } 0 < u \le v < 1;$$

$$= 4v\gamma(v)u \qquad \text{if } 0 < v < u < 1$$

and by integration, $\gamma(v) = 2/[v(4 + 3v - 4v^2)]$. The distribution function of (X, Y) is

$$F_{X,Y}(u, v) = \frac{u^2}{2}(3v - 2u + u^2) \qquad \text{if } 0 < u \leq v < 1;$$

$$= \frac{v^2}{2}(2u^2 + v - v^2) \qquad \text{if } 0 < v < u < 1$$

and the marginal distribution function of Y is

$$F_Y(v) = \frac{v^2}{2}(2 + v - v^2) \qquad \text{if } 0 < v < 1$$

$$= 1 \qquad \text{if } v \geq 1.$$

Letting $A_v = [Y \leq v]$ (where $0 < v < 1$) yields $F_X(u \mid A_v) = F_{X,Y}(u, v)/F_Y(v)$. Differentiating, the density is

$$f_X(u \mid A_v) = \frac{2u(3v - 3u + 2u^2)}{v^2(2 + v - v^2)} \qquad \text{if } 0 < u \leq v < 1;$$

$$= \frac{4u}{2 + v - v^2} \qquad \text{if } 0 < v < u < 1.$$

In the remainder of this section we explore the multiplicative relation implied by rewriting (1) as the multiplication rule

$$P[\mathbf{X} \in A, \mathbf{Y} \in B] = P(\mathbf{X} \in A \mid \mathbf{Y} \in B)P[\mathbf{Y} \in B]. \qquad (5)$$

Formula (5) is a special case of the multiplication rule (12.8).

Suppose that for every pair of rectangles $A \subset \mathcal{R}^m$ and $B \subset \mathcal{R}^n$ the values of $P(\mathbf{X} \in A \mid \mathbf{Y} \in B)$ and $P[\mathbf{Y} \in B]$ are specified. Then (5) determines the joint probability model (on $\mathcal{R}^{m+n}$) for the pair $(\mathbf{X}, \mathbf{Y})$. It often happens, as shown in Chapter 2, that the required conditional probabilities and the corresponding marginal probabilities are available from the context or presented as plausible assumptions. The multiplication rule and its extension to any number of factors embody one of the most important uses of the concept of conditioning.

The version of the multiplication rule provided by (2) is

Multiplication Rule

The joint probability function or density of the random vectors $\mathbf{X}$ and $\mathbf{Y}$ satisfies

$$f_{\mathbf{X},\mathbf{Y}}(\mathbf{u}, \mathbf{v}) = f_{\mathbf{X}}(\mathbf{u} \mid \mathbf{Y} = \mathbf{v})f_{\mathbf{Y}}(\mathbf{v}) = f_{\mathbf{Y}}(\mathbf{v} \mid \mathbf{X} = \mathbf{u})f_{\mathbf{X}}(\mathbf{u}). \qquad (6)$$

Formula (12.10) is the multiplication rule for n events. Its analog in this context is a generalization of (6):

$$f_{\mathbf{X}}(\mathbf{t}) = f_{X_1}(t_1) \prod_{i=1}^{n-1} f_{X_{i+1}}(t_{i+1} \mid (X_1, \ldots, X_i) = \mathbf{u}_i), \tag{7}$$

where $\mathbf{X} = (X_1, \ldots, X_n)$, $\mathbf{t} \in \mathcal{R}^n$, and $\mathbf{u}_i = (t_1, \ldots, t_i)$.

Example 4. Let Y be the outcome of the first roll in a game of craps and let X be the additional number of rolls required to complete the game. In Example 22.2 the conditional probability function of X given $[Y = 4]$ was derived. In a similar manner, $f_X(r \mid Y = y)$ can be obtained for every y.

$$f_X(0 \mid Y = 2) = f_X(0 \mid Y = 3) = f_X(0 \mid Y = 7)$$
$$= f_X(0 \mid Y = 11) = f_X(0 \mid Y = 12) = 1$$

$$f_X(r \mid Y = 4) = f_X(r \mid Y = 10) = \left(\frac{1}{4}\right)\left(\frac{3}{4}\right)^{r-1}$$

$$f_X(r \mid Y = 5) = f_X(r \mid Y = 9) = \left(\frac{5}{18}\right)\left(\frac{13}{18}\right)^{r-1}$$

$$f_X(r \mid Y = 6) = f_X(r \mid Y = 8) = \left(\frac{11}{36}\right)\left(\frac{25}{36}\right)^{r-1}$$

The probability function of Y is

$$f_Y(y) = \frac{6 - |7 - y|}{36} \qquad \text{if } y = 2, \ldots, 12.$$

The joint probability function of X and Y is therefore obtainable from (6) for any specified pair (r, y).

Example 5. Suppose that you arrive at a store at a time X, uniformly distributed on $(0, 1)$, and given that $X = t$, the time Y that you spend in the store is uniformly distributed on $(0, 2t)$ so that the later you arrive, the longer you are likely to stay in the store. The density of X is $f(t) = 1$ for $0 < t < 1$, and the conditional density of Y given $X = t$ is $g(u \mid t) = 1/(2t)$ for $0 < u < 2t$. Hence, for example, $P(Y > t \mid X = t) = 1/2$. The density of (X, Y) is

$$h(t, u) = g(u \mid t)f(t) = \frac{1}{2t} \qquad \text{if } 0 < u < 2t; \quad 0 < t < 1.$$

The conditional density of X given $Y = u$ is

$$f_X(t \mid u) = \frac{\gamma(u)}{t} \quad \text{if } \frac{u}{2} < t < 1; \quad 0 < u < 2,$$

from which, by integration, $\gamma(u) = \log(2/u)$.

Summary

Conditional probability function: Conditional density function of $\mathbf{Y}$ given $\mathbf{Z} = \mathbf{v}$:

$$f_\mathbf{Y}(\mathbf{u} \mid \mathbf{Z} = \mathbf{v}) = \frac{f_{\mathbf{Y},\mathbf{Z}}(\mathbf{u}, \mathbf{v})}{f_\mathbf{Z}(\mathbf{v})} \quad \text{for } f_\mathbf{Z}(\mathbf{v}) > 0.$$

For fixed $\mathbf{v}$ $f_\mathbf{Y}(\mathbf{u} \mid \mathbf{Z} = \mathbf{v})$, as a function of $\mathbf{u}$, is a probability function or a density.

Let $\mathbf{X} = (X_1, \ldots, X_n)$ be an n-dimensional random vector. Then

$$f_\mathbf{X}(\mathbf{t}) = f_{X_1}(t_1) \prod_{i=1}^{n-1} f_{X_{i+1}}(t_{i+1} \mid (X_1, \ldots, X_i) = \mathbf{u}_i),$$

where $\mathbf{u}_i = (t_1, \ldots, t_i)$.

Problems

1. (a) Among n computer chips, two are defective. Chips are selected one at a time without replacement until both defectives are found. Let X be the number of chips required to find the first defective and Y be the total number of chips selected. Find the conditional probability function of X given $Y = y$.
   (b) Repeat part (a) assuming that there are m $(2 \le m \le n)$ defective chips and Y is the number of chips selected until all defectives are found.
2. A random sample of size 2 is drawn without replacement from the integers $1, \ldots, N$. Let X_i be the ith number selected.
   (a) Find the conditional probability function of X_1 given $X_2 = t_2$.
   (b) Find the conditional probability function of X_2 given $X_1 = t_1$.
   (c) Compare your answers to parts (a) and (b) with the corresponding marginal probability functions.
3. In Problem 2, let $Y = \max(X_1, X_2)$.
   (a) Find $f_{X_1}(u \mid Y = k)$.
   (b) Repeat part (a) for sampling with replacement.
4. A random sample of size n is drawn with replacement from the integers $1, \ldots, N$. Let X_i be the ith selection and $Y = \max X_i$. Find the conditional distribution of $\mathbf{X} = (X_1, \ldots, X_n)$ given $Y = k$.

5. Suppose that the joint probability function of (X, Y) at (t, u) is either $f_1(t, u)$ or $f_2(t, u)$, where $f_i(t, u) = g_i(u)h(t, u)$ $(i = 1, 2)$ and $g_1(u) > 0$ if, and only if, $g_2(u) > 0$. Show that the conditional probability function of X given $Y = u$ does not depend on i.

6. Consider the device in Figure 1. Suppose that each component has probability p of failing. Let X be the number of failing components and A be the event that the device fails. Find the conditional probablity functions of X given A and given A^c.

7. A box has n balls with numbers $1, \ldots, n$, respectively. A sample of size $m < n$ is drawn without replacement. If ball n is drawn, stop; otherwise, draw one additional ball from those remaining in the box. Let X be the maximum of the numbers on the balls in the initial sample and Y be the maximum of the numbers on all the balls sampled. Find the joint probability function of X and Y and the marginal probability function of Y.

8. A new drug is being tested in three stages. For stage i, let N_i be the number of patients receiving the drug and let M_i be the number uncured. Assume Bernoulli trials. Suppose that $N_1 = 3$ and for $i = 2, 3$ that

$$N_i = N_{i-1} - 1 \quad \text{if } M_{i-1} = 0$$

$$= N_{i-1} \quad \text{if } M_{i-1} > 0.$$

Let $X = N_1 + N_2 + N_3$, the total number of patients tested, and let Y be the total number of successes.
(a) Find the probability function of (X, Y).
(b) Find the marginal probability functions of X and of Y.

9. In Example 2 find the conditional density of X_2 given $X_1 = t_1$.

10. In Example 3 find the conditional density of Y given $X = u$.

11. Let $\mathbf{X} = (X_1, X_2)$ be the result of choosing a point at random in the triangle in $\Re^2$ bounded by the lines $t_1 = 0$, $t_2 = 1$, and $t_1 = t_2$.
(a) Give the density of $\mathbf{X}$.
(b) Give the conditional density of X_1 given $X_2 = t_2$.
(c) Repeat part (b), interchanging X_1 and X_2.

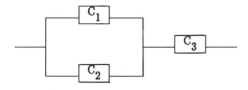

Figure 1

(d) Compare your answers in parts (b) and (c) with the corresponding marginal densities.

12. Let **X** be a two-dimensional random vector with the density

$$f(\mathbf{t}) = 2e^{-(t_1 + 2t_2)} \qquad \text{if } t_1, t_2 > 0.$$

Solve Problem 11 in this case.

13. Let $\mathbf{X} = (X_1, X_2, X_3)$ have the density

$$f(\mathbf{t}) = ct_1 t_2^{1/2} t_3^3 (1 - t_1 - t_2) \qquad \text{if } t_1, t_2, t_3 > 0; \quad t_1 + t_2 + t_3 < 1.$$

Find the conditional density of X_1 given $X_2 = t_2$ and $X_3 = t_3$.

14. Let $\mathbf{X} = (X_1, \ldots, X_n)$ have the density $f(\mathbf{t}) = 1$ for $0 < t_i < 1$ $(i = 1, \ldots, n)$.
    (a) Find the conditional density of **X** given the event $[\max X_i \le 1/2]$.
    (b) Find the conditional model for **X** given the event $[\max X_i = 1/2]$. Observe that the probability that more than one of the X_i achieves $\max X_i$ is 0.

15. Suppose that $A \subset \mathcal{R}^2$ is bounded and has positive area. A point is chosen at random from A. Let X and Y be the coordinates of this point.
    (a) Show that the conditional model of X given $Y = u$ is uniform on some $B_u \subset \mathcal{R}$ and identify B_u.
    (b) Let $C \subset A$ have positive area. Find the conditional density of (X, Y) given $[(X, Y) \in C]$.

16. Let the density of (X, Y) be

$$f(u, v) = ce^{-2uv} \qquad \text{if } u, v > 0.$$

Find the conditional density of X given $Y = v$.

17. Suppose that X has the density $f(t) = \lambda e^{-\lambda t}$ for $t > 0$ and that the conditional density of Y given $X = t$ is given by $g(u \mid t) = \gamma(t)e^{-\lambda u}$ for $u > t$.
    (a) Evaluate $\gamma(t)$.
    (b) Find the joint density of X and Y.
    (c) Find the conditional density of X given $Y = u$.

18. Let $f_{X,Y}(t, u) = ce^{-(t+u)}$ for $0 < t < u$. Find the conditional density of X given $Y = u$.

19. When a fair die is rolled n times, the probability function of $\mathbf{X} = (X_1, \ldots, X_6)$ is

$$f_{\mathbf{X}}(\mathbf{k}) = \binom{n}{k_1, \ldots, k_6} \left(\frac{1}{6}\right)^n,$$

where X_i is the number of occurrences of i. This is a *multinomial* probability function described in Section 41. Find the conditional probability function of (X_1, X_2) given $(X_5, X_6) = (u, v)$.

20.† Let (X, Y) be a two-dimensional random vector. Assume that $f_X(u \mid Y = v) = \gamma(u)$ for every (u, v) such that $f_{X,Y}(u, v) > 0$. Show that:

(a) $f_X(u \mid Y = v) \equiv f_X(u)$
(b) $f_Y(v \mid X = u) \equiv f_Y(v)$
(c) $f_{X,Y}(u, v) \equiv f_X(u)f_Y(v)$

27. INDEPENDENCE FOR RANDOM VARIABLES AND RANDOM VECTORS

In the preceding section, conditional models were described and analyzed in terms of probability functions, densities, and distribution functions. In this section we analyze, in terms of the same representations, the assumption of independence of random vectors originally defined for arbitrary experiments in Section 15.

Whatever the context, independence of two or more entities is expressible in either of two (essentially equivalent) probabilistic forms: (a) joint probabilities equal products of marginals, or (b) conditional probabilities equal marginal (unconditional) probabilities. What follows in this section are precise formulations of (a) and (b) with regard to random variables and vectors.

Independent Random Vectors

The random vectors $\mathbf{X}$ and $\mathbf{Y}$ (with marginals P_1, P_2, respectively) are called *independent* if for every pair of rectangles, $A \subset \mathfrak{R}^m$ and $B \subset \mathfrak{R}^n$, the events $[\mathbf{X} \in A]$ and $[\mathbf{Y} \in B]$ are independent, that is,

$$P[\mathbf{X} \in A, \mathbf{Y} \in B] = P(A \times B)$$
$$= P_1(A)P_2(B) = P[\mathbf{X} \in A]P[\mathbf{Y} \in B]. \tag{1}$$

Remark 1. When $m = n = 1$, relation (1) defines the independence of two rv's X and Y. The "rectangles" A and B reduce to intervals and $A \times B$ is an ordinary two-dimensional rectangle. We have chosen to define independence for random vectors of arbitrary dimension to avoid repetition of what is essentially a simple relation between joint and marginal probabilities, whatever the dimensionality may be.

Remark 2. The definition given for independence of two random vectors has a natural extension to the case of k random vectors as follows: Let $\mathbf{X}_1$, $\mathbf{X}_2, \ldots, \mathbf{X}_k$ ($k \geq 2$) be random vectors of dimensions $n_1, n_2, \ldots, n_k$, respectively, and let $A_i \subset \mathscr{R}^{n_i}$ be a rectangle for each $i = 1, 2, \ldots, k$. Then, $\mathbf{X}_1, \ldots, \mathbf{X}_k$ are *independent* if

$$P[\mathbf{X}_i \in A_i, i = 1, \ldots, k] = P\left(\underset{i=1}{\overset{k}{\times}} A_i \right)$$

$$= \prod_{i=1}^{k} P_i(A_i) = \prod_{i=1}^{k} P[\mathbf{X}_i \in A_i], \tag{2}$$

where P_i (the probability assignment to subsets of $\mathscr{R}^{n_i}$) is the marginal of $\mathbf{X}_i$. If $n_i = 1$ for $i = 1, \ldots, k$, then (2) defines independence of a set of rv's (the most commonly encountered situation).

Remark 3. The following theorems are easily understood and intuitively plausible. Their proofs are applications of (1) and (2) and some results from earlier chapters concerning Cartesian products. In particular, see GP 2.7. Following the theorems we give a general outline (without details) of some of the proofs. Detailed proofs are left as problems for the reader.

Theorem 1. *Let* $\mathbf{X}$ *and* $\mathbf{Y}$ *be m- and n-dimensional random vectors, respectively. Then* $\mathbf{X}$ *and* $\mathbf{Y}$ *are independent if, and only if,*

$$P(\mathbf{X} \in A_1 \mid \mathbf{Y} \in A_2) = P[\mathbf{X} \in A_1] \tag{3}$$

for every rectangle $A_2 \subset \mathscr{R}^n$ *such that* $P[\mathbf{Y} \in A_2] > 0$, *and for all rectangles* $A_1 \subset \mathscr{R}^m$.

Theorem 2. *For each* $i = 1, 2, \ldots, k$, *let* $\mathbf{X}_i$ *be an* n_i-*dimensional random vector. Then* $\mathbf{X}_1, \mathbf{X}_2, \ldots, \mathbf{X}_k$ *are independent if, and only if, for each* $\mathbf{t}_i \in \mathscr{R}^{n_i}$ ($i = 1, \ldots, k$)

$$F_{\mathbf{X}}(\mathbf{t}) = \prod_{i=1}^{k} F_{\mathbf{X}_i}(\mathbf{t}_i), \tag{4}$$

where $\mathbf{X} = (\mathbf{X}_1, \ldots, \mathbf{X}_k)$ *and* $\mathbf{t} = (\mathbf{t}_1, \ldots, \mathbf{t}_k)$.

Theorem 3. *Let* $\mathbf{X} = (\mathbf{X}_1, \ldots, \mathbf{X}_k)$ *and* $\mathbf{t} = (\mathbf{t}_1, \ldots, \mathbf{t}_k)$ *be as in Theorem 2. Then* $\mathbf{X}_1, \ldots, \mathbf{X}_k$ *are independent if, and only if, for every* $\mathbf{t}$,

$$f_{\mathbf{X}}(\mathbf{t}) = \prod_{i=1}^{k} f_{\mathbf{X}_i}(\mathbf{t}_i), \tag{5}$$

where f_X and the f_{X_i} are any of the following representations:

(i)   *Densities*
(ii) *Probability functions*
(iii) *Mixed*: *densities in some components, probability functions in others.*

Corollary 1.   *Let* $X = (X_1, \ldots, X_k)$ *and* $t = (t_1, \ldots, t_k)$. *Let* f_X *be as in Theorem 3. Then* $X_1, \ldots, X_k$ *are independent if, and only if, there exist nonnegative functions* $g_1, \ldots, g_k$ *such that*

$$f_X(t) = \prod_{i=1}^{k} g_i(t_i). \tag{6}$$

Theorem 4.   *If* $X_1, \ldots, X_k$ *are independent random vectors, then* (2) *remains valid if the* A_i's *are any events that can be obtained from rectangles by event operations* (*finite or countably infinite*).

Theorem 5.   *Let* X *and* Y *be random vectors and let* $U = \varphi(X)$ *and* $V = \psi(Y)$ *be arbitrary real-valued functions of* X *and* Y, *respectively. Then* X *and* Y *are independent if, and only if, for every choice of*[2] φ *and* ψ, *the rv's* U *and* V *are independent.*

Remark 4.   Theorem 1 embodies the crucial and most practical significance of the relation of independence between random vectors. The joint model for X and Y will be the product model whenever information about the value of X (or Y) provides no new (or altered) probability information about Y (or X).

Remark 5.   In theorem 2 the "only if" part is proved by noting that for each i, $F_{X_i}(t_i)$ is the probability of a rectangle in $\mathscr{R}^{n_i}$ and that $F_X(t)$ is the probability of their Cartesian product.

Remark 6.   Theorem 3 is valid (and relevant) only in those cases for which representations of type (i), (ii), and (iii) exist. To prove that independence implies (5) for case (i), differentiate both sides of (4). To prove that independence implies (5) for case (ii), note that any one point set in $\mathscr{R}^n$ is an n-dimensional rectangle.

---

[2]When referring to functions of random vectors, the restriction following (21.1) will be assumed.

Remark 7. Theorem 4 follows from the unique extension of probability beyond the class $\mathcal{C}$ of Section 5. See also GP 2.7.

Example 1. Let

$$f_{\mathbf{X}}(t_1, t_2) = 8t_1t_2 \qquad \text{if } 0 < t_1 < t_2 < 1.$$

The function $f_{\mathbf{X}}$ is a bivariate density which may appear to satisfy (6) with

$$g_1(t_1) = 8t_1 \qquad \text{if } 0 < t_1 < 1$$

and

$$g_2(t_2) = t_2 \qquad \text{if } 0 < t_2 < 1.$$

However, this argument ignores the order relation $t_1 < t_2$, which must be taken into account. For example, $f(1/2, 1/3) = 0$ and $g_1(1/2)g_2(1/3) = 4/3$. Furthermore g_1 and g_2 are not the marginals.

Example 2. Let

$$f_{\mathbf{Y}}(t_1, t_2) = 4t_1t_2 \qquad \text{if } 0 < t_1, t_2 < 1.$$

The function $f_{\mathbf{Y}}$ is a bivariate density with marginal densities $f_{Y_1}(t_1) = 2t_1$ and $f_{Y_2}(t_2) = 2t_2$ for $0 < t_1, t_2 < 1$. Thus, $f_{\mathbf{Y}}$ satisfies (5).

In fact, using indicators, the density in Example 1 can be expressed as

$$f_{\mathbf{X}}(\mathbf{t}) = 8t_1t_2I_S(t_1, t_2),$$

where $S = \{\mathbf{t}: 0 < t_1 < t_2 < 1\}$. The indicator I_S does not factor. On the other hand, the density in Example 2 can be written as

$$f_{\mathbf{Y}}(\mathbf{t}) = 4t_1t_2I_{(0,1)}(t_1)I_{(0,1)}(t_2).$$

The most thoroughly analyzed and frequently encountered models for random vectors are those for which the components are not only independent but also identically distributed. We will denote this case by the letters iid. Models of this type arise in connection with sampling from populations and in repeated trials of a fixed experiment. The representation of the model for such a case can be given by either the joint distribution function,

$$G(t_1, t_2, \ldots, t_n) = \prod_{i=1}^{n} F(t_i), \qquad (7)$$

or by the joint density or probability function,

$$g(t_1, \ldots, t_n) = \prod_{i=1}^{n} f(t_i). \qquad (8)$$

The simplicity of these expressions stems from the fact that for this case the joint model is completely determined from one, fixed, marginal.

Example 3. Let A_1, A_2, . . . , A_n denote equiprobable and mutually independent events with $P(A_i) = p$. We consider the random vector $(I_{A_1}, I_{A_2}, \ldots, I_{A_n})$. From the given assumptions the components are iid (see Problem 9) and each has probability function

$$P[I_{A_i} = x] = f(x) = q \qquad \text{if } x = 0$$
$$= p \qquad \text{if } x = 1,$$

where $q = 1 - p$. This can also be expressed as

$$f(x) = p^x q^{1-x} \qquad \text{if } x = 0, 1.$$

Therefore, the joint probability function of the n indicators is given by

$$g(x_1, \ldots, x_n) = \prod_{i=1}^{n} p^{x_i} q^{1-x_i} = p^{\Sigma x_i} q^{n - \Sigma x_i}$$

$$\text{if } x_i = 0, 1; \quad i = 1, \ldots, n. \quad (9)$$

This formula determines, for example, the probability of every member of the partition generated by the A_i's (see Section 6).

Example 4. Let X be a point selected at random in the interval $[0, 1]$ and let X_1, X_2, . . . , X_n be iid, each with the same distributions as X. Then

$$f(x) = 1 \qquad \text{if } 0 \leq x \leq 1$$

is the common marginal density of the X_i's. From (8),

$$g(x_1, x_2, \ldots, x_n) = \prod_{i=1}^{n} f(x_i)$$

$$= 1 \qquad \text{if } 0 \leq x_i \leq 1, i = 1, \ldots, n.$$

Note that g is the density for a uniform model on the n-dimensional cube determined by the conditions $0 \leq x_i \leq 1$, $i = 1, 2, \ldots, n$. Thus choosing a point at random in this cube is equivalent to n independent repetitions of the experiment of choosing a point at random in $[0, 1]$.

Note also that if X_1, . . . , X_n are independent rv's, then X_1^2, . . . , X_n^2 are independent as are $X_1^2 + X_2^2$ and $X_3^2 + X_4^2$ when $n \geq 4$.

It is possible to have $U(\mathbf{X}, \mathbf{Y})$ and $V(\mathbf{X}, \mathbf{Y})$ independent, although $\mathbf{X}$ and $\mathbf{Y}$ are not.

Example 5. Let X and Y be rv's with joint density

$$f(s, t) = e^{-t} \quad \text{if } 0 < s < t,$$

so that X and Y are not independent. The joint distribution function of $U = X$ and $V = Y - X$ is for $u, v > 0$,

$$G(u, v) = P[X \le u, Y - X \le v] = \int_0^u \int_s^{s+v} e^{-t} \, dt \, ds$$

$$= \int_0^u [e^{-s} - e^{-(s+v)}] \, ds = (1 - e^{-u})(1 - e^{-v}),$$

so that X and Z are independent and, in fact, are iid. The region of integration is shaded in Figure 1.

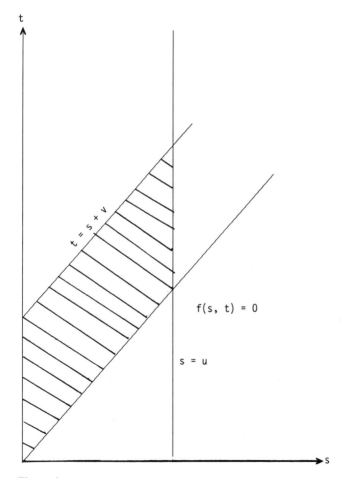

Figure 1

Summary

Independent random vectors: Random vectors $\mathbf{X}_1, \ldots, \mathbf{X}_k$, of dimensions $n_1, \ldots, n_k$, respectively, are independent if $P[\mathbf{X}_i \in A_i, i = 1, \ldots, k] = \Pi_{i=1}^{k} P[\mathbf{X}_i \in A_i]$ for all rectangles $A_i \subset R^{n_i}$.
iid: Independent and identically distributed.

Various product relations for representations of models for independent random vectors are given in Theorem 1 to 4 and Corollary 1. If $\mathbf{X}$ and $\mathbf{Y}$ are independent random vectors, then real-valued functions of $\mathbf{X}$ and $\mathbf{Y}$, respectively, are independent rv's.

Problems

1. In each of the following cases, decide whether f is the joint density of independent rv's.
   (a) $f(x, y, z) = 6e^{-(x+2y+3z)}$ if $x, y, z > 0$
   (b) $f(x, y, z) = ce^{-xy}$ if $1 < x, y < 2$
   (c) $f(x, y, z) = cxy^2z^{1/2}$ if $x, y, z > 0; x + y + z < 1$
2. Samples of students from two sections of a course are drawn simultaneously. Let X and Y be the averages of the heights of students in the two samples. Give an argument supporting the assumption that X and Y are independent rv's.
3. In Example 1, compare $P(X_1 \leq 1/2 \mid X_2 \leq 1/2)$ and $P[X_1 \leq 1/2]$.
4. Let $\mathbf{X}_1, \ldots, \mathbf{X}_k$ be independent random vectors.
   (a) Show that the random vectors in any subcollection of $\mathbf{X}_1, \ldots, \mathbf{X}_k$ are independent.
   (b) Let $\mathbf{Y} = (\mathbf{X}_1, \ldots, \mathbf{X}_l)$ and $\mathbf{Z} = (\mathbf{X}_{l+1}, \ldots, \mathbf{X}_k)$. Show that $\mathbf{Y}$ and $\mathbf{Z}$ are independent random vectors.
5. Let $A_1, \ldots, A_n$ be events. Show that $A_1, \ldots, A_n$ are independent if, and only if, $I_{A_1}, \ldots, I_{A_n}$ are independent rv's.
6. Let $X_1, \ldots, X_n$ be rv's. Set $X^{(1)} = \min(X_1, \ldots, X_n)$ and $X^{(n)} = \max(X_1, \ldots, X_n)$. Are $X^{(1)}$ and $X^{(n)}$ ever independent? If so, under what conditions?
7.† Let $X_1, \ldots, X_n$ be iid rv's with common exponential density $f(x \mid \lambda) = \lambda e^{-\lambda x}$ for $x > 0$, where $\lambda > 0$.
   (a) Find the joint density of $X_1, \ldots, X_n$.
   (b) Find the joint distribution function of $X_1 + X_2$ and X_1/X_2. *Hint*: Sketch the region $X_1 + X_2 \leq u$, $X_1/X_2 \leq v$.
   (c) Are $X_1 + X_2$ and X_1/X_2 independent rv's?
8. Let the marginal probability function or density of $\mathbf{X}$ be g and the conditional probability function or density of $\mathbf{X}$ given $\mathbf{Y} = \mathbf{u}$ be $h(\mathbf{t} \mid \mathbf{u})$. Show that $\mathbf{X}$ and $\mathbf{Y}$ are independent if, and only if,

$h(t \mid u) = g(t)$ for all t and u for which $h(t \mid u)$ is defined. See Problem 26.5.

9. Let X, Y, and Z be rv's. Suppose that X and Y are independent, X and Z are independent, and Y and Z are independent. Does this imply that X, Y, and Z are independent? Prove or give a counter-example.

10. Let $\mathcal{S}$ be a finite subset of $\mathcal{R}^n$ and choose an element of $\mathcal{S}$ at random. Let $\mathbf{X} = (X_1, \ldots, X_n)$ be the resulting n-dimensional random vector.
    (a) What are the conditions on $\mathcal{S}$ that make $X_1, \ldots, X_n$ independent?
    (b) Under the conditions in part (a), what are the marginal probability functions of the X_i?

11. Let $\mathcal{E}_1, \ldots, \mathcal{E}_n$ be independent experiments and, for each $i = 1, \ldots, n$, let $\mathbf{X}_i$ be a random vector defined on $\mathcal{E}_i$. Show that $\mathbf{X}_1, \ldots, \mathbf{X}_n$ are independent.

12. (a) Prove that independence implies that

$$F_{\mathbf{X}_1,\ldots,\mathbf{X}_k}(\mathbf{t}_1, \ldots, \mathbf{t}_k) = \prod_{i=1}^{k} F_{\mathbf{X}_i}(\mathbf{t}_i).$$

    (b) Prove the converse of part (a) for $k = 2$ and rv's X_1 and X_2.

13. Prove Theorem 1.
14. Prove Theorem 3 for cases (i) and (ii) only.
15. Prove Corollary 1.
16. Prove Theorem 4.
17. Prove Theorem 5.
18. Let $\mathbf{X}_1, \ldots, \mathbf{X}_k$ be random vectors and let $\mathbf{U}_i = \varphi_i(\mathbf{X}_i)$ where the functions $\varphi_1, \ldots, \varphi_k$ take values in $\mathcal{R}^{m_1}, \ldots, \mathcal{R}^{m_k}$, respectively, and are subject to the conditions given in Section 21. State and prove the extension of Theorem 5 to this case.

28. TOTAL PROBABILITY AND BAYES' THEOREM

The multiplication rule

$$f_{\mathbf{X},\mathbf{Y}}(\mathbf{u}, \mathbf{v}) = f_{\mathbf{Y}}(\mathbf{v})f_{\mathbf{X}}(\mathbf{u} \mid \mathbf{Y} = \mathbf{v}) \tag{1}$$

is valid independently of the types of X and Y (even if they are different) as long as the functions in (1) are appropriate representations. The extended multiplication rule (26.7) also applies.

Example 1. In Example 25.3, let X be the continuous rv and Y be the discrete rv. The representation

$$f(u) = 1 - u_1 \quad \text{if } 0 < u_1 < 1; \quad u_2 = 0;$$
$$= u_1 \quad \text{if } 0 < u_1 < 1; \quad u_2 = 1.$$

was found for the joint model for X and Y. The marginal probability function of X was found to be $f_Y(u_1) = 1$ for $0 < u_1 < 1$, so that the conditional probability function of Y given $X = u_1$ is

$$f_Y(u_2 \mid X = u_1) = 1 - u_1 \quad \text{if } u_2 = 0; \quad 0 < u_1 < 1;$$
$$= u_1 \quad \text{if } u_2 = 1; \quad 0 < u_1 < 1.$$

Let A be the event that a coin toss results in a head and let $Y = I_A$. Thus as noted previously, the model results from an experiment in which a number X is chosen at random in $(0, 1)$ and then a coin, for which the probability of a head is X, is tossed.

The version of the total probability law (12.12) that follows from (1) is

$$f_X(u) = \sum_v f_{X,Y}(u, v) = \sum_v f_X(u \mid Y = v)f_Y(v) \qquad (2)$$

for Y discrete and

$$f_X(u) = \int_{\Re k} f_X(u \mid Y = v)f_Y(v) \, dv \qquad (3)$$

for Y with a density.

Theorem 1. (Another Bayes' theorem) *Let X and Y be random vectors. Then*

$$f_Y(v \mid X = u) = \frac{f_Y(v)f_X(u \mid Y = v)}{f_X(u)}. \qquad (4)$$

Proof. The denominator in (4) is obtained from (2) or (3). The theorem then follows from (1).

Example 2. Suppose that the coin of Example 1 is tossed until a head occurs and let Z be the number of tosses required. The conditional probability function of Z given that $Y = v$ is the geometric $h(u \mid v) = v(1 - v)^{u-1}$ for $u = 1, 2, \ldots$ The joint model for Z and Y is represented by

$$f(u, v) = v(1 - v)^{u-1} \quad \text{if } 0 < v < 1; \quad u = 1, 2, \ldots$$

The conditional density of Y, given $Z = u$ is, then,

$$g(v \mid u) = \gamma(u)v(1 - v)^{u-1} \qquad \text{if } 0 < v < 1.$$

It is easy to verify that $\gamma(u) = u(u + 1)$.

In Example 2, replace v by p. Then $g(u \mid p)$ is a probability function for each $p \in [0, 1]$. As p varies over $[0, 1]$, these probability functions identify a family of probability models on $\mathcal{R}$. In such cases we refer to p as the *parameter* of the family. In applications we are often willing to assume that a random variable or vector has a probability model that belongs to a family which is indexed by one or more parameters as in Example 2. Specification of the parameters amounts to complete identification of the model.

In Bayes' theorem, when $f_X(u \mid v)$ is a probability function or density with a parameter v and f_Y is the probability function or density of this parameter, we use the following terminology: f_Y is called the *prior probability function* or *density* of the parameter and $f_Y(v \mid u)$ is called the *posterior probability function* or *density* of the parameter given $X = u$. (See Section 13 for the introduction of prior and posterior probabilities.) This terminology suggests, as is often the case, that f_Y represents the uncertainty about the parameter prior to experimentation and $f_Y(v \mid u)$ the revised uncertainty about the parameter if the result of the experiment is $X = u$. Similar remarks apply if v is a vector of parameters.

Notation. Although parameters with prior and posterior distributions are considered to be rv's, they are not necessarily capitalized.

Example 3. Suppose that the density of X is exponential, that is,

$$f_X(t \mid \lambda) = \lambda e^{-\lambda t} \qquad \text{if } t > 0,$$

where $\lambda > 0$. Suppose that the prior density of λ is uniform on $(0, 1)$. For $t > 0$, the posterior density of λ given $X = t$ is

$$h(z \mid t) = \frac{ze^{-zt}}{\int_0^1 ue^{-ut}\, du} = \frac{t^2 ze^{-zt}}{1 - e^{-t}(1 + t)} \qquad \text{if } 0 < z < 1.$$

Example 4. A shipment of N identical components has an unknown number D of defectives. A sample of n components is to be chosen at random from the shipment, so that if X is the number of defectives in the sample, then

$$f_X(k \mid D = d) = \frac{\binom{d}{k}\binom{N-d}{n-k}}{\binom{N}{n}}$$

$$= \frac{\binom{n}{k}\binom{N-n}{d-k}}{\binom{N}{d}} \qquad \text{if } k = 0, 1, \ldots, n,$$

which is a hypergeometric. Assume that the prior probability function of D is the binomial

$$f_D(d) = \binom{N}{d} p^d q^{N-d},$$

where $0 < p < 1$ is fixed and $q = 1 - p$. The posterior probability function of D given $X = k$ is

$$f_D(d \mid k) = \gamma(k) \binom{N-n}{d-k} p^d q^{N-d} \qquad \text{if } d - k = 0, \ldots, N - n.$$

From the form of the binomial distribution (Section 22), $\gamma(k) = 1/(p^k q^{n-k})$. Hence

$$f_D(d \mid k) = \binom{N-n}{d-k} p^{d-k} q^{(N-n)-(d-k)}$$

$$\text{if } d = k, k + 1, \ldots, k + N - n.$$

It can be seen that the model for $D - k$ given $X = k$ is binomial.

Let F be a mixture (see Section 25) of any distribution functions, G and H [i.e., $F(\mathbf{u}) = \alpha G(\mathbf{u}) + (1 - \alpha)H(\mathbf{u})$ for some $\alpha \in (0, 1)$]. Then F is produced by the following experiment:

1. There is an event A with $P(A) = \alpha$.
2. $G(\mathbf{u}) = F_X(\mathbf{u} \mid A)$.
3. $H(\mathbf{u}) = F_X(\mathbf{u} \mid A^c)$.

Applying the total probability law (12.12), $\mathbf{X} = \mathbf{X}I_A + \mathbf{X}I_{A^c}$ has distribution function F.

Every countable mixture of distribution functions has a similar experimental interpretation.

Example 5.† In Example 25.1 we considered the distribution function

$$G(y) = p + (1 - p)(1 - e^{-y/10}) \quad \text{if } 0 \le y \le 8;$$
$$= 1 \quad \text{if } y > 8$$

and expressed it as a mixture of a discrete and a continuous distribution function. The following describes an experiment that leads to an expression for G as a mixture of one continuous and two degenerate distribution functions.

The device is turned on at time 0 and may fail (the event A) with probability p. The device has lifetime given by the continuous distribution function $F(y) = 1 - e^{-y/10}$ for $y > 0$, so that it may fail during $(0, 8)$ (the event B). At time 8, if the device has survived (the event C), it is turned off. The time the device is on is $X = 0 \cdot I_A + XI_B + 8I_C$, where it was found that

$$F_X(x \mid B) = H_2(x) = \frac{1 - e^{-x/10}}{1 - e^{-4/5}} \quad \text{if } 0 \le x \le 8;$$
$$= 1 \quad \text{if } x > 8.$$

The distribution function of X is

$$G(x) = pK_0(x) + (1 - \alpha)H_2(x) + \beta K_8(x),$$

where K_t is the distribution function of a rv degenerate at t and $\alpha = 1 - (1 - p)(1 - e^{-4/5})$. Since $G(8) = 1$, it follows that $\beta = 1 - (p + \alpha) = (1 - p)e^{-4/5}$.

Summary

The formula in Section 26 for conditional probability functions and densities is valid when marginal models are of different types. The same is true of the multiplication rule.

Bayes' Theorem:

$$f_Y(v \mid u) = \frac{f_Y(v)f_X(u \mid v)}{f_X(u)}.$$

Problems

1. For $i = 1, 2, \ldots$, let

$$F_i(x) = \frac{x}{i} \quad \text{if } 0 < x < i;$$
$$= 1 \quad \text{if } x \ge i$$

and $\alpha_i = i(1/2)^{i+1}$. Set

$$F(x) = \sum_{i=1}^{\infty} \alpha_i F_i(x).$$

(a) Show that

$$F(x) = 1 - \frac{i + 1 - x}{2^i} \qquad \text{if } i - 1 \le x < i, \quad i = 1, 2, \dots.$$

(b) Give an experiment in which a rv with distribution function F will result.

2. Two machines, 1 and 2, each produce 10 mousetraps per hour. Suppose that during one hour machine 1 has produced two defective traps and machine 2 has produced three defective traps. A machine is selected at random, and from the production of that machine, we select two traps at random. Let X be the number of defective traps selected and Y be the number of defective traps produced by the selected machine. Find the conditional probability function of Y given $X = t$ for each t.

3. Let the marginal density of X be

$$f(t) = 2t \qquad \text{if } 0 < t < 1.$$

Assume that the conditional density of Y given $X = t$ is $g(u \mid t) = t^2 u e^{-tu}$ for $u > 0$. Find the conditional density of X given $Y = u$.

4. A family has income Y and owns Z cars. Assume that the marginal probability function of Z is

$$f(v) = \frac{1}{10} \qquad \text{if } v = 0;$$

$$= \frac{4}{10} \qquad \text{if } v = 1;$$

$$= \frac{3}{10} \qquad \text{if } v = 2;$$

$$= \frac{2}{10} \qquad \text{if } v = 3$$

and that the conditional density of Y given $Z = v$ is

$$g(u \mid v) = \frac{1}{v + 1} e^{-u/(v+1)} \qquad \text{if } u > 0.$$

Find the conditional probability function of Z given $Y = u$.

5. In two courses (1 and 2) the numbers of students with each class code (1 = freshman, etc.) is given in Table 1. Suppose that a course is chosen at random and a student is chosen at random from that course. Let Z be the course chosen and Y be the class code of the student chosen. For each possible value u of Y, find the conditional probability function of Z given $Y = u$.

6. Let X have the probability function $f(k \mid \lambda) = (\lambda^k/k!)e^{-\lambda}$ for $k = 0$, $1, \ldots$ and assume that the prior density of λ is $g(u) = e^{-u}$ for $u > 0$. As $\lambda > 0$ varies, these probability functions represent the members of the family of Poisson models. The density g is an exponential density. Find the posterior density of λ.

7. Repeat Problem 6 assuming that X has the density

$$f(t \mid \lambda) = \lambda^2 t e^{-\lambda t} \qquad \text{if } t > 0.$$

8.† Suppose that the probability function of Y is $g(k \mid p) = \binom{4}{k} p^k(1 - p)^{4-k}$ for $k = 0, 1, 2, 3,$ or 4 and that the prior distribution of p is uniform on $(0, 1)$. Find the posterior density of p.

9. The U.S. Senate (100 members) is about to vote on ratification of a major arms control treaty. A random sample of 15 senators is taken and 9 in the sample favor ratification. Assume that the prior distribution of the number of senators favoring ratification is binomial with some specified p.

   (a) What is the posterior distribution of the number of senators favoring ratification?

   (b) What is the posterior probability that a senator not in the sample favors ratification?

   (c) Comment on the plausibility of the assumed prior.

Table 1

Code	Course 1	Course 2
1	2	0
2	15	4
3	10	7
4	8	7
5	5	2
Total	40	20

10. Show for random vectors $\mathbf{X}$, $\mathbf{Y}$, and $\mathbf{Z}$ that

$$f_{\mathbf{X}}(\mathbf{u} \mid \mathbf{Y} = \mathbf{v}, \mathbf{Z} = \mathbf{w}) = \frac{f_{\mathbf{X,Y}}(\mathbf{u}, \mathbf{v} \mid \mathbf{Z} = \mathbf{w})}{f_{\mathbf{Y}}(\mathbf{v} \mid \mathbf{Z} = \mathbf{w})}.$$

This states that "today's posterior is tomorrow's prior." Note the analogy to Problem 12.4.

11. Verify (4).
12. Find the posterior models in Example 1 and verify that the posterior model given that $Z = 1$ is the same as the posterior model in Example 2 given that $Y = 1$.
13. Find the marginal probability function of Y in Example 3.

29. ILLUSTRATIONS

In this section examples using conditioning and independence are considered.

A. Clinical Trials

Two frequently studied procedures for comparing two drugs or treatments are:

(1) *Pair at a time*: Patients are considered in pairs. One member of each pair receives drug 1 and the other receives drug 2.
(2) *Play the winner*: Patients are considered individually and sequentially. If a drug succeeds on a patient, the next patient gets the same one; otherwise, the next patient is switched to the other drug. The first patient is randomly assigned one of the two drugs.

A complete description of the procedure requires a rule for terminating the experiment.

The following describes a class of hybrid procedures along with assumptions and notation to be used.

(i) Pair at a time experimentation is used until, for the first time, one drug succeeds and the other fails.
(ii) Testing proceeds using play the winner starting with the last successful drug in (i).
(iii) The stopping rule for the second stage guarantees that there will be at least one switch (e.g., stop the second stage when each drug has failed one additional time).
(iv) $p_i = 1 - q_i$ denotes the probability of success for drug i, and without loss of generality, $p_1 \geq p_2$ (drug 1 is "better" than drug 2).

Consider two questions:

1. What is the probability of the event A that the play the winner phase begins with the inferior drug?
2. What is the distribution of X, the number of patients treated with the inferior drug, before the first switch in the play the winner phase?

The probability that in stage (i), the drugs give different results within any pair is $\alpha = p_1 q_2 + p_2 q_1$. Then $P(A) = p_2 q_1 / \alpha$.

Let Y be the number of pairs of patients in stage (i) and Z be the number of patients in stage (ii), so that $X = Y + I_A Z$. Then

$$f_Y(r) = \alpha(1 - \alpha)^{r-1} \qquad \text{if } r = 1, 2, \ldots$$

and

$$f_Z(k \mid A) = q_2 p_2^{k-1} \qquad \text{if } k = 1, 2, \ldots.$$

Thus

$$f_X(k \mid Y = r, A^c) = 1 \qquad \text{if } k = r;$$
$$f_X(k \mid Y = r, A) = q_2 p_2^{k-r-1} \qquad \text{if } k = r + 1, r + 2, \ldots$$

and hence

$$f_X(k \mid Y = r) = P(A^c) f_X(k \mid Y = r, A^c) + P(A) f_X(k \mid Y = r, A)$$

$$= \frac{p_1 q_2}{\alpha} \qquad \text{if } k = r;$$

$$= \frac{q_2 q_1 p_2^{k-r}}{\alpha} \qquad \text{if } k = r + 1, r + 2, \ldots.$$

For $k = 1, 2, \ldots$,

$$f_X(k) = \sum_{r=1}^{k} f_Y(r) f_X(k \mid Y = r)$$

$$= \sum_{r=1}^{k-1} q_2 q_1 p_2^{k-r} (1 - \alpha)^{r-1} + p_1 q_2 (1 - \alpha)^{k-1}$$

$$= q_2 \left[p_2^{k-1} q_1 \sum_{m=0}^{k-2} \left(\frac{1 - \alpha}{p_2} \right)^m + p_1 (1 - \alpha)^{k-1} \right].$$

Now $1 - \alpha = p_1 p_2 + q_1 q_2$. It is easy to show that $p_2 = 1 - \alpha$ if, and only if, $p_2 = 1/2$, and in this case,

$$f_X(k) = (1/2)^k [(k - 1) q_1 + p_1] \qquad \text{if } k = 1, 2, \ldots.$$

When $p_2 \neq 1/2$ and for $k = 1, 2, \ldots,$

$$f_X(k) = q_2 \left[p_2^{k-1} q_1 \frac{1 - [(1 - \alpha)/p_2]^{k-1}}{1 - [(1 - \alpha)/p_2]} + p_1(1 - \alpha)^{k-1} \right]$$

$$= \frac{q_2[p_2^k - \alpha(1 - \alpha)^{k-1}]}{p_2 - q_2}.$$

The last equality uses the easily verified relation $p_2 - (1 - \alpha) = q_1(p_2 - q_2)$. Note that the numerator and denominator have the same sign, positive or negative.

B. Reliability

When a system of components breaks down it becomes necessary to seek and replace failed components. If the cost of examination is the same for each component, the search should begin with the one most likely to have caused the failure at that particular time (i.e., the one with the largest conditional probability of failure given the time of system failure). It may also be desirable to find and replace components that failed but were *not* the cause of a system failure (e.g., spare components in a parallel subsystem).

The problems to be solved are combinations and extensions of the following. Let C_1 and C_2 be components whose respective lives are X_1 and X_2 and consider $U = \min(X_1, X_2)$ and $V = \max(X_1, X_2)$. Given that X_1 and X_2 are independent and have respective densities f_1 and f_2 and distribution functions F_1 and F_2, determine:

(i) $P(X_1 = U \mid U = t)$, the conditional probability that C_1 failed at time t given that at least one of the components failed at time t.

(ii) $P(X_1 < t \mid V \geq t)$, the conditional probability that C_1 failed by time t given that at least one of the components survived to time t.

The rv U is the lifetime of a series system and V is the lifetime of a parallel system.

The distribution function of U is

$$F_U(t) = P[U \leq t] = 1 - P[U > t]$$

$$= 1 - P[X_1 > t, X_2 > t] = 1 - \overline{F}_1(t)\overline{F}_2(t)$$

where, as in Section 24B, $\overline{F} = 1 - F$. It follows that

$$f_U(t) = f_1(t)\overline{F}_2(t) + f_2(t)\overline{F}_1(t). \tag{1}$$

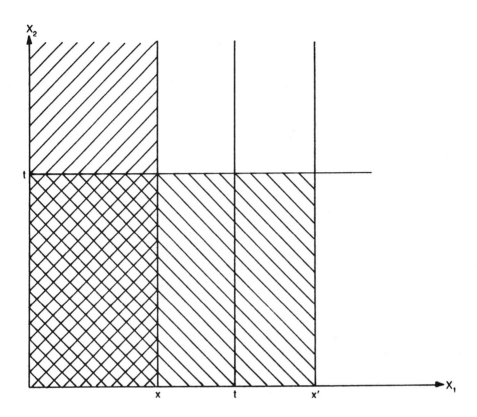

Figure 1

As can be seen from Figure 1, for $x, t > 0$,

$$P[X_1 \leq x, U \leq t] = F_1(x) \qquad\qquad \text{if } x \leq t;$$
$$= F_1(x)F_2(t) + F_1(t)\overline{F}_2(t) \qquad \text{if } x > t. \qquad (2)$$

The joint distribution function given by (2) represents a model in which:

(i) $P[X_1 = U]$, the probability that C_1 fails first, is positive.
(ii) In the region $x > t$, the derivative of the joint distribution function exists and is given by $h(x, t) = f_1(x)f_2(t)$.

Thus the conditional probabilities for X_1 given $U = t$ for $x > t$ are obtained by integrating $f_1(x)f_2(t)/g(t)$. This is,

$$P(X_1 > t \mid U = t) = \int_t^\infty \frac{f_1(x)f_2(t)}{g(t)} \, dx = \frac{f_2(t)\overline{F}_1(t)}{f_1(t)\overline{F}_2(t) + f_2(t)\overline{F}_1(t)},$$

and hence

$$P(X_1 = t \mid U = t) = 1 - P(X_1 > t \mid Y = t)$$

$$= \frac{f_1(t)\overline{F}_2(t)}{f_1(t)\overline{F}_2(t) + f_2(t)\overline{F}_1(t)} = \frac{r_1(t)}{r_1(t) + r_2(t)}, \qquad (3)$$

where $r_i(t)$ is the failure rate of F_i. If X_1 and X_2 are identically distributed, then $f_1 = f_2$, so that $F_1 = F_2$ and hence $P(X_1 = t \mid U = t) = 1/2$.

In the exponential case $r_i(t) = \lambda_i$, the conditional probability that C_1 fails at time t given that the system fails at time t [given in (3)] reduces to

$$P(X_1 = t \mid U = t) = \frac{\lambda_1}{\lambda_1 + \lambda_2}.$$

This probability does not depend on t. Pairs of distributions for X_1 and X_2 which share this property can easily be characterized in terms of the r_i (see Problem 2).

The distribution function of V is

$$F_V(t) = P[X_1 \le t, X_2 \le t] = F_1(t)F_2(t),$$

so that

$$f_V(t) = f_1(t)F_2(t) + f_2(t)F_1(t).$$

Furthermore,

$$P[X_1 < t, V \ge t] = P[X_1 < t, X_2 \ge t] = F_1(t)\overline{F}_2(t).$$

Hence

$$P(X_1 < t \mid V \ge t) = \frac{F_1(t)\overline{F}_2(t)}{\overline{F}_V(t)} = \frac{F_1(t)\overline{F}_2(t)}{1 - F_1(t)F_2(t)}. \qquad (4)$$

If X_1 and X_2 are identically distributed so that $F_1 = F_2 = F$, then (4) simplifies to

$$P(X_1 < t \mid V \ge t) = \frac{F(t)}{2 - \overline{F}(t)} = \frac{F(t)}{1 + F(t)}.$$

In the exponential case with $\lambda_1 = \lambda_2 = \lambda$,

$$P(X_1 < t \mid V \ge t) = \frac{1 - e^{-\lambda t}}{2 - e^{-\lambda t}}.$$

Problems

1. In the procedure in Subsection A, suppose that the experiment stops when one drug has had M more success than the other. Let W be the number of patients treated with the inferior drug until the first switch or termination of the experiment, whichever is earlier. Find f_W.

2. Suppose that the clinical trials of Subsection A terminate at the second failure in the play the winner stage and let U be the number of times the inferior drug is used. Find f_U.

3. Find the relationship between F_1 and F_2 which guarantees that the probability in (3) does not depend on t.

4. Suppose that a system consists of three components, C_1, C_2, and C_3. The lifetimes of the components are independent rv's with densities. Consider two cases:

   (i) C_1 and C_2 are in series and the resulting subsystem is in parallel with C_3.

   (ii) C_1 and C_2 are in parallel and the resulting subsystem is in series with C_3.

   Let L be the lifetime of the system. In each case:

   (a) Find f_L.

   (b) Given that the system fails at time t, what is the probability that $C_i(i = 1, 2, 3)$ has failed?

   (c) Given that the system is functioning at time t, what is the probability that $C_i(i = 1, 2, 3)$ has failed?

   (d) Specialize the answers in (a), (b), and (c) to the case in which the lifetimes of the C_i are iid.

   (e) Specialize the answers in part (d) to the case of an exponential distribution.

5.† The baseball World Series is a sequence of games, none of which can end in a tie, played until one team wins four games, thus winning the series. Suppose that the games are a sequence of Bernoulli trials in which team A has probability p of winning. Let X be the number of games in the series.

   (a) Find the probability function of X.

   (b) Find the conditional probability function of X given that team A wins the first game.

   (c) If team A won the series, what is the conditional probability that they won the first game?

6. In a population of size N, the individuals have been assigned numbers serially from 1 to N. For example, these might be licence plate numbers. A sample of size n is taken with replacement. Let X_i be the number assigned to the ith individual in the sample and $Y = \max X_i$.

   (a) Find the probability function of Y.

   (b) For $c > 1$, find $P[Y \le N \le cY]$.

7. There are 518 consecutively numbered tickets (from 1 to 518) in a drawing for a state millionaire lottery. There are three urns. Urn 1 contains digits 0 to 5 and urns 2 and 3 each contain digits 0 to 9. The winning number is $Y = 100X_1 + 10X_2 + X_3$, where X_i is randomly selected from urn i with the following rules:
    (i) If X_2 is inadmissible ($X_1 = 5$ and $X_2 \geq 2$), a new ball is drawn from urn 2 and the procedure is repeated until an admissible value of X_2 is obtained.
    (ii) If X_3 inadmissible given X_1 and X_2, the same procedure is followed.
    (a) Find the probability function of $X = (X_1, X_2, X_3)$.
    (b) Find the probability function of Y.
    (c) Comment on the fairness of this method. If you conclude that this method is unfair:
        (i) What ticket(s) would you prefer to hold?
        (ii) Devise a fair method based on selections from these urns.

GENERAL PROBLEMS

1.† The random vectors $\mathbf{X}$ and $\mathbf{Y}$ are *conditionally independent given the event B* if $\mathbf{X}$ and $\mathbf{Y}$ are independent in the conditional model for $(\mathbf{X}, \mathbf{Y})$ given B. If $\mathbf{X}$ and $\mathbf{Y}$ are conditionally independent given $[\mathbf{Z} = \mathbf{t}]$ for each possible value $\mathbf{t}$ of $\mathbf{Z}$, then $\mathbf{X}$ and $\mathbf{Y}$ are *conditionally independent given* $\mathbf{Z}$. Let $\mathbf{X} = (\mathbf{X}_1, \mathbf{X}_2)$ and $\mathbf{Y} = (\mathbf{Y}_1, \mathbf{Y}_2)$ be independent. Show that $\mathbf{X}_1$ and $\mathbf{Y}_1$ are conditionally independent given $(\mathbf{X}_2, \mathbf{Y}_2)$.

2.† Let A, B, and C be events. Assume that I_A and I_B are conditionally independent given I_C. Show that A and B are independent if, and only if, A and C are independent or B and C are independent.

3. Show that the following condition is equivalent to $\mathbf{X}$ and $\mathbf{Y}$ being independent: For each possible value, $\mathbf{t}$, the conditional model for $\mathbf{X}$ given $\mathbf{Y} = \mathbf{t}$ does not depend on $\mathbf{t}$.

4. In Problem 29.17, let λ have exponential prior density $g(\lambda) = e^{-\lambda}$ for $\lambda > 0$. Hence the distribution in Problem 29.17 is conditional given λ. Solve each of (a), (b), and (c) using the marginal distribution of $(X_1, \ldots, X_n)$.

5. Let X be uniformly distributed on $(0, 1)$ and Y be uniformly distributed on the integers $0, \ldots, n$. Find the distribution of $X + Y$. *Hint*: Solve for $n = 1, 2$.

6. See the definition of odds in GP 1.11. Let A and A^c each represent a model on $\mathfrak{R}^n$ with probability functions or densities $f(\mathbf{x} \mid A)$ and $f(\mathbf{x} \mid A^c)$, respectively. The *likelihood ratio* is $\lambda(\mathbf{x}) = f(\mathbf{x} \mid A^c)/f(\mathbf{x} \mid A)$. Express the posterior odds (odds using posterior probabilities) against A in terms of prior odds (odds using prior probabilities) and $\lambda(\mathbf{x})$.

REFERENCES

DeGroot, Morris H. (1986). *Probability and Statistics*, 2nd ed. Reading, Mass.: Addison-Wesley Publishing Co., Inc. Exposition and application of Bayes' theorem.

Hoel, Paul G., Port, Sidney C., and Stone, Charles J. (1971). *Introduction to Probability Theory*. Boston: Houghton Mifflin Company.

Woodroofe, Michael (1975). *Probability with Applications*. New York: McGraw-Hill Book Company.

6.

Expectation and Other Characteristics of Distributions

30. EXPECTATION

The uncertainty about a random variable or vector is completely described by its distribution, which may be very complicated and detailed. In practice, however, it often happens that only a few summary attributes of the distribution are of genuine interest. For instance, in choosing various investments, the choice may depend only on such things as (1) the probability of making a profit of a specified size, (2) the expected profit, or (3) some entirely different but reasonably simple criterion.

In this chapter we describe and define some properties of random variables and vectors that have been particularly useful. Many are defined in terms of "expected value," a concept that was first introduced in Section 7 for the special case of simple rv's. This section consists of three subsections. In Subsection A the expected value, $E(X)$, is defined for discrete and continuous rv's; Subsection B contains a formula for $E(X)$ in terms of tail probabilities; Subsection C contains a general definition and related theorems.

A. Definitions of $E(X)$, Discrete and Continuous Cases

If $X = \sum_{i=1}^{n} a_i I_{A_i}$, then X is a simple rv and

$$E(X) = \sum_{i=1}^{n} a_i P(A_i). \tag{1}$$

This definition was discussed in some detail in Section 7, where it was shown that no matter what (finite) linear combination of indicators is used to represent X, the value of $E(X)$ is unchanged. As often happens in mathematics, going from the finite to the infinite introduces a variety of complications. Analysis of these complications will be deferred to Subsection C. Here we present the most useful forms of the definitions based on two sets of assumptions. The notation for expected value is $E(X)$, as before, or, for convenience, μ_X.

If $\mathbf{X} = (X_1, \ldots, X_n)$ is a random vector, the notation "$E(X)$" stands for the vector whose ith component is $E(X_i)$ $(i = 1, \ldots, n)$.

Expected Value

The *expected value*, $E(X)$ (or μ_X), of a rv is a number defined as follows:

(A)   (i) X is discrete and (ii) $\sum_{r \in \mathcal{S}(X)} |r| \, f_X(r) < \infty$. Then

$$E(X) = \sum_{r \in \mathcal{S}(X)} r f_X(r). \tag{2}$$

[Recall that $\mathcal{S}(X)$ is a countable set containing all possible values of X.]

(B)   (i) X has a density and (ii) $\int_{-\infty}^{\infty} |x| \, f_X(x) \, dx < \infty$. Then

$$E(X) = \int_{-\infty}^{\infty} x f_X(x) \, dx. \tag{3}$$

If the absolute convergence assumption is not satisfied, $E(X)$ *does not exist*. When "$E(X)$" is used in an expression, it will be assumed that the absolute convergence condition is satisfied.

As was shown in the discussion of histograms (Section 23B), for every $\varepsilon > 0$, a rv X with density f_X can be approximated by a simple rv Y_ε and, from (2),

$$E(Y_\varepsilon) = \sum_i m_i \eta(m_i),$$

where $\varepsilon(m_i) = h(m_i)\Delta_i$, h is the histogram, Δ_i is the ith interval width, and ε measures the closeness of the approximation. By an appropriate limiting argument (Problem 19), it can be shown that

$$E(Y_\varepsilon) \longrightarrow \int_{-\infty}^{\infty} x f_X(x) \, dx = E(x)$$

as $\varepsilon \to 0$ provided that (ii) holds.

To illustrate the computation of $E(X) = \mu_X$ for the two types of models, here are some examples.

Example 1. Consider the infinite set $\mathcal{S}(X) = \{x_1, -x_1, x_2, -x_2, \ldots\}$, where $0 < x_i$, $i = 1, 2, \ldots$. If the rv X has probability function

$$f(x_n) = f(-x_n) = \left(\frac{1}{2}\right)^{n+1}, \qquad n = 1, 2, \ldots,$$

then under assumption (A)(ii), $E(X) = 0$ since

$$E(X) = x_1 \left(\frac{1}{4}\right) - x_1 \left(\frac{1}{4}\right) + x_2 \left(\frac{1}{8}\right) - x_2 \left(\frac{1}{8}\right) + \cdots .$$

Assumption (A)(ii) will be satisified for some x_n's but not for others.

(i)   If $x_n = n$, then

$$\sum_x |x| f(x) = \sum_{n=1}^{\infty} nf(n) + \sum_{n=1}^{\infty} |-n| f(-n) = 2 \sum_{n=1}^{\infty} nf(n)$$

$$= 2 \sum_{n=1}^{\infty} n \left(\frac{1}{2}\right)^{n+1} = 2.$$

(ii)   If $x_n = 2^n$, then

$$\sum_x |x| f(x) = 2 \sum_{n=1}^{\infty} 2^n \left(\frac{1}{2}\right)^{n+1} = \infty.$$

When the same probabilities are attached to a different set of x's, it can make a big difference.

Example 2. Arrivals at a certain telephone booth occur at an average rate of 10 per hours. In 1 hour, therefore, we "expect" 10 arrivals and it seems reasonable to "expect" to wait 6 minutes (.1 hour) for each arrival. Let N denote the actual number of arrivals in 1 hour and let W denote the waiting time between arrivals. The stated conditions suggest that $E(N) = 10$ and $E(W) = .1$. Under assumptions often invoked for such phenomena, the resulting model for N is Poisson and the corresponding model for W is exponential. That is,

$$f_N(k) = \frac{10^k e^{-10}}{k!} \qquad \text{if } k = 0, 1, 2, \ldots$$

$$f_W(t) = 10e^{-10t} \qquad \text{if } t > 0.$$

These models have expectations

$$E(N) = \sum_{k=0}^{\infty} k \frac{10^k e^{-10}}{k!} = 10 \sum_{k=1}^{\infty} \frac{10^{k-1} e^{-10}}{(k-1)!} = 10$$

$$E(W) = \int_0^{\infty} t \cdot 10 e^{-10t} \, dt = .1 \int_0^{\infty} u e^{-u} \, du = .1.$$

We now augment and generalize the interpretations of $E(X)$ given in Section 7.

Case 1: *Population mean.* A population consists of N individuals and the ith individual is assigned the number x_i. For example, in a population of buildings, x_i might be the age of or the number of rooms in the ith building. Let $X = \sum_{i=1}^{N} x_i I_{A_i}$, where A_i designates the ith individual. Then the prevalence of A_i is $P(A_i) = 1/N$ $(i = 1, \ldots, N)$, so that $E(X) = \sum_{i=1}^{N} x_i/N$, the arithmetic mean of $x_1, \ldots, x_N$. Note that if each x_i is either 0 or 1, then $E(X)$ is the population proportion of those individual assigned value 1.

Case 2: *Average per trial value.* Suppose that you play a sequence of identical games in which the amount you win in the ith game is the rv X_i. Since the games are identical, so are the values of $E(X_i)$, and this common value is the amount you expect to win per game. If you play a long sequence of such games, the actual amount you win per game should be close to this expectation. The last statement is a consequence of the analogous statement for relative frequencies of events in repeated trials and applies to all such experiments.

Case 3: *Subjective expectation.* You hold a ticket in a football pool and X, your payoff, depends on the number of games for which you pick the correct winner. If the probabilities of various results of the games are subjective, so is $E(X)$, the worth of the ticket. If the ticket cost you C, then $E(X) - C$ is your expected net income from the pool. Another purchaser of a ticket in the pool, even one making the same choices as you, may have a different value for $E(X)$ due to different subjective probabilities.

There is a useful physical analog of expected value. Let $r_1, r_2, \ldots$ be the possible values of the rv X. On a weightless rod with a number scale, attach the mass $f_X(r_i)$ at each point r_i. Then the center of gravity (balance point) of this system is $E(X)$. This physical analog applies to rv's which are not discrete and to random vectors.

In view of these interpretations, the alternative names for $E(X)$ that are sometimes used are *mean of X*, *average (value) of X*, and *expectation of X*.

B. Use of Tail Events

An integer-valued rv X can be expressed in terms of tail events as follows (see Problem 21.13):

$$X = \sum_{k=0}^{\infty} I_{C_k} - \sum_{k=1}^{\infty} I_{D_k}, \tag{4}$$

where $C_k = [X > k]$ and $D_k = [X \le -k]$. In this subsection it will be shown that existence of $E(X)$ means the finiteness of $\sum_{k=0}^{\infty} P(C_k)$ and $\sum_{k=1}^{\infty} P(D_k)$. By analogy with the case of simple rv's,

$$E(X) = \sum_{k=0}^{\infty} P(C_k) - \sum_{k=1}^{\infty} P(D_k) = \sum_{k=0}^{\infty} [1 - F(k)] - \sum_{k=1}^{\infty} F(-k), \tag{5}$$

where F is the distribution function of X.

As it stands, (5) already provides a useful alternative for calculating $E(X)$ but only for integer-valued X. A modification of (5), however, produces an expression that can be used to define $E(X)$ for arbitrary rv's. For the integer-valued case, since F is constant between successive values of X (integer valued), $F(-k) = \int_{-k}^{-k+1} F(t)\, dt$ and $[1 - F(k)] = \int_{k}^{k+1} [1 - F(t)]\, dt$. Hence, from (5),

$$E(X) = \int_{0}^{\infty} P[X > t]\, dt - \int_{-\infty}^{\infty} P[X \le t]\, dt$$

$$= \int_{0}^{\infty} [1 - F(t)]\, dt - \int_{-\infty}^{0} F(t)\, dt. \tag{6}$$

Figure 1 illustrates the computation of each term in (6) for the discrete case. The shaded area to the right of 0 is the first term, while the shaded area to the left of 0 is the second term.

While (6) can be used as a general definition of $E(X)$, the following analysis applies only to the discrete and continuous cases.

For the two lemmas that follow, we adopt the special notation $h(u, v) = I_{[0,v)}(u) = 1$ for $0 \le u < v$ and 0 otherwise. Then $x = \int_0^\infty h(t, x)\, dt$ for $x > 0$.

Lemma 1.   *Let $x_i \in [0, \infty)$ for $i = 1, 2, \ldots$ and let $g \ge 0$ with $\sum_{i=1}^{\infty} x_i g(x_i) < \infty$. Then*

$$\sum_{i=1}^{\infty} x_i g(x_i) = \int_{0}^{\infty} \sum_{x_i > t} g(x_i)\, dt.$$

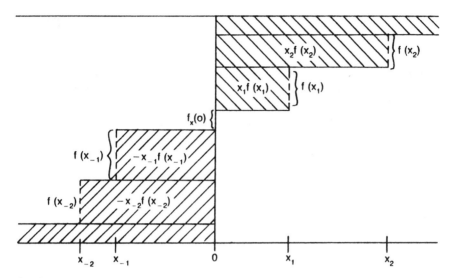

Figure 1

Proof. With h defined as above,

$$\sum_{i=1}^{\infty} x_i g(x_i) = \sum_{i=1}^{\infty} \left[\int_0^{\infty} h(t, x_i) \, dt \right] g(x_i). \tag{7}$$

Since the functions on the right side of (7) are nonnegative, the operations Σ and $\int$ can be interchanged so that

$$\sum_{i=1}^{\infty} x_i g(x_i) = \int_0^{\infty} \left[\sum_{i=1}^{\infty} h(t, x_i) g(x_i) \right] dt.$$

But $\sum_{i=1}^{\infty} h(t, x_i) g(x_i) = \sum_{x_i > t} g(x_i)$, completing the proof.

Lemma 2.   *Let $g \geq 0$ and $xg(x)$ both be integrable on $[0, \infty)$. Then $\int_0^{\infty} xg(x) \, dx = \int_0^{\infty} \int_t^{\infty} g(x) \, dx \, dt$.*

Proof.   For $x > 0$, replacing x by $\int_0^x dt$ yields

$$\int_0^{\infty} xg(x) \, dx = \int_0^{\infty} \int_0^x g(x) \, dt \, dx = \int_0^{\infty} \int_t^{\infty} g(x) \, dx \, dt. \tag{8}$$

The last equality in (8) follows by changing the order of integration and is justified by the nonnegativity of g.

Theorem 1. *For any discrete or continuous rv X, $E(X)$ exists if, and only if, $\int_0^\infty [1 - F(t)] \, dt < \infty$ and $\int_{-\infty}^0 F(t) \, dt < \infty$ and in that case, $E(X)$ is given by (6).*

Proof. The proof in the continuous case, using Lemma 2, is given below. The proof in the discrete case, using Lemma 1, is similar and has been left as a problem.

Let f be the density of X. Then, by Lemma 2,

$$\int_0^\infty xf(x) \, dx = \int_0^\infty \int_t^\infty f(x) \, dx \, dt = \int_0^\infty [1 - F(t)] \, dt. \qquad (9)$$

Also,

$$\int_{-\infty}^0 xf(x) \, dx = -\int_0^\infty xf(-x) \, dx = -\int_0^\infty \int_t^\infty f(-x) \, dx \, dt$$

$$= -\int_0^\infty \left[\int_{-\infty}^{-t} f(x) \, dx \right] dt = -\int_0^\infty F(-t) \, dt = -\int_{-\infty}^0 F(t) \, dt. \qquad (10)$$

Existence of $E(X)$ means that $\int_{-\infty}^{+\infty} |x| \, f(x) \, dx < \infty$, which is equivalent to

$$\int_0^\infty xf(x) \, dx < \infty \quad \text{and} \quad \int_{-\infty}^0 xf(x) \, dx > -\infty. \qquad (11)$$

If (11) is true, then

$$E(X) = \int_0^\infty xf(x) \, dx + \int_{-\infty}^0 xf(x) \, dx. \qquad (12)$$

Using (9) and (10), it follows that (11) is equivalent to

$$\int_0^\infty [1 - F(t)] \, dt < \infty, \qquad \int_{-\infty}^0 F(t) \, dt < \infty$$

and, finally, that (10) is equivalent to $E(X) = \int_0^\infty [1 - F(t)] \, dt - \int_{-\infty}^0 F(t) \, dt$, as was to be shown.

Remark 1. Note that $E(X)$ only depends on the model for X and is usually calculated through one of the model's representations. Therefore, we call the number $E(X)$ (or μ_X) the expected value of the model or the expected value of any of its representations. The notation μ_F where F is a distribution function is, we hope, self explanatory.

Remark 2. It can be shown that (6) is equivalent to the general definition of $E(X)$ given in Subsection C. The proof is beyond the level of this text.

Example 3†   Let $0 \le p \le 1$, $q = 1 - p$, and T be a rv with distribution function $F(x) = 1 - q^n$ when $n \le x < n + 1$ and $n = 0, 1, \ldots$. Then T is a nonnegative, integer-valued rv, and using (6), we obtain

$$E(T) = \sum_{n=0}^{\infty} q^n = \frac{1}{1 - q} = \frac{1}{p}.$$

The probability function of T is a geometric:

$$f(n) = (1 - q^n) - (1 - q^{n-1}) = pq^{n-1} \qquad \text{if } n = 1, 2, \ldots .$$

Example 4†   Let W be a rv with distribution function $F(t) = 1 - e^{-10t}$ for $t > 0$. This is the distribution function of the exponential model given in Example 2. Using (6), $E(W) = \int_0^{\infty} e^{-10t} \, dt = .1$ as before.

Example 5.   For $a, b > 0$ let

$$f_X(t) = \frac{a}{2} e^{at} \qquad \text{if } t < 0;$$

$$= \frac{b}{2} e^{-bt} \qquad \text{if } t \ge 0.$$

The corresponding distribution function is

$$F_X(t) = \frac{1}{2} e^{at} \qquad \text{if } t < 0;$$

$$= 1 - \frac{1}{2} e^{-bt} \qquad \text{if } t \ge 0.$$

Thus $\int_{-\infty}^{0} F_X(t) \, dt = (1/2) \int_{-\infty}^{0} e^{at} \, dt = 1/(2a)$ and $\int_0^{\infty} [1 - F_X(t)] \, dt = (1/2) \int_0^{\infty} e^{-bt} \, dt = 1/(2b)$, so that $E(X) = 1/(2b) - 1/(2a)$.

C. A General Definition; Consequences

The purpose of this subsection is to present and examine a definition of expected value that generalizes (1), the definition of $E(X)$ for simple rv's. The main results are embodied in Theorem 2 and Corollary 1, which are generalized versions of similar theorems proved in Sections 5 and 7. Although the proofs of these results have been included and are not, we believe, beyond the assumed level of prospective readers, they are of a decidedly mathematical character and may be skipped without adversely affecting the reader's understanding of the remainder of the text.

Every rv X, of whatever type, can be expressed as

$$X = \sum_{i=1}^{\infty} a_i I_{A_i} \tag{13}$$

for suitably chosen numbers a_i and events A_i. The A_i may or may not be disjoint even when X is discrete. [Unless the a_i and A_i satisfy certain conditions, the right side of (13) will not determine a rv.]

One way to represent an arbitrary rv X, in the manner of (13) is to adopt the methods of Example 21.9 and Problem 21.15 (binary expansion). Admittedly, indicator representations of rv's are not necessarily natural or desirable. If X is discrete, however, there are several useful and natural versions of (13). Tail events, for example, have been used previously. Although most applications of such representations in this text are to the discrete case, the results are as easily obtained in general and provide a general tool.

Expected Value. (C):

If in (13) the a_i and A_i can be chosen so that

$$\sum_{i=1}^{\infty} |a_i| P(A_i) < \infty, \tag{14}$$

then the *expected value of X* exists and is defined to be

$$\sum_{i=1}^{\infty} a_i P(A_i). \tag{15}$$

This latest definition of expected value will now be justified and discussed. The first step is to prove an analog to Theorems 5.1 and 7.2.

Theorem 2. *Let $a_i \geq 0$, $b_j \geq 0$ and assume that $\sum_{i=1}^{\infty} a_i I_{A_i} \geq \sum_{j=1}^{\infty} b_j I_{B_j}$. Then*

$$\sum_{i=1}^{\infty} a_i P(A_i) \geq \sum_{j=1}^{\infty} b_j P(B_j). \tag{16}$$

Proof. Let J be an arbitrary, but fixed positive integer and let $K = \sum_{j=1}^{J} b_j I_{B_j}$. If it can be shown that

$$\sum_{i=1}^{\infty} a_i P(A_i) \geq \sum_{j=1}^{J} b_j P(B_j), \tag{17}$$

then (16) will follow by letting $J \to \infty$. To prove (17), define

$\mathfrak{D}$: The partition induced by $\{B_1, \ldots, B_J\}$: $\mathfrak{D} = \{D_1, \ldots, D_{2^J}\}$.
$\mathfrak{D}_n$: The partition induced by $\{A_1, \ldots, A_n\}$.
$\mathfrak{D}_n^*$: The common refinement of $\mathfrak{D}$ and $\mathfrak{D}_n$ (see Problems 4.14 and 4.15),
$L_n = \Sigma_{i=1}^n a_i I_{A_i}$, and $L = \Sigma_{i=1}^\infty a_i I_{A_i}$.

Since $a_i \geq 0$ for every i, $L_n \leq L_{n+1}$ and it follows that $E(L_n) \uparrow$ $\Sigma_{i=1}^\infty a_i P(A_i)$.

Let $\varepsilon \geq 0$ be arbitrary and let $\varphi = \max K$. Now $K = \Sigma_{k=1}^{2^J} d_k I_{D_k}$, so that $\varphi = \max\{d_k : D_k \in \mathfrak{D}, D_k \neq \phi\}$. Consider the simple rv $K - L_n$ (its values are defined on $\mathfrak{D}_n^*$). There is an event Δ_n (a union of members of $\mathfrak{D}_n^*$) such that

$$(K - L_n)I_{\Delta_n} > \varepsilon I_{\Delta_n}$$

$$(K - L_n)I_{\Delta_n^c} \leq \varepsilon I_{\Delta_n^c}.$$

Since $L_n \uparrow L \geq K$ and $\Delta_n = [K - L_n > \varepsilon]$, it follows that $\Delta_n \downarrow \phi$ as $n \to \infty$. Hence, by Theorem 6.3, $P(\Delta_n) \to 0$.

Finally, define $M_n = \varphi I_{\Delta_n} + \varepsilon I_{\Delta_n^c}$. The rv M_n is simple and $M_n \geq K - L_n$, which implies that

$$E(K) - E(L_n) \leq E(M_n) = \varphi P(\Delta_n) + \varepsilon P(\Delta_n^c).$$

Since $E(L_n) \uparrow \Sigma_{i=1}^\infty a_i P(A_i)$ and $P(\Delta_n) \downarrow 0$, it follows that

$$E(K) = \sum_{j=1}^J b_j P(B_j) \leq \sum_{i=1}^\infty a_i P(A_i),$$

as was to be shown.

Corollary 1. *Let $\Sigma_{i=1}^\infty a_i I_{A_i} \equiv 0$ and assume that $\Sigma_{i=1}^\infty |a_i| P(A_i) < \infty$. Then $\Sigma_{i=1}^\infty a_i P(A_i) = 0$.*

Proof. Let $a_i^+ = a_i I_{[a_i > 0]}$ and $a_i^- = -a_i I_{[a_i < 0]}$. Then

$$\sum_{i=1}^\infty a_i I_{A_i} = \sum_{i=1}^\infty a_i^+ I_{A_i} - \sum_{i=1}^\infty a_i^- I_{A_i} \equiv 0$$

and since $a_i^+, a_i^- \geq 0$,

$$\sum_{i=1}^n a_i^+ I_{A_i} - \sum_{i=1}^\infty a_i^- I_{A_i} \leq 0 \leq \sum_{i=1}^\infty a_i^+ I_{A_i} - \sum_{i=1}^n a_i^- I_{A_i} \qquad (18)$$

for every n. Applying Theorem 2 to (18),

$$\sum_{i=1}^{n} a_i^{+} P(A_i) \le \sum_{i=1}^{\infty} a_i^{-} P(A_i)$$

$$\sum_{i=1}^{n} a_i^{-} P(A_i) \le \sum_{i=1}^{\infty} a_i^{+} P(A_i).$$

Letting $n \to \infty$ yields

$$\sum_{i=1}^{\infty} a_i^{+} P(A_i) = \sum_{i=1}^{\infty} a_i^{-} P(A_i).$$

Since $\sum_{i=1}^{\infty} |a_i| P(A_i) < \infty$, the theorem follows.

Corollary 2. Let $X = \sum_{i=1}^{\infty} a_i I_{A_i} = \sum_{j=1}^{\infty} b_j I_{B_j}$ be such that $\sum_{i=1}^{\infty} |a_i| P(A_i) < \infty$ and $\sum_{j=1}^{\infty} |b_j| P(B_j) < \infty$. Then $\sum_{i=1}^{\infty} a_i P(A_i) = \sum_{j=1}^{\infty} b_j P(B_j)$.

Proof. The conclusion follows from Corollary 1 by noting that

$$\sum_{i=1}^{\infty} a_i I_{A_i} - \sum_{j=1}^{\infty} b_j I_{B_j} \equiv 0$$

and the absolute convergence condition is satisfied.

Corollary 2 shows that the definition of expected values, case (C), produces the same value for $E(X)$, if it exists, regardless of which linear combination of indicators satisfying (14) is used to represent X.

To relate case (C) to case (B), the continuous case [or to the general definition suggested by (7)], and to show that no conflicting values of $E(X)$ will be obtained requires detailed analysis of the definition of the integral and will not be attempted here.

Case (A), the discrete case, is, however, covered by the following theorem.

Theorem 3. Let X be a discrete rv with probability function f_X. If for some numbers a_i and events A_i, it is the case that $X = \sum_i a_i I_{A_i}$ and $\sum_i |a_i| P(A_i) < \infty$, then

$$\sum_{r \in \mathcal{S}(X)} |r| f_X(r) < \infty \tag{19}$$

and

$$E(X) = \sum_i a_i P(A_i) = \sum_{r \in \mathcal{S}(X)} r f_X(r). \tag{20}$$

Proof. Let $\{[X = r_i]: i = 1, 2, \ldots\}$ denote the partition induced by X. Then

$$X = \sum_i r_i I_{[X=r_i]} = \sum_i a_i I_{A_i}$$

and since the $[X = r_i]$ form a partition,

$$\sum_i |r_i|\, I_{[X=r_i]} = \left|\sum_i r_i I_{[X=r_i]}\right| = \left|\sum_i a_i I_{A_i}\right| \leq \sum_i |a_i|\, I_{A_i}.$$

Hence, by Theorem 2,

$$\sum_i |r_i|\, P[X = r_i] \leq \sum_i |a_i|\, P(A_i) < \infty. \tag{21}$$

The left side of (21) is $\sum_{r \in \mathcal{S}(X)} |r|\, f_X(r)$ and (19) is proved. Equation (20) then follows from Corollary 2.

Example 6. Let $X = 2 \sum_{n=1}^{\infty} (1/3)^n I_{A_n}$, where $P(A_n) = 1/2$. Then

$$E(X) = 2 \sum_{n=1}^{\infty} \left(\frac{1}{3}\right)^n \cdot \left(\frac{1}{2}\right) = \frac{1}{2}.$$

Now consider two cases. At one extreme all the A_n are the same event, so that X is the simple rv taking values 0 and 1, each with probability 1/2. In this case we could have used (1) to calculate $E(X)$.

At the other extreme, if the A_n are independent, the distribution of X is called the Cantor distribution. It is the example of a singular distribution alluded to in Section 25.

The theorem that follows will be useful in finding expectations of functions of random vectors (Section 31).

Theorem 4. *Let $\{A_1, A_2, \ldots\}$ and $\{B_1, B_2, \ldots\}$ be partitions. Assume that $\sum_i a_i I_{A_i} = \sum_j b_j I_{B_j}$. Then $\sum_i |a_i|\, P(A_i) = \sum_j |b_j|\, P(B_j)$ and hence $\sum_i |a_i|\, P(A_i) < \infty$ implies that $\sum_j |b_j|\, P(B_j) < \infty$.*

Proof. Since $\sum_i |a_i|\, I_{A_i} = |\sum_i a_i I_{A_i}|$ for any partition $\{A_1, A_2, \ldots\}$, the result follows easily.

Summary

Expectation of a discrete rv: $E(X) = \sum_{r \in \mathcal{S}(X)} r f_X(r)$ *provided that* $\sum_{r \in \mathcal{S}(X)} |r|\, f_X(r) < \infty.$

Expectation of a rv X with a density; $E(X) = \int_{-\infty}^{\infty} x f_X(x)\, dx$ *provided that* $\int_{-\infty}^{\infty} |x|\, f_X(x)\, dx.$

Expectation of a rv $X = \sum_{k=1}^{\infty} a_k I_{A_k}$: $E(X) = \sum_{k=1}^{\infty} a_k P(A_k)$ provided that $\sum_{k=1}^{\infty} |a_k| P(A_k) < \infty$.

$E(X) = \int_0^{\infty} [1 - F_X(t)] \, dt - \int_{-\infty}^0 F_X(t) \, dt$ provided that both integrals are finite. In the integer-valued case, this becomes

$$E(X) = \sum_{k=1}^{\infty} [1 - F_X(k)] - \sum_{k=0}^{\infty} F_X(-k).$$

Problems

1.† Let X be a random variable with density $f(t) = (1/2)e^{-|t|}$. This is the *standard double exponential density*. Verify that f is a density, show that $E(X)$ exists, and find $E(X)$.

2. A fair die is rolled twice. Let X be the result of the first roll, Y be the minimum of the two rolls, and $Z = X + Y$. Find $E(X)$, $E(Y)$, and $E(Z)$. What relationship do you observe between these expectations? Do you expect this to be generally true?

3. For fixed $\alpha > 1$, let

$$f(t) = (\alpha - 1)t^{-\alpha} \qquad \text{if } t > 1.$$

Verify that f is a density. Let X be an rv with density f. For what values of α does $E(X)$ exist? Evaluate $E(X)$ in those cases.

4. Let X be an rv with density

$$f(x) = c(1 - x^2) \qquad \text{if } |x| < 1.$$

Find $E(X)$.

5. Let $f_X(k) = 1/[k(k + 1)(k + 2)]$ for $k = 1, 2, \ldots$. Find $E(X)$.

6. Let $f_X(t) = (3/8)t^2$ for $0 < t < 2$. Find $E(X)$.

7. Let $f_X(t) = (1/2)t^{-4}e^{-1/t}$ for $t > 0$. Find $E(X)$. *Hint:* When integrating, use the transformation $t = 1/x$ and then integrate by parts.

8.† Let the distribution X be uniform on the interval (a, b). Show that $E(X) = (a + b)/2$. Solve this problem in at least two ways.

9. Let $f(x) = (1/\pi) \cdot [1/(1 + x^2)]$, the *standard Cauchy density*. Show that this density has no mean.

10. The distribution of an rv X, is given by $F(t) = 1 - 1/[k(k + 1)]$ for $k \le t < k + 1$; $k = 1, 2, \ldots$.
    (a) Determine the probability function of X.
    (b) Determine $P[X > 3.5]$.
    (c) Does $E(X)$ exist? If so, find it.

11. Let X have the distribution function

$$F(t) = \frac{2}{3} e^{2t} \qquad \text{if } t < 0;$$

$$= 1 - \frac{1}{4} e^{-3t} \qquad \text{if } t \geq 0.$$

Find $E(X)$.
12. Let F be the distribution function corresponding to the density in Problem 3.
    (a) Let Y have distribution function F^2. For what α does $E(Y)$ exist? Generalize to the distribution function F^n $(n \geq 1)$.
    (b) Let Z have distribution function $1 - (1 - F)^n$ $(n \geq 1)$. For what α does $E(Z)$ exist?
    (c) Evaluate $E(Y)$ and $E(Z)$ when they exist.
13. In a sequence of Bernoulli trials with success probability p, let X be the number of trials until at least one success and at least one failure are achieved.
    (a) Find the right-tail probabilities for X.
    (b) Find $E(X)$.
14. Let f be a density on $\mathcal{R}$ for which $\int_{-\infty}^{\infty} |x| f(x) \, dx < \infty$, and for $\varepsilon > 0$, let R be such that $\int_{|x|>R} |x| f(x) \, dx < \varepsilon$. Let Y_ε be simple rv such that $|Y_\varepsilon| \leq R$ and the probability function of Y_ε is obtained as usual from a histogram approximating f. Show that for $\max_i \Delta_i$ sufficiently small, $|E(Y_\varepsilon) - \int_{-\infty}^{\infty} xf(x) \, dx| < 2\varepsilon$. Recall that $\eta(m_i) = \int_{a_{i-1}}^{a_i} f(x) \, dx$ unless $a_{i-1} = -R$ or $a_i = R$.
15. In Example 1(ii) give an ordering of the terms that results in $\Sigma_x \, xf(x) \neq 0$.
16. Prove Theorem 1 in the discrete case.
17. Let X be a rv with values $0 = a_0 < a_1 < a_2 < \cdots$ and $0 = b_0 > b_1 > b_2 > \cdots$. Show that

$$E(X) = \sum_{k=1}^{\infty} (a_{k+1} - a_k)[1 - F(a_k)] + \sum_{k=0}^{\infty} (b_k - b_{k+1})F(b_{k+1})$$

provided that $E(X)$ exists (see Problem 21.15).
18. Show that in the evaluation of $E(X)$ by integrating tail probabilities as in (6) the inclusion or exclusion of endpoints of the tails in immaterial.
19. Consider the mixture $F = \alpha G + (1 - \alpha)H(0 \leq \alpha \leq 1)$ of distribution functions. Show that if μ_G and μ_H exist, so does μ_F and $\mu_F = \alpha \mu_G + (1 - \alpha)\mu_H$.

20. Recall the distribution function G of Example 25.1 (a mixture of a discrete and a continuous distribution function). See Example 28.5 for a continuation. Find μ_G in as many ways as you can.
21. Let $X = \Sigma_{k=1}^{\infty} a_k Y_k$ where each Y_k is a simple rv. Show that under suitable conditions, $E(X) = \Sigma_{k=1}^{\infty} a_k E(Y_k)$. State these conditions.

31. PROPERTIES OF EXPECTATION

In this section we investigate the computation of expectations of functions of random vectors and associated results. Some of these were presented in Section 7 for simple rv's. In particular, Theorem 1 generalizes (7.8), which can now be stated in terms of probability functions.

Theorem 1. *Let* $\mathbf{X}$ *be an n-dimensional random vector and g be a real-valued function[1] on* $\mathfrak{R}^n$. *Then*

(i) $\mathbf{X}$ *discrete*: $E(g(\mathbf{X}))$ *exists if, and only if,* $\Sigma_t |g(\mathbf{t})| f_{\mathbf{X}}(\mathbf{t}) < \infty$ *and then*

$$E(g(\mathbf{X})) = \sum_t g(\mathbf{t}) f_{\mathbf{X}}(\mathbf{t}). \qquad (1)$$

(ii) $\mathbf{X}$ *continuous*: $E(g(\mathbf{X}))$ *exists if, and only if,* $\int_{\mathfrak{R}^n} |g(\mathbf{t})| f_{\mathbf{X}}(\mathbf{t}) \, d\mathbf{t} < \infty$ *and then*

$$E(g(\mathbf{X})) = \int_{\mathfrak{R}^n} g(\mathbf{t}) f_X(\mathbf{t}) \, d\mathbf{t}. \qquad (2)$$

Proof. In the discrete case, let $A_t = [\mathbf{X} = \mathbf{t}]$ [the A_t are the events of the partition induced by $\mathbf{X}$ and $f_{\mathbf{X}}(\mathbf{t}) = P(A_t)$] and assume that $\mathbf{X}$ is discrete. Then $g(\mathbf{X}) = \Sigma_t \, g(\mathbf{t}) I_{A_t}$, so that (1) follows from Theorem 30.3.

Assume that X is continuous and $g > 0$. (The proof for other g is left as a problem.) By (30.5),

$$E(g(\mathbf{X})) = \int_0^{\infty} P[g(\mathbf{X}) > u] \, du = \int_0^{\infty} \left[\int_{g(t)>u} f_{\mathbf{X}}(\mathbf{t}) \, d\mathbf{t} \right] du$$

$$= \int_{\mathfrak{R}^n} f_{\mathbf{X}}(\mathbf{t}) \left[\int_0^{g(t)} du \right] d\mathbf{t} = \int_{\mathfrak{R}^n} g(\mathbf{t}) f_{\mathbf{X}}(\mathbf{t}) \, d\mathbf{t}.$$

---

[1]The function g in the statement of this theorem must satisfy the conditions of Section 21.

Example 1. Let X be a rv with density $f(t) = c(t + 1)^2$ for $|t| < 1$. Then $c = 3/8$ and

(a) $E\left(\dfrac{1}{X + 1}\right) = \dfrac{3}{8}\displaystyle\int_{-1}^{1}\dfrac{1}{t+1}\cdot(t + 1)^2\,dt = \dfrac{3}{8}\int_{-1}^{1}(t + 1)\,dt = \dfrac{3}{4};$

(b) $E(|X|) = \dfrac{3}{8}\displaystyle\int_{-1}^{1}|t|\,(t + 1)^2\,dt$

$= \dfrac{3}{8}\left[\displaystyle\int_{0}^{1}t(1 + t)^2\,dt + \int_{0}^{1}t(1 - t)^2\,dt\right] = \dfrac{9}{16}.$

Remark 1. Theorem 1 provides the most commonly used method of computing $E(g(X))$. Sometimes, as in Example 2, it is simpler to determine a representation of the model for $g(X)$ and apply an appropriate formula.

Example 2. Let $\mathbf{X}$ be the three-dimensional random vector that results when a point is chosen at random from the interior of the sphere of radius 1, centered at the origin. Let $Y = g(\mathbf{X}) = (X_1^2 + X_2^2 + X_3^2)^{1/2}$, the distance of $\mathbf{X}$ from the origin. Evaluation of $E(Y)$ using (2) involves a tedious three-fold integration. However, $F_Y(r) = P[Y \le r] = r^3$ for $0 < r < 1$ and hence

$$E(Y) = \int_{0}^{1}(1 - r^3)\,dr = \dfrac{3}{4}.$$

Corollaries 1 and 2 follow from linearity properties of sums and integrals. Corollary 2 generalizes (7.9) and should be intuitively clear.

Corollary 1. *Let a and b be any constants. Then*

$$E(bX + a) = bE(X) + a. \tag{3}$$

Corollary 1 is a special case of Corollary 2 if the constant a is interpreted as a degenerate rv.

Corollary 2. *Let $\mu_i = E(X_i)$ for $i = 1, \ldots, n$ and let $a_1, \ldots, a_n$ be any constants. Then*

$$E\left(\sum_{i=1}^{n}a_iX_i\right) = \sum_{i=1}^{n}a_i\mu_i. \tag{4}$$

Proof. Equation (4) will be proved for $n = 2$ in the continuous case. The discrete case can be handled similarly or as in Problem 7.3. See also Problem 15, which deals with the general case.

Let the joint density of X_1 and X_2 be f. Using (2) with $g(\mathbf{X}) = a_1 X_1 + a_2 X_2$ yields

$$E(a_1 X_1 + a_2 X_2) = \int_{-\infty}^{\infty} \int_{-\infty}^{\infty} (a_1 t_1 + a_2 t_2) f(t_1, t_2) \, dt_1 \, dt_2$$

$$= a_1 \int_{-\infty}^{\infty} t_1 \left[\int_{-\infty}^{\infty} f(t_1, t_2) \, dt_2 \right] dt_1 \tag{5}$$

$$+ a_2 \int_{-\infty}^{\infty} t_2 \left[\int_{-\infty}^{\infty} f(t_1, t_2) \, dt_1 \right] dt_2.$$

But, $\int_{-\infty}^{\infty} f(t_1, t_2) \, dt_2$ is the marginal density of X_1 so that the first term on the right side of (5) is $a_1 E(X_1)$. Similarly, the second term is $a_2 E(X_2)$.

Existence of the terms on the right side of (4) implies the existence of the left side. This is proved by using the inequality $|a_1 t_1 + a_2 t_2| \leq |a_1| \, |t_1| + |a_2| \, |t_2|$ and making appropriate modifications in (5). The proof of the theorem for $n = 2$ in the discrete case is similar and the result for arbitrary n follows by induction.

Since $E(X_1 + X_2) = E(X_1) + E(X_2)$, we see that $E(X_1 + X_2)$ depends only on the marginal models and not on the joint model.

Suppose that $A_1, \ldots, A_n$ are events and $X_i = I_{A_i}$. Observe that Corollary 2 is a generalization of a similar property of simple rv's (Theorem 7.3).

Example 3. A machine has just failed. There are three costs incurred while the machine is idle:

(a) A \$25 fixed repair cost.
(b) A cost of \$15 per hour for time the machine is idle.
(c) A lost profit of \$2 per item that could have been produced during the idle time.

The total cost of the breakdown is $Z = 25 + 15X + 2Y$. Assume that X has the exponential density $f_X(x) = e^{-x}$ for $x > 0$ and that Y has the geometric probability function $f_Y(y) = (.9)^{y-1}(.1)$ for $y = 1, 2, \ldots$. Then $E(X) = \int_{-\infty}^{\infty} xe^{-x} \, dx = 1$ and $E(Y) = \sum_{y=1}^{\infty} y(.9)^{y-1}(.1) = 10$, so that $E(Z) = 25 + (15)(1) + (2)(10). = \$60.$

The next corollary is an important result for independent rv's.

Corollary 3. If $X_1, \ldots, X_n$ are independent rv's, then

$$E\left(\prod_{i=1}^{n} X_i\right) = \prod_{i=1}^{n} E(X_i) \tag{6}$$

provided that each $E(X_i)$ exists.

Proof. We prove (6) for the continuous case. The proof in the discrete case is similar. A more complicated proof is required in general.

Let the marginal densities of X_1 and X_2 be f and g, respectively, so that the joint density is $f(t)g(u)$. Then

$$E(X_1X_2) = \int_{-\infty}^{\infty} \int_{-\infty}^{\infty} tuf(t)g(u)\, dt\, du$$

$$= \left[\int_{-\infty}^{\infty} tf(t)\, dt\right]\left[\int_{-\infty}^{\infty} ug(u)\, du\right]. \tag{7}$$

The first factor on the right side of (7) is $E(X_1)$ and the second is $E(X_2)$. Hence $E(X_1X_2) = E(X_1)E(X_2)$ and the case of arbitrary n follows by induction.

Problem 14 asks the reader to investigate questions of existence.

Example 4. A container with nominal dimensions 10 cm $\times$ 5 cm $\times$ 3 cm is made by some process. However, the ith dimension is in fact a rv $Y_i = c_i + X_i$, where c_i is the nominal dimension and the X_i are iid rv's with density

$$f(x) = c(.1 - |x - .05|) \text{ if } -.05 < x < .15.$$

By symmetry, $E(X_i) = .05$. Since X_i are independent, the expected volume of the container is

$$E[(10 + X_1)(5 + X_2)(3 + X_3)] = (10.05)(5.05)(3.05) = 161.72625.$$

The converse to Corollary 3 is false, as demonstrated by the following example.

Example 5† Let the joint probability function of X and Y be

$$f(u, v) = p \qquad \text{if } u = \pm 1, v = 0;$$
$$= 1 - 2p \qquad \text{if } u = 0, v = 1,$$

where $0 < p < 1/2$. Clearly, X and Y are *not* independent, $E(X) = 0$, and XY is degenerate at 0, so that $E(XY) = 0 = E(X)E(Y)$.

If, as above, the joint probability function (or density) is symmetric about any value of u or any value or v (any vertical or horizontal line), then $E(XY) = E(X)E(Y)$ even if X and Y are *not* independent.

Theorem 2 states the fact that averages are always between the bounds of the values being averaged. These results and Corollary 4 generalize Theorm 7.3 and Corollary 7.3. If X is a bounded rv, then $E(X)$ exists, as can easily be verified.

Theorem 2. If $P[a < X < b] = 1$, then $a < E(X) < b$. [*If $E(X)$ exists, then the statement remains true if $a = -\infty$ or $b = \infty$.*]

Proof. Left as a problem.

Corollary 4. If $P[X \le Y] = 1$, then $E(X) \le E(Y)$.

Proof. Left as a problem.

Moments

The expectations of certain functions are called *moments* of the rv X. Let $\alpha > 0$. For g as defined below, $E(g(X))$ is called:

1. *αth moment of X about a:* $g(t) = (t - a)^\alpha$, α an integer or $X \ge a$.
2. *αth absolute moment of X about a:* $g(t) = |t - a|^\alpha$.

In these cases, when $a = 0$ we delete "about a" and when $a = E(X)$ we use the term *central*.

3. *αth factorial moment of X:* $g(t) = t(t - 1) \cdots (t - \alpha + 1) = (t)_\alpha$, α an integer.
4. *αth reverse factorial moment of X:* $g(t) = t(t + 1) \cdots (t + \alpha - 1) = (t + \alpha - 1)_\alpha$, α an integer.

The first moment is then the mean. The second central moment is called the *variance*, discussion of which begins in the next section. Our interest will generally be confined to small integer values of α.

By definition $E(X)$, the first moment, exists if, and only if, $E(|X|)$, the first absolute moment, exists.

Theorem 3. Let $0 < \beta < \alpha$. If the αth absolute moment of X exists, so does the βth absolute moment of X.

Proof. Left as a problem.

Let α be an integer. By Theorem 3 and Corollary 2, if any one of the following exists, so do the others:

(i) αth moment
(ii) αth absolute moment
(iii) αth factorial moment
(iv) αth reverse factorial moment
(v) αth moment about a for some a

For α a noninteger some of these equivalences are still valid in special cases; for example, for $X \geq 0$, (i) and (ii) are still equivalent.

Two useful formulas are

$$E(X^2) = E(X(X - 1)) + E(X) \tag{8}$$

and

$$E(X^2) = E(X(X + 1)) - E(X), \tag{9}$$

relating the second moment of X to the first and second factorial and reverse factorial moments of X.

Example 6.   Let the probability function of X be

$$f(k) = \frac{18}{k(k + 1)(k + 2)(k + 3)} \quad \text{if } k = 1, 2, \ldots .$$

Then the first two reverse factorial moments are

$$E(X) = 18 \sum_{k=1}^{\infty} \frac{1}{(k + 1)(k + 2)(k + 3)}$$

$$= 9 \sum_{k=1}^{\infty} \left[\left(\frac{1}{k + 1} - \frac{1}{k + 2} \right) - \left(\frac{1}{k + 2} - \frac{1}{k + 3} \right) \right] = \frac{3}{2}.$$

and

$$E(X(X + 1)) = 18 \sum_{k=1}^{\infty} \frac{1}{(k + 2)(k + 3)}$$

$$= 18 \sum_{k=1}^{\infty} \left(\frac{1}{k + 2} - \frac{1}{k + 3} \right) = 6,$$

so that $E(X^2) = 6 - 3/2 = 9/2$.

The third moment of X does not exist.

Example 7. Consider the binomial probability function

$$f_X(k) = \binom{n}{k} p^n q^{n-k} \qquad \text{if } k = 0, 1, \ldots, n.$$

The αth factorial moment of X is 0 if $\alpha > n$ and, otherwise, is

$$E((X)_\alpha) = \sum_{k=\alpha}^{n} \frac{k!}{(k-\alpha)!} \frac{n!}{(n-k)!k!} p^n q^{n-k}$$

$$= (n)_\alpha p^\alpha \sum_{k=\alpha}^{n} \frac{(n-\alpha)!}{(k-\alpha)!(n-k)!} p^{n-\alpha} q^{n-k}$$

$$= (n)_\alpha p^\alpha \sum_{m=0}^{n-\alpha} \binom{n-\alpha}{m} p^m q^{(n-\alpha)-m} = (n)_\alpha p^\alpha.$$

In particular, $E(X) = np$ and $E(X(X-1)) = n(n-1)p^2$, so that $E(X^2) = np[(n-1)p + 1]$.

Summary

Let **X** be a random vector with probability function or density f and let g be a real-valued function. Then

$$E(g(X)) = \sum_{\mathbf{t}} g(\mathbf{t})f(\mathbf{t})$$

in the discrete case and

$$E(g(X)) = \int_{-\infty}^{\infty} g(\mathbf{t})f(\mathbf{t}) \, d\mathbf{t}$$

in the continuous case. Important consequences of this are

$$E(bX + a) = bE(X) + a,$$

$$E\left(\sum_{i=1}^{n} a_i X_i\right) = \sum_{i=1}^{n} a_i E(X_i)$$

and when $X_1, \ldots, X_n$ are independent,

$$E\left(\prod_{i=1}^{n} X_i\right) = \prod_{i=1}^{n} E(X_i).$$

These equations are valid for all types of rv's, not just discrete and continuous.

Expectations of X^α, $|X|^\alpha$, $(X)_\alpha$, and $(X + \alpha - 1)_\alpha$ are called moments of various types.

Problems

1. A fair die is rolled and X is the face that results. Find $E(X)$ and $E(2X - 5)$.
2. Let X have the exponential density

$$f(t) = \frac{1}{\beta} e^{-t/\beta} \qquad \text{if } t > 0.$$

Find $E(\sin X)$.

3. A machine produces 10% defective items. Suppose that a day's production consists of 35 items and let X be the number of defectives produced in a day. Find $E(X)$. *Hint*: Let B_i be the event that the ith item is defective.
4. A fair die is rolled n times. Let X_i be the result of the ith roll. Find $E(\Sigma_{i=1}^{n} X_i)$ and $E(\Pi_{i=1}^{n} X_i)$.
5. In Problem 4, let Y_1 be the smaller and Y_2 be the larger of X_1 and X_2. Find $E(Y_1 Y_2)$, $E(Y_1 + Y_2)$, and $E(Y_2 - Y_1)$.
6. Let the two-dimensional random vector X be uniformly distributed on the triangle with vertices $(0, 0)$, $(0, 1)$, and $(1, 0)$. Find $E(X_1)$, $E(X_2)$, and $E(X_1 X_2)$.
7. Suppose that the daily demand for a certain commodity has the exponential density

$$f(x) = \frac{1}{100} e^{-x/100} \qquad \text{if } x > 0.$$

(a) Find the mean daily demand.
(b) A retailer will buy the commodity for $50 per unit and sell at $57 per unit. Any left at the end of the day will be sold for $47 per unit. How much should the retailer buy each day?

8. Complete the proof of Theorem 1.
9. Prove Corollary 2 by expressing each X_i as in (30.13).
10. Prove Theorem 2.
11. Prove Corollary 4.
12. Let λ_α and ρ_α be the αth factorial moment of $-X$ and the αth reverse factorial moment of X, respectively. Show that $\rho_\alpha = (-1)^\alpha \lambda_\alpha$.
13. Prove Theorem 3. *Hint*: Show that $|t|^\beta \le 1 + |t|^\alpha$.
14. Prove or give a counterexample to each of the following statements:
    (a) $E(X)$ and $E(Y)$ exist implies that $E(XY)$ exists.
    (b) $E(XY)$ exists implies that $E(X)$ exists.
    (c) $E(X + Y)$ exists implies that $E(X)$ and $E(Y)$ exist.
15.† For $i = 1, \ldots, n$, let $E(X_i) = \mu$. Set $\overline{X}_n = (1/n) \Sigma_{i=1}^{n} X_i$. Show that $E(\overline{X}_n) = \mu$.

16. Let X and Y be m- and n-dimensional random vectors, respectively, and g and h be real-valued functions on $\mathcal{R}^m$ and $\mathcal{R}^n$, respectively.
    (a) Show that X and Y independent implies that

$$E(g(X)h(Y)) = E(g(X))E(h(Y)). \qquad (10)$$

   (b) Show that if (10) is true for all choices of g and h, then X and Y are independent.

17.† Show that $|E(X)| \le E(|X|)$. *Hint*: Use Corollary 4.

32. VARIANCE AND STANDARD DEVIATION

One of the characteristics of the distribution of a rv is its mean. If X is degenerate at c, not only is $E(X) = c$, but we can be certain that $X = c$. On the other hand, if $P[Y = c + 100] = P[Y = c - 100] = 1/2$, the $E(Y) = c$, but we can be certain that $Y \ne c$ and, in fact, that Y must be 100 units aways from c. Predicting the value of X is clearly an easier task than predicting the value of Y. Generally, greater concentration of mass near some value such as the mean makes prediction easier. In this section a measure of variability is examined which reflects this.

Suppose that you have to predict the value of some rv X and that if $X = t$ and you predict $X = a$, you suffer a penalty depending on the error of prediction, $t - a$. The measure of variability to be discussed arises from consideration of the penalty $(t - a)^2$.

Assume that $E(X^2)$ exists and that you wish to choose a so as to minimize your average loss, $E(X - a)^2$. It will be shown that you can achieve this minimum by setting a $= E(X) = \mu_X$.

First observe that for any number t,

$$(t - a)^2 = [(t - \mu_X) + (\mu_X - a)]^2$$
$$= (t - \mu_X)^2 + 2(\mu_X - a)(t - \mu_X) + (\mu_X - a)^2.$$

Then, using the linearity of E,

$$E(X - a)^2 = E(X - \mu_X)^2 + 2(\mu_X - a)E(X - \mu_X) + (\mu_X - a)^2. \qquad (1)$$

Since $E(X - \mu_X) = \mu_X - \mu_X = 0$ it follows that

$$E(X - a)^2 = E(X - \mu_X)^2 + (\mu_X - a)^2, \qquad (2)$$

which is minimized by setting $a = \mu_X$. Thus

$$\min_a E(X - a)^2 = E(X - \mu_X)^2.$$

When $E(X - \mu_X)^2$ is small, the distribution of X is highly concentrated near μ_X, its mean, and prediction of X should be fairly accurate. Conversely, if the distribution of X is substantially spread out, then $E(X - \mu_X)^2$ will be large.

Variance, Standard Deviation

If X is a rv for which $E(X^2)$ exists, its *variance* is

$$\text{Var}(X) = E(X - \mu_X)^2, \tag{3}$$

the second central moment of X. The *standard deviation* of X is $\text{SD}(X) = \sqrt{\text{Var}(X)}$. Sometimes $\text{Var}(X)$ will be denoted by σ_X^2, so that $\text{SD}(X) = \sigma_X$. In keeping with our convention about expectations, when a variance appears in an expression, its existence is assumed. Notation such as σ_F^2 or σ_F is self-explanatory.

Note that $\text{Var}(X)$ is measured in squared X units, while $\text{SD}(X)$ has the same unit as X. For example, if X is a measurement in meters, then $\text{Var}(X)$ is measured in squared meters while $\text{SD}(X)$ is in meters.

The right side of (3) is often inconvenient for calculating the variance. Setting $a = 0$ in (2) yields the usual computational formula

$$\text{Var}(X) = E(X^2) - \mu_X^2. \tag{4}$$

Theorem 1. *Let a and b be any constants. Then* $\text{Var}(bX + a) = b^2 \text{Var}(X)$ *so that* $\text{SD}(bX + a) = |b|\text{SD}(X)$. *In particular* $(b = 0)$, *the variance of any degenerate rv is* 0.

Proof. Since $(bX + a) - E(bX + a) = b(X - \mu_X)$, application of (3) yields

$$\text{Var}(bX + a) = E[b(X - \mu_X)]^2 = b^2 E(X - \mu_X)^2$$

$$= b^2 \text{Var}(X).$$

Example 1. For any event A, $E(I_A) = P(A)$. Futhermore, $I_A^2 = I_A$, so that

$$\text{Var}(I_A) = P(A) - [P(A)]^2 = P(A)P(A^c).$$

Example 2. Suppose that you have a choice between two investments, say A and B. Let X_A and X_B be the profits, in units of \$1000, for A and

Table 1

r	$f_A(r)$	r	$f_B(r)$
$-.1$	.45	-1	.8
1.9	.55	9	.2

B, respectively, and assume the probability functions are f_A and f_B, respectively, given in Table 1. It is easy to verify that $E(X_A) = E(X_B) = 1$, $\mathrm{Var}(X_A) = .99$, and $\mathrm{Var}(X_B) = 16$.

Which investment would you choose? Both investments provide expected profits of \$1000. The fact that A is less risky than B is reflected in its smaller variance. However, despite its greater risk, B may be more attractive to you because it provides an opportunity for a far greater gain.

Example 3.   Consider the family of densities given by

$$f_{a,b}(x) = \frac{2\,|x - a|}{b^2} \qquad \text{if } 0 \le \frac{x - a}{b} \le 1,$$

where $b \ne 0$. Thus $f_{a,b}(a) = 0$, $f_{a,b}(a + b) = 2/|b|$, and $f_{a,b}$ is linear between those points. If X has density $f_{a,b}$, then

$$E(X) = \frac{2b}{3} + a$$

$$E(X^2) = \frac{b^2}{2} + \frac{4ab}{3} + a^2 \tag{5}$$

Thus

$$\mathrm{Var}(X) = \frac{b^2}{2} + \frac{4ab}{3} + a^2 - \left(\frac{2b}{3} + a\right)^2 = \frac{b^2}{18} \tag{6}$$

and $\mathrm{SD}(X) = |b|/(3\sqrt{2})$.

Let Y have the density $f_{0,1}$. Then $f_{a,b}$ is the density of $bY + a$ and it follows from previous results that $E(X)$ and $\mathrm{Var}(X)$ are as in (5) and (6).

The discussion of variances of linear combinations is deferred to Section 35.

Summary

Variance of a rv X: $\mathrm{Var}(X) = E(X - \mu_x)^2$.
Standard deviation of a rv X: $\mathrm{SD}(X) = \sqrt{\mathrm{Var}(X)}$.

An alternative to the definition is $\text{Var}(X) = E(X^2) - \mu_X^2$. Furthermore, for constants a and b, $\text{Var}(bX + a) = b^2 \text{Var}(X)$, $\text{SD}(bX + a) = |b|\text{SD}(X)$.

Problems

1. A die is rolled twice. Let X be the result of the first roll, Y be the result of the second roll, and $Z = X + Y$. Find $\text{Var}(X)$, $\text{Var}(Y)$, and $\text{Var}(Z)$. What relationship do you observe between these variances? Do you think this is a general relationship? Also find $\text{Var}(2X - 5)$.
2. For fixed $\alpha > 1$, let

$$f_X(t) = (\alpha - 1)t^{-\alpha} \qquad \text{if } t > 1.$$

For what values of α does $\text{Var}(X)$ exist? Evaluate $\text{Var}(X)$ for such α. Compare your answer with the answer to Problem 30.3.
3. Let X have density

$$f(x) = c(x - \mu)^s \qquad \text{if } \mu < x < \mu + a,$$

where $s > 0$ and c depends on s, a, and μ.
   (a) Why does $\text{Var}(X)$ not depend on μ?
   (b) Find $\text{Var}(X)$ and give its numerical value for $s = 3$, $a = 1$.
4. A point is chosen at random on the perimeter of the circle centered at the origin and with radius 1. The probability that the selected point is in a specified arc is proportional to the length of the arc. Find $E(X^2 + Y^2)$ and $\text{Var}(X^2 + Y^2)$. Does the solution depend on the specific distribution given?
5. Among 10 items, 3 are defective. The items are inspected successively (without replacement) and at random until a good one is found. Let T be the number of trials required.
   (a) Find the probability function of T.
   (b) Determine $E(T)$ and $\text{SD}(T)$.

33. LOCATION AND SCALE PARAMETERS; SYMMETRY

The linearity property $E(bX + a) = bE(X) + a$ identifies E as a *measure of location*. The corresponding property $\text{SD}(bX + a) = |b|\text{SD}(X)$ identifies SD as a *measure of spread (dispersion)*. In this section other measures with these properties are discussed.

Let X be a rv with distribution function F. For $b > 0$, let $G(t \mid a, b)$ be the distribution function of the rv $Y = bX + a$. Then

$$G(t \mid a, b) = P[bX + a \le t] = P\left[X \le \frac{t - a}{b}\right] = F\left(\frac{t - a}{b}\right). \qquad (1)$$

Location and Scale Family

The rv Y is obtained from X by change of location (addition of a) and of scale (multiplication by $b > 0$). Thus the family of distribution function $\{G(t \mid a, b): b > 0\}$ defined in (1) is called a *location and scale family*. We call a the *location parameter* and b the *scale parameter* of the family. The distribution function F, obtained by setting $a = 0$ and $b = 1$ in (1), *generates* the family and is called the *standard member of the family*. The subfamily obtained by holding b constant is a *location family*, while that obtained by holding $a = 0$ is a *scale family*.

Any member of a location and scale family can be used as the standard member by suitably redefining a and b. The particular choice of a standard member is a matter of convenience.

In the discrete case, let f be the probability function of X and $g(t \mid a, b)$ be the probability function of Y. Replacing inequalities in (1) by equalities yields

$$g(t \mid a, b) = f\left(\frac{t - a}{b}\right). \tag{2}$$

In the continuous case, let f be the density of X and $g(t \mid a, b)$ be the density of Y. Differentiating in (1) gives

$$g(t \mid a, b) = \frac{1}{b} f\left(\frac{t - a}{b}\right). \tag{3}$$

For location and scale families with second moments, the standard member is often chosen so that its mean is 0 and its standard deviation is 1.

Standardization

Suppose that $E(Y) = \mu_Y$ and $SD(Y) = \sigma_Y$. The *standardization* of Y is $Y^* = (Y - \mu_Y)/\sigma_Y$. Note that $E(Y^*) = 0$ and $SD(Y^*) = 1$.

The location and scale family obtained from the distribution function of Y^* is parametrized so that $a = E(Y)$ and $b = SD(Y)$. With this parametrization, the distribution function of Y^* is the standard member of the family.

Example 1. Let $F_1(t) = 1 - 1/(t + 1)^2$ if $t > 0$ and $F_2(t) = 1 - 4/t^2$ if $t > 2$. Both F_1 and F_2 generate the same family, but F_1 is simpler. Note that $F_2(t) = F_1((t - 2)/2)$.

Example 2. Consider the density

$$g(t \mid a, b) = \frac{1}{b} \qquad \text{if } a < t < a + b$$

when $b > 0$. As a and b vary, we obtain the family of uniform densities discussed in Section 43. This is a location and scale family whose standard member is the uniform distribution on $(0, 1)$. This model has mean $\mu = 1/2$ and standard deviation $\sigma = 1/2\sqrt{3}$. Thus, in general, $\mu = b/2 + a$ and $\sigma = b/(2\sqrt{3})$. The member with $a = -\sqrt{3}$ and $b = 2\sqrt{3}$ has mean 0 and standard deviation 1. Now consider the densities

$$h(t \mid a, b) = \frac{1}{2\sqrt{3b}} \qquad \text{if } a - \sqrt{3b} \leq t \leq a + \sqrt{3b}.$$

The family of h's is the same as the family of g's. But in the new parametrization a is the mean and b is the standard deviation of the member.

Because of its simplicity, the uniform distribution on $(0, 1)$ is usually chosen as the standard member of the family in Example 2.

Parametrization with $a = E(X)$ and $b = SD(X)$ requires the first two moments. When $E(X)$ does not exist, there are alternative measures of location and/or scale that can be substituted.

Quantiles, Percentiles

An αth quantile (100αth percentile) of the rv X is any number (not necessarily unique), denoted by t_α, satisfying $P[X \leq t_\alpha] \geq \alpha$ and $P[X \geq t_\alpha] \geq 1 - \alpha$. In terms of the distribution function, the condition is $F_X(t_\alpha -) \leq \alpha \leq F_X(t_\alpha)$.

Any solution of $F(t) = \alpha$ is an αth quantile of F. If there is no solution, the unique αth quantile is the smallest value of t satisfying $F(t) > \alpha$. In each case the set of αth quantiles is a closed interval that reduces to a single point when t_α is unqiue.

We call any $t_{.5}$ a *median*, any $t_{.25}$ a *first quartile*, and any $t_{.75}$ a *third quartile*. The quantity $t_{.75} - t_{.25}$ is called the *interquartile range* with the understanding that the quantiles are the midpoints of their intervals. The interquartile range, and indeed, any difference between quantiles, is a scale parameter. Any quantile is a location parameter in the sense that if t_α is any αth quantile of X, then $a + bt_\alpha$ is an αth quantile of $a + bX$.

Example 3. Consider the density $f(t) = t^{-2}$ if $t > 1$, for which the mean does not exist. The corresponding distribution function is $F(t) = 1 - t^{-1}$

if $t > 1$. For this distribution function, $t_\alpha = 1/(1 - \alpha)$, so that, in particular, $t_{.25} = 4/3$, $t_{.5} = 2$, and $t_{.75} = 4$. Thus the interquartile range is $t_{.75} - t_{.25} = 8/3$. Consider the location and scale family $(G(t \mid a, b) = F((t - a)/b)$. A member of this family has $t_\alpha = b/(1 - \alpha) + a$. In particular, $t_{.5} = 2b + a$ and $t_{.75} - t_{.25} = 8b/3$. The member with $a = -3/4$ and $b = 3/8$ has median 0 and interquartile range 1.

Symmetry

A rv X is *symmetric (about* 0) if X and $-X$ have the same distribution. A rv X is *symmetric about* a if $X - a$ is symmetric. Any representation of a model for a symmetric rv (or for a rv symmetric about a) will be called *symmetric* (or *symmetric about a*).

Remark 1.   Symmetry of X means that for every $t \in \mathcal{R}$, $P[X \le t] = P[X \ge -t]$. Hence, in terms of the usual representations:

(a)   A continuous distribution function F is symmetric if, and only if,

$$F(t) = 1 - F(-t) \qquad \text{for every } t \in \mathcal{R}. \qquad (4)$$

(b)   If X has a probability function or a density, then the symmetry of X is equivalent to

$$f_X(t) = f_X(-t) = 0 \qquad \text{for every } t \in \mathcal{R}. \qquad (5)$$

Remark 2.   If X is symmetric about a, the implications noted in Remark 1 remain valid with X replaced by $X - a$. Thus (4) becomes

$$F(a + t) = 1 - F(a - t) \qquad \text{for every } t \in \mathcal{R} \qquad (4')$$

and (5) becomes

$$f_X(a + t) = f_X(a - t) \qquad \text{for every } t \in \mathcal{R}. \qquad (5')$$

Remark 3.   If X is symmetric, 0 is a median and, if it exists, $\mu_X = 0$. Also, for any odd integer k, X^k is symmetric and $Y = bX + a$ is symmetric about a.

Example 4.   If X is uniform on $\{0, 1, \ldots, n\}$, then $Y = X - n/2$ is a symmetric rv, $E(Y) = 0$, and $E(X) = n/2$ (X is symmetric about $n/2$). Similarly, if X is uniform on the interval (α, β), then X is symmetric about $a = (\alpha + \beta)/2$, its mean and median.

Means and medians describe "typical" or "central" values of rv's. In the same vein, modes describe their "most probable" values.

Mode (Discrete Case)

In the discrete case, the number m is a *mode of the rv X* (or of F_X) if

$$P[X = m] = \max_t P[X = t]. \tag{6}$$

If only one number m satisfies (6), then X is called *unimodal*. In that case m is that t which maximizes $f_X(t)$ or, equivalently, the location of the maximum jump of F_X.

Mode (Continuous Case)

Let X have a density f_X.[2] Then m is a *mode* of X (f_X or F_X) if (1) for all $\varepsilon > 0$, f_X is unbounded in $(m - \varepsilon, m + \varepsilon)$, or (2) for some interval $J = (a, b)$ such that $m \in J$,

$$0 < f_X(m) = \max_{t \in J} f_X(t), \tag{7}$$

$f_X(m) > f_X(a)$, and $f_X(m) > f_X(b)$. Thus modes are located at the global (and some local) maxima of f_X. If $f_X(a) \le f_X(b) \le f_X(m)$ whenever $a < b < m$ or $m < b < a$, then X is *unimodal at* m. A unimodal density can have many modes, but they must form an interval. See Figure 1 for examples.

Example 5. (a) The gamma densities $f(x) = cx^{\alpha-1}e^{-x}$ for $x > 0$ ($\alpha > 0$) are all unimodal. For $\alpha < 1$, the density is unbounded at 0. For $\alpha > 1$, by differentiation it can be seen that the mode is a solution of $cx^{\alpha-2}e^{-x}(\alpha - 1 - x) = 0$, $x > 0$. The unique mode in this case is therefore $\alpha - 1$.
    (b) Let

$$F(x) = \frac{1}{2}x^2 \qquad\qquad \text{if } 0 < x \le 1$$

$$= \frac{1}{2}[2 - (2 - x)^3] \qquad \text{if } 1 \le x < 2.$$

---

[2]For the purpose of identifying a density uniquely, assume that $f_X(x) = F'_X(x)$, where this derivative exists and is the larger of the right- and left-handed derivatives (assumed finite) otherwise.

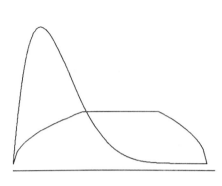

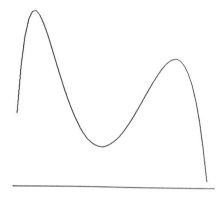

(a)   Unimodal   (b)   Bimodal

Figure 1

The density corresponding to this distribution function is

$$f(x) = x \qquad\qquad \text{if } 0 < x < 1$$
$$= \frac{3}{2}(2 - x)^2 \qquad \text{if } 1 \le x < 2,$$

which is unimodal at 1.

(c) For $a > 0$, let

$$f(x) = c[e^{-(x+a)^2/2} + e^{-(x-a)^2/2}],$$

a mixture of two normal densities. The following behavior can be proved:

(i) For $a \le 1$, the density is unimodal at 0.

(ii) For $a > 1$, the density has two modes, at some m $(0 < m < a)$ and at $-m$.

Certain measures of location and scale are naturally related as indicated in Table 1.

Table 1

Location	Scale	Reference
$E(X)$	$SD(X)$	Section 32
$t_{.5}$	$E(\lvert X - t_{.5}\rvert)$	Problem 8
t_α	$E(c_1(t_\alpha - X)I_{[X<t_\alpha]} + c_2(X - t_\alpha I_{[X>t_\alpha]}))$	Problem 9
	$\left(\alpha = \dfrac{c_2}{c_1 + c_2}\right)$	
Mode	$\min\limits_{x} \dfrac{1}{f(x)}$ if f is a bounded density	Problem 13

Summary

Location and scale family: A family $\{G(t \mid a, b): b > 0\}$ of distribution
functions where $G(t \mid a, b) = F((t - a)/b)$. Then a is called the *location
parameter* and b is called the *scale parameter*.
Standard member: The member of a location and scale family with a =
0, $b = 1$.
αth quantile of X: Any t_α satisfying $P[X \le t_\alpha] \ge \alpha$ and $P[X \ge t_\alpha] \ge
1 - \alpha$.
Median: Any value of $t_{.5}$.
Symmetric rv X: Distribution of X and $-X$ are the same.
Rv X symmetric about a: $X - a$ is symmetric.
Mode of a probability function f: Any m satisfying $f(m) = \max_t f(t)$.
Mode of a density: Local maximum.
Unimodal density: $f(a) \le f(b) \le f(m)$ whenever $a < b < m$ or $m <
b < a$.

Means, quantiles, and modes are location parameters. The standard
deviation, and any differences of quantiles are scale parameters. If Y has
distribution function $G(t \mid 0, 1) = F(t)$, the standard member of a location
and scale family, then $bY + a$ has distribution function $G(t \mid a, b) =
F((t - a)/b)$.

Problems

1. Consider the family

$$g(t \mid a, b) = \frac{1}{b} e^{-(x-a)/b} \qquad \text{if } x > a.$$

   (a) Determine the mean and variance of the standard member.
   (b) Determine the mean and variance of an arbitrary member.
   (c) Reparametrize this family so that μ_Y and σ_Y are the parameters.
2. Consider the location-scale family for which the density of one member is

$$g(t) = \frac{1}{\pi} \cdot \frac{1}{1 + t^2}$$

   parametrize the family so that the location parameter is $t_{.5}$ and the
   scale parameter is $t_{.75} - t_{.25}$. This family is called the *Cauchy family*.
3.† Let X be a rv with standardization X^* and let $\beta \neq 0$. Show that the
   standardization of $\beta X + \alpha$ is $(\beta/|\beta|)X^*$.
4. Let X and Y be iid rv's. Show that $g(X, Y)$ and $g(Y, X)$ have the
   same distribution for any function g and hence that $X - Y$ is symmetric.

5. Let X have the binomial probability function

$$f(k) = \binom{n}{k}\left(\frac{1}{2}\right)^n \quad \text{if } k = 0, 1, \ldots, n.$$

   (a) Find the mode of X.
   (b) Is X symmetric about some a? If so, give a.

6. Show that for a symmetric, unimodal distribution 0 is a mode.

7. Let X be a rv symmetric about a and assume that $E(X)$ exists. Show that $E(X) = a$.

8. (a) Let θ be any location parameter for the distribution of X. Show that $d(\theta) = E(|X - \theta|)$ (called the *mean absolute deviation about* θ) is a scale parameter. *Note*: The usual choice is $\theta = t_{.5}$.
   (b) Show that $d(a)$ is minimized by setting $a = t_{.5}$ *Hint*: Evaluate $\int_0^\infty P[|X - a| > t]\, dt$.

9. Consider the family of densities of the form $f(x \mid a, b) = c(x - a)$ for $a < x < b$.
   (a) Find the mean, standard deviation, median m, and $d(m)$.
   (b) Parametrize so that the location and scale parameters are the mean and standard deviation, respectively.
   (c) Parametrize so that the location and scale parameters are m and d, respectively.

10. Let $c_1, c_2 > 0$. Show that

$$E(c_1(a - X)I_{[X<a]} + c_2(X - a)I_{[X>a]})$$

   is minimized by setting $a = t_\alpha$, where $\alpha = c_2/(c_1 + c_2)$.

11. Consider the *standard double exponential* density, $f(t) = (1/2)e^{-|t|}$.
    (a) Give the location and scale family generated by f.
    (b) Find the mean and variance for a member of this family.

12. Find the mean and variance of the density $f(t) = c(t - 7)^\pi$ for $7 < t < 10$.

13. Let $G_1(t \mid a, b)$ and $G_2(t \mid a, b)$ be members of the family generated by F_1 and F_2 in Example 1. How are they related?

14. (a) Show that if m is a mode of X, then $bm + a$ is a mode of $bX + a$, so that a mode is a location parameter.
    (b) Let f be a bounded density and show that $\min_x 1/f(x)$ is a scale parameter with value $1/f(m)$ for some mode m.

15. Let X be symmetric and h be an odd function. Show that $h(X)$ is symmetric.

16. Let $\alpha < \beta$ and assume that the quantiles t_α and t_β are the centers of their appropriate quantile intervals. Show that t_α is a location parameter and that $t_\beta - t_\alpha$ is a scale parameter.

17. Give the formulas for densities other than those in Example 5 satisfying each condition below. Sketch each choice.
    (a) Unimodal
    (b) Bimodal
    (c) Symmetric about some point
18. Find all the modes of the density $f(x) = ce^{-x} \cos^2 x$ for $x > 0$.

Problems 19 and 20 are of interest in statistical applications.

19. For each of the following models, determine (i) the shortest interval $[a, b]$ such that $P[a \le X \le b] = .75$, and (ii) the value of d such that $P[d \le X \le d + 1]$ is a maximum.
    (a) $f_X(x) = cx^2 e^{-x/2}$ for $x > 0$
    (b) $f_X(x) = ce^{-|x-3|/2}$
20. Let f_X be a unimodal density and let $0 < \gamma < 1$.
    (a) Show that among all intervals $[a, b]$ such that $P[a \le X \le b] = \gamma$, the one that minimizes $b - a$ satisfies $f_X(b) = f_X(a)$.
    (b) Let $L(t)$ be a strictly increasing, differentiable function of t. Among all intervals $[a, b]$ such that $P[a \le X \le b] = \gamma$, characterize those that minimize $L(b) - L(a)$.

34. CONDITIONAL EXPECTATION

In Section 12 the conditional expectation of a simple rv X given an event B was defined. The definition can now be extended using the more general concepts of expectation from Section 30. Like conditional probability, conditional expectation provides a useful technique for obtaining unconditional values.

Let X be a rv and B be an event. Like any other model, a conditional model for X given B may or may not have a mean.

Conditional Expectation

If the conditional model for X given B has a mean, it is called the *conditional expectation of X given B*, denoted by $E(X \mid B)$.

With minor adjustments, all the results of Sections 30 and 31 are valid for conditional expectations if both sides of any equation are conditioned by the same event. Thus the result of Corollary 31.2 becomes

$$E\left(\sum_{i=1}^{n} a_i X_i \;\middle|\; B\right) = \sum_{i=1}^{n} a_i E(X_i \mid B). \tag{1}$$

Let $F(t \mid B)$ be the conditional distribution function of X given B. Then (30.6) becomes

$$E(X \mid B) = \int_0^\infty [1 - F(t \mid B)] \, dt - \int_{-\infty}^0 F(t \mid B) \, dt. \tag{2}$$

The following theorem is the appropriate reformulation of Corollary 31.3 concerning products of independent rv's.

Theorem 1. *Let the rv's $X_1, \ldots, X_n$ be conditionally independent given the event B. Then*

$$E\left(\prod_{i=1}^n X_i \;\middle|\; B\right) = \prod_{i=1}^n E(X_i \mid B). \tag{3}$$

See GP 5.1 for the definition of conditional independence. Rv's can be independent without being conditionally independent given B and can be conditionally independent given B without being independent.

When $P(B) > 0$ we have the following useful formula.

Theorem 2. *Let $P(B) > 0$. Then if $E(X)$ exists, so does $E(X \mid B)$ and*

$$E(X \mid B) = \frac{E(XI_B)}{P(B)}. \tag{4}$$

Proof. Note that $P([X > t]B) = P[XI_B > t]$ and $P([X \le t]B) = P[XI_B \le t]$. Then

$$E(X \mid B) = \int_0^\infty P(X > t \mid B) \, dt - \int_{-\infty}^0 P(X \le t \mid B) \, dt$$

$$= \int_0^\infty \frac{P[XI_B > t]}{P(B)} \, dt - \int_{-\infty}^0 \frac{P[XI_B \le t]}{P(B)} \, dt = \frac{E(XI_B)}{P(B)}.$$

Existence of the right side follows from existence of $E(X)$.

By setting $X = I_A$ it can be seen that (4) generalizes (12.7), the definition of $P(A \mid B)$.

If $B = [\mathbf{Y} = \mathbf{u}]$ for some random vector $\mathbf{Y}$, the notation used is $E(X \mid B) = E(X \mid \mathbf{Y} = \mathbf{u})$. Suppose that $g(\mathbf{u}) = E(X \mid \mathbf{Y} = \mathbf{u})$ exists for all values $\mathbf{u}$ of $\mathbf{Y}$. The rv $g(\mathbf{Y})$ is denoted by $E(X \mid \mathbf{Y})$. The notation $\mathrm{Var}(X \mid \mathbf{Y})$ and other generalizations have obvious meanings.

Example 1. Let X and Y be rv's with

$$f_{X,Y}(t, u) = 2 \quad \text{if } t, u > 0; t + u < 1.$$

Then

$$f_X(t \mid u) = \frac{1}{1 - u} \quad \text{if } 0 < t < 1 - u < 1.$$

Thus the function g on $(0, 1)$ is $g(u) = (1 - u)/2$, the mean of the conditional density given $Y = u$. Clearly X and Y have the same marginal densities and $E(X) = E(Y) = 1/3$.

Then $g(Y) = E(X \mid Y) = (1 - Y)/2$. Now $f_Y(u) = 2(1 - u)$ for $0 < u < 1$, so that

$$E(g(Y)) = E(E(X \mid Y)) = \frac{1}{3} = E(X).$$

The relation $E(E(X \mid Y)) = E(X)$ holds in general when $E(X)$ exists. The following theorems give two versions of this relationship.

Theorem 3. *Assume that $E(X)$ exits and let $\{B_1, B_2, \ldots ,\}$ be a partition with each $P(B_i) > 0$. Define $a_i = E(X \mid B_i)$ and let $Z = \sum_{i=1}^{\infty} a_i I_{B_i}$. Then $\sum_{i=1}^{\infty} |a_i| P(B_i) < \infty$, and $E(X) = E(Z)$.*

Proof. Since $E(X)$ exists and $P(B_i) > 0$, then $E(X I_{B_i}) = a_i P(B_i)$ by (4). Furthermore, $X = \sum_{i=1}^{\infty} X I_{B_i}$ so that

$$E(X) = \sum_{i=1}^{\infty} E(X I_{B_i}) = \sum_{i=1}^{\infty} a_i P(B_i) = E(Z).$$

Note that $|a_i| \leq E(|X| \mid B_i)$ (Problem 31.7). Repeating the previous argument with X replaced by $|X|$ proves that $\sum_{i=1}^{\infty} |a_i| P(B_i) < \infty$.

Theorem 3 generalizes (12.12), the theorem of total probability, and has been previously proved when X is a simple rv [see (12.13)].

In Theorem 3, let $\mathbf{Y}$ be any random vector that induces the partition $\{B_1, B_2, \ldots\}$. Then $Z = E(X \mid \mathbf{Y})$, so that the result of the theorem becomes $E(X) = E(E(X \mid \mathbf{Y}))$ for any discrete random vector $\mathbf{Y}$.

Theorem 4. *Let X be a rv and $\mathbf{Y}$ be a random vector. If $E(X)$ exists, then*

$$E(X) = E(E(X \mid \mathbf{Y})). \tag{5}$$

Proof. The proof when $\mathbf{Y}$ is discrete has already been given. We now consider the case of continuous $\mathbf{Y}$.

For any event A, using (28.3) yields

$$\int_{\mathfrak{R}^n} P(A \mid \mathbf{Y} = \mathbf{u}) f_{\mathbf{Y}}(\mathbf{u}) \, d\mathbf{u}$$

$$= \int_{\mathfrak{R}^n} f_{I_A} (1 \mid \mathbf{Y} = \mathbf{u}) f_{\mathbf{Y}}(\mathbf{u}) \, d\mathbf{u} = f_{I_A}(1) = P(A). \tag{6}$$

Applying (6) to tail events,

$$E(E(X \mid \mathbf{Y})) = \int_{\mathfrak{R}^n} \left[\int_0^{\infty} P(X > t \mid \mathbf{Y} = \mathbf{u}) \, dt \right.$$

$$\left. - \int_{-\infty}^0 P(X \le t \mid \mathbf{Y} = \mathbf{u}) \, dt \right] f_{\mathbf{Y}}(\mathbf{u}) \, d\mathbf{u}$$

$$= \int_0^{\infty} \left[\int_{\mathfrak{R}^n} P(X > t \mid \mathbf{Y} = \mathbf{u}) f_{\mathbf{Y}}(\mathbf{u}) \, d\mathbf{u} \right] dt \tag{7}$$

$$- \int_{-\infty}^0 \left[\int_{\mathfrak{R}^n} P(X \le t \mid \mathbf{Y} = \mathbf{u}) f_{\mathbf{Y}}(\mathbf{u}) \, d\mathbf{u} \right] dt.$$

The proof is now completed in two steps: (i) interchange the order of each integration on the right side of (7) and (ii) apply (6) to the result. Step (i) is valid by nonnegativity of the integrands and the finiteness of each term is guaranteed by the existence of $E(X)$.

The proof for more general $\mathbf{Y}$ will be omitted.

Recall that $E(X)$ can be interpreted as a poplulation mean (see Section 30). Suppose that the population is partitioned into subpopulations by the value of $\mathbf{Y}$. Then (5) states that the overall population mean is a suitably weighted average of the subpopulation means.

Let $\mathbf{X}$ and $\mathbf{Y}$ be m- and n-dimensional random vectors, respectively, and let g be a real-valued function on $\mathfrak{R}^{m+n}$. The conditional expectation of $g(\mathbf{X}, \mathbf{Y})$ given $\mathbf{Y} = \mathbf{u}$ can be calculated by replacing $\mathbf{Y}$ by the constant $\mathbf{u}$. Thus

$$E(g(\mathbf{X}, \mathbf{Y}) \mid \mathbf{Y} = \mathbf{u}) = E(g(\mathbf{X}, \mathbf{u}) \mid \mathbf{Y} = \mathbf{u}). \tag{8}$$

A useful application of (8) is in the calculation of $E(XY)$ for rv's X and Y. By (8),

$$E(XY \mid Y = u) = E(Xu \mid Y = u) = uE(X \mid Y = u),$$

so that

$$E(XY \mid Y) = YE(X \mid Y). \tag{9}$$

Applying (5) to (9) yields

$$E(XY) = E[YE(X \mid Y)]. \tag{10}$$

Example 2.† Let (X, Y) be a point chosen at random in C, the interior of the unit circle centered at the origin. The area of the rectangle with corners $(\pm X, \pm Y)$ is $Z = 4|XY|$ and $E(Z)$ is to be found. By symmetry, the problem can be solved with the simplifying assumption that (X, Y) is chosen at random from the portion of C in the first quadrant.

The conditional distribution of X given Y is uniform on $(0, [1 - Y^2]^{1/2})$, so that $E(X \mid Y) = [1 - Y^2]^{1/2}/2$. Furthermore,

$$f_Y(t) = \frac{4}{\pi}[1 - t^2]^{1/2} \qquad \text{if } 0 < t < 1.$$

Hence, using (10), we obtain

$$E(XY) = E\left(Y\frac{[1 - Y^2]^{1/2}}{2}\right) = \frac{2}{\pi}\int_0^1 t(1 - t^2)\, dt = \frac{1}{2\pi}$$

so that $E(Z) = 4E(|XY|) = 2/\pi$.

Theorems 3 and 4 provide a very useful alternative to direct calculation of $E(X)$. The technique suggested is to replace the problem of finding $E(X)$ by that of finding $E(E(X \mid Y))$ where Y is "suitable" rv. The challenge is to find the Y that leads to simple (or interesting) results. This will be illustrated in the remaining examples of this section.

Example 3. Suppose that you are driving to keep an appointment. Within walking distance of your appointment there are five street parking spaces and a parking lot that is never full. Number the street spaces 1 to 5, with 5 the closest to your destination, 4 the next closest, and so on. If you park in space n, the time you spend walking will cost $\$(5 - n)$, while the price of the lot is $\$5$ with no walking cost. Beginning with space 1, you can only see one at a time. You may not return to a previously examined space. Call "space empty" a success and assume that the five trials form a sequence of Bernoulli trials with unknown success probability p. Suppose that your prior distribution for p is uniform on $(0, 1)$.

Let A_i be the event "success on the ith trial" and set $S_n = \sum_{i=1}^n I_{A_i}$. It will be shown in Section 44 that the posterior distribution of p given $S_n = k$ is beta with parameters $\alpha = k + 1$ and $\beta = n + 1 - k$ and therefore that $E(p \mid S_n = k) = (k + 1)/(n + 2)$. Let X be the cost. The object is to find the strategy that minimizes $E(X)$.

If you are now looking at space 5, the only relevant information is the value of I_{A_5}. Clearly, $E(X \mid I_{A_5} = 0) = 5$, and for the optimal strategy, $E(X \mid I_{A_5} = 1) = 0$. When looking at spaces $n = 2, 3,$ and 4, the relevant information is the value of (S_{n-1}, I_{A_n}).

Let $\gamma_{n,k} = E(X \mid S_{n-1} = k, I_{A_n} = 0)$, so that $\gamma_{5,k} = 5$. Denote expectation with respect to the conditional distribution of $I_{A_{n+1}}$ given S_n by E_{n+1}. Then

$$\lambda_{n,k} = E_{n+1}(E(X \mid S_n = k))$$

$$= \frac{k+1}{n+2} E(X \mid S_n = k, I_{A_{n+1}} = 1) + \frac{n+1-k}{n+2} \lambda_{n+1,k}.$$

If you behave optimally beginning with the $(n+1)$st space, then

$$E(X \mid S_{n-1} = k - 1, I_{A_n} = 1, \text{"don't park"}) = \lambda_{n,k}$$

$$E(X \mid S_{n-1} = k - 1, I_{A_n} = 1, \text{"park"}) = 5 - n.$$

For the optimal strategy,

$$E(X \mid S_{n-1} = k - 1, I_{A_n} = 1) = \min(\lambda_{n,k}, 5 - n).$$

Now $\lambda_{4,k} = [(k+1)/6](0) + [(5-k)/6](5) = 5(5-k)/6$, so that

$$E(X \mid S_3 = k - 1, I_{A_4} = 1) = 1 \qquad \text{if } k = 1, 2, 3$$

$$= \frac{5}{6} \qquad \text{if } k = 4$$

and you should park if $k = 1, 2, 3$ and not park if $k = 4$. Hence

$$\lambda_{3,k} = \frac{k+1}{5} \cdot 1 + \frac{4-k}{5} \cdot \frac{5(5-k)}{6} \qquad \text{if } k = 0, 1, 2$$

$$= \frac{4}{5} \cdot \frac{5}{6} + \frac{1}{5} \cdot \frac{10}{6} = 1 \qquad \text{if } k = 3$$

so that

$$E(X \mid S_2 = k - 1, I_{A_3} = 1) = 2 \qquad \text{if } k = 1 \text{ (park)}$$

$$= \frac{8}{5} \qquad \text{if } k = 2 \text{ (don't park)}$$

$$= 1 \qquad \text{if } k = 3 \text{ (don't park)}.$$

Continuing these calculations (Problem 9), the optimal strategy is:

Don't park in spaces 1 or 2.
Park in space 3 only if it is the first one vacant.
Park in space 4 unless all spaces have been vacant.
Park in space 5 if it is vacant.

The expected cost of this strategy is 131/60.

The method used in Example 3 of first optimizing the last action taken, then the next to last, and so on, is called *dynamic programming*.

Example 4. In Example 30.3 tail probabilities were used to show that $E(T) = 1/p$, where T is the waiting time for one success in a sequence of Bernoulli trials with $P(S) = p > 0$. By conditioning on the outcome of the first trial, one obtains this result quite simply (Problem 10). Now (for the same Bernoulli sequence) consider the more complicated rv X, the waiting time for the first occurrence of the pattern SSS. The method of calculation of $E(X)$ is illustrated in the tree diagram of Figure 1. As the conditioning rv take Y, the trial number on which the first failure (F) occurs. Then

$$E(X \mid Y = 1) = 1 + E(X_1) = 1 + E(X)$$
$$E(X \mid Y = 2) = 2 + E(X_2) = 2 + E(X) \qquad (11)$$
$$E(X \mid Y = 3) = 3 + E(X_3) = 3 + E(X)$$
$$E(X \mid Y > 3) = 3$$

where X_1, X_2, X_3 are rv's with exactly the same distribution as X. From (11),

$$E(X) = q[1 + E(X)] + pq[2 + E(X)] + p^2q[3 + E(X)] + 3p^3$$

and it follows that

$$(1 - q - pq - p^2q)E(X) = q + 2pq + 3p^2$$

since $E(X)$ exists. Finally, by algebraic manipulation,

$$E(X) = \frac{1}{p} + \frac{1}{p^2} + \frac{1}{p^3}.$$

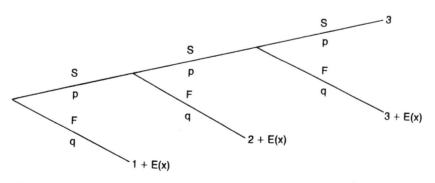

Figure 1

Summary

Conditional expectation of X given B, E(X | B): mean of the conditional model for the rv X given the event B.

Conditional expectations satisfy the properties of expectation obtained in Sections 30 and 31. The translation of Corollary 31.3 requires that the X_i be conditionally independent given B. For **Y** a random vector and $B = [\mathbf{Y} = \mathbf{u}]$, set $E(X \mid B) = E(X \mid \mathbf{Y} = \mathbf{u})$. If $g(\mathbf{u}) = E(X \mid \mathbf{Y} = \mathbf{u})$, we denote $g(\mathbf{Y})$ by $E(X \mid \mathbf{Y})$ and obtain $E(X) = E(E(X \mid \mathbf{Y}))$. As a result, $E(XY) = E[YE(X \mid Y)]$.

Problems

1. Let $f_{X,Y}(t, u) = ce^{-2u}$ for $0 < t < u$.
   (a) Find $E(X \mid Y)$.
   (b) Find $E(X)$.
   (c) Find $E(XY)$.

2. Let N be the number of patients to be tested for a certain disease. Group testing is to be used (see GP 3.1) and M is the number of tests performed. Let $f_N(k) = (\lambda^k/k!)e^{-\lambda}$ for $k = 0, 1, \ldots$, a Poisson probability function. Assume that the probability that an individual has the disease is p.
   (a) Find $E(M \mid N)$.
   (b) Find $E(M)$.
   (c) Suppose that (p, λ) has a prior distribution in which p and λ are independent, p is distributed uniformly on $(0, 1)$ and λ has prior density

$$g(t) = \frac{1}{100} te^{-t/10} \qquad \text{if } t > 0,$$

   which is a gamma density. Find $E(M)$ in this case.

3. Suppose that team A has won the first game in a World Series (see Problem 29.5). What is the conditional mean of the length of the series?

4.† Suppose that the random vector **Y** is observed and the value of the rv X is to be predicted with loss $(X - a)^2$ if the prediction is a. Find the function h to minimize the expected loss, $E(X - h(\mathbf{Y}))^2$. What expectations do you assume exist?

5. A fair coin is tossed until a head results. If k tosses are required, X has the conditional density

$$f_X(t \mid k) = \frac{1}{k} e^{-t/k} \qquad \text{if } t > 0,$$

an exponential density. Find $E(X)$ and $E(X^2)$.

6. Let X be a rv and Y and Z be random vectors. Set $g(Y, Z) = E(X \mid Y, Z)$ and show that $E(X \mid Y) = E(g(Y, Z) \mid Y)$.
7. Obtain an expression analogous to (10) for $E(XYZ)$.
8. (a) Why is (5) a generalization of the total probability laws [(28.2) and (28.3)]?
   (b) State (28.2) and (28.3) in the form of (5).
9. Complete the calculations in Example 3 and, for the optimal stategy, find the probability of finding street parking.
10. In a sequence of Bernoulli trials, let T_n be the waiting time for n successes in a row.
    (a) Use the method of Example 4 to find $E(T_1)$.
    (b) Find $E(T_2)$.
    (c) Find $E(T_n)$.

35. COVARIANCE AND CORRELATION; VARIANCES OF SUMS

The relationship between a given pair of random vectors, **X** and **Y**, can be described in many interesting and useful ways. The joint distribution, which gives the complete story, often gives many more details then one needs or can assimilate. Thus we look for various methods of summarizing essential features of the relation. One such summary, the conditional mean of **X** given **Y**, was the subject of the preceding section. Independence of **X** and **Y** was the subject of another.

The subject of this section is "covariance," which in a standardized form called "correlation," measures the extent to which two rv's are linearly related. Covariance also comes up in calculating variances and in higher dimensions may be used to measure the spread of a vector about its mean.

A. Pairs of Random Variables

We begin with the following definition:

Covariance

The *covariance* of the rv's X and Y is

$$\text{Cov}(X, Y) = E((X - \mu_X)(Y - \mu_Y)) \tag{1}$$

provided that the expectations exist. Alternatively,

$$\text{Cov}(X, Y) = E(XY) - \mu_X \mu_Y. \tag{2}$$

The proof that (1) and (2) are equivalent is left to the reader (Problem 1).

Correlation

Let X and Y be rv's with finite, positive variances, and let X^* and Y^* be their respective standardizations. The *correlation* between X and Y is

$$\rho(X, Y) = \text{Cov}(X^*, Y^*). \tag{3}$$

From linearity properties of expectation,

$$\rho(X, Y) = E(X^*Y^*) = \frac{\text{Cov}(X, Y)}{\sigma_X \sigma_Y}. \tag{4}$$

The last expression in (4) is the usual definition of $\rho(X, Y)$ and the straightforward way of calculating it.

Remark 1. Observe that $\text{Cov}(X, X) = \text{Var}(X)$. Thus covariance is a generalization of variance.

Remark 2. Formally, no second moments are needed for the existence of $\text{Cov}(X, Y)$, yet they are for $\rho(X, Y)$. However, $\text{Cov}(X, Y)$ is not of much interest when either $\text{Var}(X)$ or $\text{Var}(Y)$ fails to exist.

Remark 3. It should be clear by inspection of (1) that the sign of $\text{Cov}(X, Y)$ [the same as the sign of $\rho(X, Y)$] reflects the "average" sign of $(X - \mu_X)(Y - \mu_Y)$. Thus if X and Y tend to be large together and small together, then $\text{Cov}(X, Y) > 0$. If on the other hand, X is large when Y is small (and vice versa), then $\text{Cov}(X, Y) < 0$. By applying this form of gross reasoning to real quantities, we can often guess the signs of their correlations.

For example, height and weight will have positive correlation (or covariance) in most populations, whereas outdoor temperature and heating expenditures will have negative correlation. The following terminology is used to distinguish various cases: X and Y are called:

positively correlated   if $\text{Cov}(X, Y) > 0$;
negatively correlated   if $\text{Cov}(X, Y) < 0$;
uncorrelated   if $\text{Cov}(X, Y) = 0$.

An immediate consequence of Corollary 31.3 is that independent rv's with moments are uncorrelated [since $E(XY) = \mu_X \mu_Y$]. This is not the only case of uncorrelated rv's, however, as Example 31.5 demonstrates. Example 3 below provides an additional illustration of this fact. Note that a degenerate rv is uncorrelated with (in fact, independent of) any rv.

The proofs of the following propositions are based on linearity properties of expectations.

Theorem 1. *Let a, b, c, and d be any numbers. Then*

$$\text{Cov}(bX + a, dY + c) = bd \, \text{Cov}(X, Y). \tag{5}$$

Proof. The proof is similar to that of Theorem 32.2 and is left as a problem (Problem 2).

Corollary 1. *If bd $\neq$ 0, then*

$$\rho(bX + a, dY + c) = \frac{bd}{|bd|} \rho(X, Y). \tag{6}$$

Proof. Problem 7.

Example 1.† Suppose that the ith member of a population of size M has value v_i ($i = 1, \ldots, M$). Two members are randomly selected without replacement. Let X and Y be the values of the first and second, respectively. It has been shown that X and Y are identically distributed with mean $\mu = \sum_{i=1}^{M} v_i/M$ and variance $\sigma^2 = \sum_{i=1}^{M} (v_i - \mu)^2/M$. Furthermore,

$$\sum_{i=1}^{M} \sum_{j \neq i} v_i v_j = \sum_{i=1}^{M} \sum_{j=1}^{M} v_i v_j - \sum_{i=1}^{M} v_i^2$$

$$= M^2 \mu^2 - \sum_{i=1}^{M} v_i^2.$$

Hence

$$E(XY) = \frac{1}{M(M-1)} \sum_{i=1}^{M} \sum_{j \neq i} v_i v_j = \frac{M\mu^2}{M-1} - \frac{1}{M(M-1)} \sum_{i=1}^{M} v_i^2.$$

Since $\sigma^2 = \sum_{i=1}^{M} v_i^2/M - \mu^2$, it follows that

$$E(XY) = \frac{M\mu^2}{M-1} - \frac{1}{M-1}(\sigma^2 + \mu^2) = \mu^2 - \frac{\sigma^2}{M-1}$$

so that

$$\text{Cov}(X, Y) = -\frac{\sigma^2}{M-1} < 0.$$

The negativity of $\text{Cov}(X, Y)$ could have been predicted. A large value of X amounts to removing a large value from the population, increasing the conditional probability that Y will be small.

Example 2. Let (X, Y) have the density
$$f(t, u) = 2 \qquad \text{if } 0 < u < t < 1.$$
Then
$$f_X(t) = 2t \qquad\qquad \text{if } 0 < t < 1;$$
$$f_Y(u) = 2(1 - u) \qquad \text{if } 0 < u < 1$$
and

$$E(X) = \int_0^1 2t^2 \, dt = \frac{2}{3} \qquad\qquad E(X^2) = \int_0^1 2t^3 \, dt = \frac{1}{2}$$

$$E(Y) = \int_0^1 2u(1 - u) \, du = \frac{1}{3} \qquad E(Y^2) = \int_0^1 2u^2(1 - u) \, du = \frac{1}{6}$$

$$E(XY) = \int_0^1 \left[\int_0^1 2u \, du \right] t \, dt = \frac{1}{4} \qquad \sigma_X^2 = \frac{1}{18} = \sigma_Y^2.$$

Hence

$$\text{Cov}(X, Y) = \frac{1}{4} - \frac{2}{3} \cdot \frac{1}{3} = \frac{1}{36};$$

$$\rho(X, Y) = \frac{1/36}{1/18} = .5.$$

Here X and Y are positively correlated.

Example 3. Let X be any symmetric rv, let $Y = X^2$ and assume appropriate moments. Then, since $XY = X^3$ is also symmetric, $\text{Cov}(X, Y) = \rho(X, Y) = 0$. That is, X and Y are uncorrelated although Y is actually a function of X (not, however, a linear function).

Example 4. Suppose that X and Y are jointly distributed as indicated in Table 1 which gives $P[X = t, Y = u]$. Then
$$E(X) = 2 = EY, \qquad E(X^2) = 5, \qquad E(Y^2) = 8;$$
$$E(XY) = 2, \qquad\qquad \sigma_X = 1, \qquad\qquad \sigma_Y = 2.$$
Hence
$$\text{Cov}(X, Y) = -2 \quad \text{and} \quad \rho(X, Y) = -1.$$
Checking Table 1, it can be seen that, in fact,
$$Y = 6 - 2X.$$

Table 1

		u		
t	0	2	4	6
0	0	0	0	.1	.1
1	0	0	.2	0	.2
2	0	.3	0	0	.3
3	.4	0	0	0	.4
	.4	.3	.2	.1	

That is, every pair (t, u) to which a positive probability has been assigned lies on a line.

From (6) it follows that

$$\rho(X, bX + a) = \frac{b}{|b|} \rho(X, X) = \frac{b}{|b|} = \pm 1,$$

the sign depending on whether $b > 0$ or $b < 0$. It will be shown below that $|\rho(X, Y)| = 1$ implies that the joint distribution of X and Y is concentrated on a line and that in all other cases, $|\rho(X, Y)| < 1$. First we give the main results concerning covariance and its use in calculating variance.

Theorem 2. *Let* **X** *and* **Y** *be m- and n-dimensional random vectors. Then*

$$\text{Cov}\left(\sum_{i=1}^{m} a_i X_i, \sum_{j=1}^{n} b_j Y_j\right) = \sum_{i=1}^{m} \sum_{j=1}^{n} a_i b_j \text{ Cov}(X_i, Y_j) \tag{7}$$

for all constants $a_1, \ldots, a_m, b_1, \ldots, b_n$.

Proof. By Theorem 1, it is sufficient to prove (7) for $a_1 = \cdots = a_m = b_1 = \cdots = b_n = 1$. We will prove (7) only for $m = 2$ and $n = 1$. The more general case follows by induction and the fact that $\text{Cov}(X, Y) = \text{Cov}(Y, X)$. By Theorem 1, without loss of generality, assume that $E(X) = E(Y) = 0$. Then

$$\text{Cov}(X_1 + X_2, Y_1) = E((X_1 + X_2)Y_1) = E(X_1 Y_1) + E(X_2 Y_1)$$

$$= \text{Cov}(X_1, Y_1) + \text{Cov}(X_2, Y_1).$$

At present, our interest is in applying Theorem 2 to the formula for $\text{Var}(\sum_{i=1}^{n} a_i X_i)$ given in the following corollary.

Corollary 2. *Let* **X** *be an n-dimensional random vector. Let* $a_1, \ldots, a_n$ *be any constants. Then*

$$\text{Var} \left(\sum_{i=1}^{n} a_i X_i \right) = \sum_{i=1}^{n} a_i^2 \text{ Var}(X_i) + 2 \sum_{i<j} a_i a_j \text{ Cov}(X_i, X_j). \tag{8}$$

In particular, if $X_1, \ldots, X_n$ *are uncorrelated, then*

$$\text{Var} \left(\sum_{i=1}^{n} a_i X_i \right) = \sum_{i=1}^{n} a_i^2 \text{ Var}(X_i). \tag{9}$$

Proof. Recall that $\text{Var}(Y) = \text{Cov}(Y, Y)$ and let $Y = \sum_{i=1}^{n} a_i X_i$. Then, from Theorem 2,

$$\text{Var} \left(\sum_{i=1}^{n} a_i X_i \right) = \sum_{i=1}^{n} \sum_{j=1}^{n} a_i a_j \text{ Cov}(X_i, X_j)$$

$$= \sum_{i=1}^{n} a_i^2 \text{ Cov}(X_i, X_i) + \sum_{i \neq j} a_i a_j \text{ Cov}(X_i, X_j)$$

$$= \sum_{i=1}^{n} a_i^2 \text{ Var}(X_i) + 2 \sum_{i<j} a_i a_j \text{ Cov}(X_i, X_j).$$

As a special case of Corollary 2,

$$\text{Var}(aX + bY) = a^2 \text{ Var}(X) + b^2 \text{ Var}(Y) + 2ab \text{ Cov}(X, Y). \tag{10}$$

Example 5. From Example 2,

$$\text{Var}(2X + 3Y) = 4 \left(\frac{1}{18} \right) + 9 \left(\frac{1}{18} \right) + 12 \left(\frac{1}{36} \right) = \frac{19}{18}$$

and $\text{Var}(2X - 3Y) = 7/18$.

Theorem 3. *Let X and Y be rv's with correlation* $\rho(X, Y)$. *Then* $|\rho(X, Y)| \leq 1$. *Furthermore,* $|\rho(X, Y)| = 1$ *if, and only if,* $Y = bX + a$ *with* $b \neq 0$.

Proof. By (3), $\rho(X^*, Y^*) = \rho(X, Y) = \text{Cov}(X^*, Y^*)$. Thus since variances are nonnegative,

$$\text{Var}(X^* \pm Y^*) = \text{Var}(X^*) + \text{Var}(Y^*) \pm 2 \text{ Cov}(X^*, Y^*)$$
$$= 2[1 \pm \rho(X, Y)] \geq 0.$$

The inequality implies that $|\rho(X, Y)| \leq 1$. Furthermore, $\rho(X, Y) = \pm 1$ if, and only if, $\text{Var}(X^* \mp Y^*) = 0$ and then $Y^* = \pm X^*$, which is equivalent to $Y = bX + a$ with $b \neq 0$. Here $b > 0$ if, and only if, $Y^* = +X^*$ (see Problem 18).

Example 6. Let X and Z be independent rv's and set $Y = a(X + Z) + b$ for some constants, a and b. We will compute $\rho(X, Y)$. Without loss of generality, assume that $E(X) = E(Z) = b = 0$, so that $E(Y) = 0$. Then $\sigma_Y^2 = a^2(\sigma_X^2 + \sigma_Z^2)$. Furthermore, $\text{Cov}(X, Y) = E(XY) = a\sigma_X^2$. Hence

$$\rho(X, Y) = \frac{a\sigma_X^2}{\sqrt{\sigma_X^2 a^2(\sigma_X^2 + \sigma_Z^2)}} = \frac{a}{|a|} \cdot \frac{1}{\sqrt{1 + \sigma_Z^2/\sigma_X^2}}.$$

If $\sigma_Z^2 = 0$, then $Y = aX$, and in that case, $\rho(X, Y) = \pm 1$, the sign being that of a.

Thus $\sigma_Z^2 = 0$ implies a direct liner relationship between X and Y as in Example 4. As σ_Z^2 grows, the values of (X, Y) will increasingly tend to depart from that line and $|\rho(X, Y)|$ will decrease.

The following results lead to a useful formula relating $\text{Var}(X)$ to $\text{Var}(X \mid \mathbf{Y})$ and $E(X \mid \mathbf{Y})$.

Theorem 4. If $E(Z \mid \mathbf{Y}) = 0$, then Z and $g(\mathbf{Y})$ are uncorrelated for any g for which $E(g(\mathbf{Y}))$ and $E(Zg(\mathbf{Y}))$ exist.

Proof. Since $E(Z) = E[E(Z \mid \mathbf{Y})] = 0$, it follows that $\text{Cov}[Z, g(\mathbf{Y})] = E(Zg(\mathbf{Y}))$. Then

$$E(Zg(\mathbf{Y}) \mid \mathbf{Y}) = g(\mathbf{Y})E(Z \mid \mathbf{Y}) = 0.$$

Hence $E(Zg(\mathbf{Y})) = 0$.

Corollary 3. If X is any rv and $\mathbf{Y}$ is a random vector, then $E(X \mid \mathbf{Y})$ and $Z = X - E(X \mid \mathbf{Y})$ are uncorrelated.

Proof. Since $E(Z \mid \mathbf{Y}) = 0$, the result follows from Theorem 4.

Corollary 4. If X is a rv and $\mathbf{Y}$ is a random vector, then

$$\text{Var}(X) = E(\text{Var}(X \mid \mathbf{Y})) + \text{Var}(E(X \mid \mathbf{Y})). \tag{11}$$

Proof. Since $X = [X - E(X \mid \mathbf{Y})] + E(X \mid \mathbf{Y})$, by Corollary 3 and (9),

$$\text{Var}(X) = \text{Var}(X - E(X \mid \mathbf{Y})) + \text{Var}(E(X \mid \mathbf{Y})).$$

But since $E(X - E(X \mid \mathbf{Y})) = 0$,

$$\text{Var}(X - E(X \mid \mathbf{Y})) = E(X - E(X \mid \mathbf{Y}))^2$$
$$= E(E((X - E(X \mid \mathbf{Y}))^2 \mid \mathbf{Y})) = E(\text{Var}(X \mid \mathbf{Y})),$$

completing the proof.

Example 7. Let $A_1, \ldots, A_n$ be events with $P(A_i) = p$ and set $X = \sum_{i=1}^{n} I_{A_i}$. Let the prior density of p be uniform on $(0, 1)$. Thus $E(p) = 1/2$ and $\text{Var}(p) = 1/12$. Assume that the events $A_1, \ldots, A_n$, and hence $I_{A_1}, \ldots, I_{A_n}$, are conditionally independent given p. This does not imply that $I_{A_1}, \ldots, I_{A_n}, p$ are independent, and it is not true that $\text{Var}(X) = \sum_{i=1}^{n} \text{Var}(I_{A_i})$. However, from conditional independence given p it follows that $\text{Var}(X \mid p) = \sum_{i=1}^{n} \text{Var}(I_{A_i} \mid P)$.

From indicator properties previously given, $E(I_{A_i} \mid p) = p$ and $\text{Var}(I_{A_i} \mid p) = p(1 - p)$. Hence $E(X \mid p) = np$ by Corollary 31.2 and $\text{Var}(X \mid p) = np(1 - p)$. Hence $\text{Var}(E(X \mid p)) = n^2/12$ and $E(\text{Var}(X \mid p)) = n/6$. Thus (11) yields $\text{Var}(X) = n/6 + n^2/12$. Also, $E(X) = n/2$.

In this example direct computation of $\text{Var}(X)$ is more difficult.

Section 32 examined the problem of predicting the value of a random variable X with loss $(X - a)^2$ when the predicted value is a. Suppose that a is allowed to depend on another rv Y. Extension of the result in Section 32 yields the fact that the best predictor of X based on Y is $E(X \mid Y)$ (see Problem 34.4). Another predictor of great importance is often used because it is best within a restricted, but simple class.

A predictor of X of the form $a(Y) = \alpha + \beta Y$ is called *linear*. If α and β are chosen so that $E(X - \alpha - \beta Y)^2$ is minimized, then $a(Y)$ is called the *best* (or *least square*) *linear predictor of X based on Y*.

Theorem 5. *Provided that the appropriate expectations exist, the best linear predictor of X based on Y is $a(Y) = \alpha + \beta Y$, where*

$$\beta = \frac{\text{Cov}(X, Y)}{\sigma_Y^2} = \frac{\sigma_X}{\sigma_Y} \rho(X, Y) \qquad (12)$$

and

$$\alpha = \mu_X - \beta \mu_Y. \qquad (13)$$

Proof. Consider first the case of standardized variables X^* and Y^* and denote the predictor of X^* by $\gamma Y^* + \delta$. Then, letting $\rho = \rho(X^*, Y^*)$,

$$E(X^* - (\gamma Y^* + \delta))^2 = 1 - 2\gamma \rho + \gamma^2 + \delta^2$$

$$= 1 - \rho^2 + (\rho - \gamma)^2 + \delta^2,$$

which is clearly minimized by setting $\delta = 0$ and $\gamma = \rho$. The proof is then completed by expressing X^* and Y^* in terms of X and Y (Problem 5).

B. Covariance Matrices

In the remainder of this section, familiarity with basic matrix algebra is assumed. To begin with, we extend the definition of expectation to random matrices.

Expectation of a Random Matrix

If $\mathbf{Z}$ is an $m \times n$ matrix with random entries Z_{ij} for which the $E(Z_{ij})$ exist, then the *expectation* of $\mathbf{Z}$ is the $m \times n$ matrix $E(\mathbf{Z})$ with entries $E(Z_{ij})$.

Let $\mathbf{X}$ be an n-dimensional random vector. Using the notation of matrix algebra, take $\mathbf{X}$ to be a column vector so that $\mathbf{X}'$, its transpose, is a row vector. Then $\mathbf{XX}'$ is an $n \times n$ matrix whose (i, j) entry is the rv $X_i X_j$.

Covariance matrix

Let $\mathbf{X}$ be an n-dimensional random vector with $E(\mathbf{X}) = \boldsymbol{\mu}$. The *covariance matrix* of $\mathbf{X}$ is the $n \times n$ matrix

$$\mathbf{C}(\mathbf{X}) = E[(\mathbf{X} - \boldsymbol{\mu})(\mathbf{X} - \boldsymbol{\mu})'] = E(\mathbf{XX}') - \boldsymbol{\mu}\boldsymbol{\mu}'. \tag{14}$$

The entries in $\mathbf{C}(\mathbf{X})$ are the $\mathrm{Cov}(X_i, X_j)$, so that the main diagonal entries are the $\mathrm{Var}(X_i)$. The equality of the two expressions for $\mathbf{C}(\mathbf{X})$ in (13) is a restatement of (1).

If $\mathbf{X}$ is an n-dimensional random vector, $\mathbf{A}$ is an $m \times n$ fixed matrix, and b is a fixed m-dimensional vector, then

$$\mathbf{C}(\mathbf{AX} + \mathbf{b}) = \mathbf{AC}(\mathbf{X})\mathbf{A}', \tag{15}$$

which extends Theorem 32.1 to higher dimensions. Furthermore, Corollary 31.1 may be restated as

$$E(\mathbf{AX} + \mathbf{b}) = \mathbf{A}E(\mathbf{X}) + \mathbf{b}. \tag{16}$$

If $\mathbf{a}$ is a vector and $\mathbf{A} = \mathbf{a}'$, then (15) yields the variance of the rv $\mathbf{a}'\mathbf{X} + \mathbf{b}$.

An n-dimensional random vector $\mathbf{X}$ has *essential dimension* k if there exists a set H of dimension k for which $P[\mathbf{X} \in H] = 1$ and no lower-dimensional subset of $\mathcal{R}^n$ has this property. If the essential dimension of $\mathbf{X}$ is $k < n$, there exist numbers $a_1, \ldots, a_n$, not all 0, such that $\sum_{i=1}^{n} a_i X_i$ is a degenerate rv. Whenever this occurs, at least one X_i can be expressed as a linear combination of the others (possibly with a constant term) and we choose it at our convenience.

Theorem 6. *A matrix is a covariance matrix if, and only if, it is symmetric and nonnegative definite.*

Proof. Symmetry of a covariance matrix follows from $\text{Cov}(X_i, X_j) = \text{Cov}(X_j, X_i)$. Let $\mathbf{X}$ be an n-dimensional random vector and $\mathbf{a} \in \mathcal{R}^n$. Then $\text{Var}(\mathbf{a}'\mathbf{X}) = \mathbf{a}'\mathbf{C}(\mathbf{X})\mathbf{a} \geq 0$. Let $\mathbf{M}$ be an $n \times n$ symmetric, nonnegative definite matrix and $\mathbf{X} = (X_1, \ldots, X_n)'$, where the X_i are iid with unit variance. Then there exists a matrix $\mathbf{A}$ such that $\mathbf{M} = \mathbf{A}\mathbf{A}'$ and $\mathbf{C}(\mathbf{A}\mathbf{X}) = \mathbf{M}$.

Theorem 7. *Let $\mathbf{X}$ be an n-dimensional random vector with covariance matrix $\mathbf{C}(\mathbf{X})$. The rank of $\mathbf{C}(\mathbf{X})$ is the same as the essential dimension of $\mathbf{X}$.*

Proof. Suppose that the essential dimension of $\mathbf{X}$ is $k < n$ and assume, without loss of generality, that $X_n = \sum_{i=1}^{n-1} a_i X_i$. Then, by Theorem 2, $\text{Cov}(X_n, X_j) = \sum_{i=1}^{n-1} a_i \text{Cov}(X_i, X_j)$ $(j = 1, \ldots, n)$ and hence $\mathbf{C}(\mathbf{X})$ is singular.

Now suppose that $\mathbf{C}(\mathbf{X})$ is singular, so there exist $a_1, \ldots, a_n$, not all zero, such that $\sum_{i=1}^{n} a_i \text{Cov}(X_i, X_j) = 0$ $(j = 1, \ldots, n)$ Thus $\sum_{i=1}^{n} \sum_{i=1}^{n} a_i a_j \text{Cov}(X_i, X_j) = 0$. From Corollary 2, it follows that $\text{Var}(\sum_{i=1}^{n} a_i X_i) = 0$, so that $\sum_{i=1}^{n} a_i X_i$ is degenerate.

The result is obtained by reducing dimensionality until a nonsingular covariance matrix is obtained.

The full-rank case is left as a problem.

Summary

Covariance of X and Y: $\quad \text{Cov}(X, Y) = E((X - \mu_X)(Y - \mu_Y)) = E(XY) - \mu_X\mu_Y$.

Correlation of X and Y: $\quad \rho(X, Y) = \text{Cov}(X^*, Y^*)$, where X^* and Y^* are the standardized versions of X and Y, respectively.

Uncorrelated rv's X and Y: $\quad \text{Cov}(X, Y) = 0$.

Covariance matrix: If $\mathbf{X}$ is a random column vector, then $\mathbf{C}(\mathbf{X}) = E[(\mathbf{X} - \mu_\mathbf{X})(\mathbf{X} - \mu_\mathbf{X})'] = E(\mathbf{X}\mathbf{X}') - \mu_\mathbf{X}\mu_\mathbf{X}'$.

The usual formula for computing $\rho(X, Y)$ is $\rho(X, Y) = \text{Cov}(X, Y)/(\sigma_X\sigma_Y)$. Independent rv's are uncorrelated. Some properties are:

(i) $\text{Cov}(bX + a, dY + c) = bd\,\text{Cov}(X, Y)$

(ii) $\rho(bX + a, dY + c) = (bd/|bd|)\rho(X, Y) = \pm\rho(X, Y)$

(iii) $\text{Cov}(\sum_{i=1}^{m} a_i X_i, \sum_{j=1}^{n} b_j Y_j) = \sum_{i=1}^{m} \sum_{j=1}^{n} a_i b_j \text{Cov}(X_i, Y_j)$

(iv) $\text{Var}(\sum_{i=1}^{m} a_i X_i) = \sum_{i=1}^{m} a_i^2 \text{Var}(X_i) + 2 \sum_{i<j} a_i a_j \text{Cov}(X_i, X_j)$

(v)   If $X_1, \ldots, X_n$ are uncorrelated, then $\text{Var}(\Sigma_{i=1}^{n} a_i X_i) = \Sigma_{i=1}^{n} a_i^2$
      $\text{Var}(X_i)$

(vi) $|\rho(X, Y)| \leq 1$

(vii) $|\rho(X, Y)| = 1$ if, and only if, X and Y are linearly related

A useful formula is

$$\text{Var}(X) = E(\text{Var}(X \mid Y)) + \text{Var}(E(X \mid Y)).$$

An extension of (i) to random vectors is $\mathbf{C}(\mathbf{AX} + \mathbf{b}) = \mathbf{AC}(\mathbf{X})\mathbf{A}'$. A matrix is a covariance matrix if, and only if, it is nonnegative definite and symmetric.

Problems

1. For $x \in \mathcal{R}^2$, let $f(x) = 2$ if x is in the triangle with vertices $(0, 0)$, $(0, 1)$, and $(1, 0)$. Let X be the random vector with density f. Find $\text{Cov}(X_1, X_2)$, $\rho(X_1, X_2)$, $\text{Var}(X_1 + X_2)$, and $\text{Var}(X_1 - X_2)$.

2. Repeat Problem 1 for the triangle with vertices $(1, 1)$, $(2, 2)$, and $(1, 2)$.

3. Let X be the outcome when a fair die is rolled. If $X = k$, a coin with probability $1/k$ of producing a head is tossed until a head results. Let Y be the number of tosses required. Find $\text{Cov}(X, Y)$, $\rho(X, Y)$, $\text{Cov}(X, X + Y)$, and $\text{Var}(5X - 3Y)$.

4. A box has n fuses, of which two are defective. Fuses are drawn one at a time at random and without replacement. Let X and Y be the trial numbers on which the first and the second defective fuses, respectively, are drawn. Find $\text{Cov}(X,Y)$, $\rho(X, Y)$, $\text{Cov}(X, Y - X)$, and $\rho(X, Y - X)$.

5. Prove equality of the expressions in (1) and (2) for $\text{Cov}(X, Y)$.

6. Prove Theorem 1.

7. Prove Corollary 1.

8. Let $\mathbf{X}$ be a two-dimensional random vector with probability function or density f.

    (a) Suppose that $f(-t_1, t_2) = f(t_1, t_2)$ for all t_1, t_2, that is, f is symmetric about the line $t_1 = 0$. Show that $\text{Cov}(X_1, X_2) = 0$.

    (b) Suppose that $f(t)$ is symmetric about the line $t_1 = a$, that is, $f(a - u, t_2) = f(a + u, t_2)$ for all t_2, u. Show that $\text{Cov}(X_1, X_2) = 0$.

    (c) Show that the result in part (b) holds if symmetry is about the line $t_2 = b$.

9.† For $i = 1, \ldots, n$ let $\text{Var}(X_i) = \sigma^2$. Assume that the X_i are independent and set $\overline{X}_n = \Sigma_{i=1}^{n} X_i/n$. Show that $\text{Var}(\overline{X}_n) = \sigma^2/n$.

10. (a) Find $\text{Cov}(X + Y, X - Y)$ in terms of σ_X^2, σ_Y^2, and $\text{Cov}(X, Y)$.
    (b) When are $X + Y$ and $X - Y$ uncorrelated?
11. From (3), the definition of correlation, prove that $\rho(X, Y) = E(X^*Y^*) = \text{Cov}(X, Y)/(\sigma_X\sigma_Y)$.
12. Let A and B be events. Show that:
    (a) $\text{Cov}(I_A, I_B) = P(AB) - P(A)P(B)$.
    (b) The following three statements are equivalent.
        (i) A and B are independent events.
        (ii) I_A and I_B are independent rv's.
        (iii) I_A and I_B are uncorrelated.
13. Recall that $|\rho(X, Y)| = 1$ implies that there exist constants a and b such that $Y = a + bX$ (Theorem 3). Identify a and b in terms of moments.
14. Show that the conclusion of Theorem 4 is valid if $E(Z \mid Y)$ is any constant.
15. Complete the proof of Theorem 5.
16. Assume that $E(X \mid Y)$ is linear in Y and prove that $\text{Var}(E(X \mid Y)) = \sigma_X^2\rho^2(X, Y)$.
17. Let X and Y be rv's with $E(Y \mid X) = bX + a$. Use Theorem 5 to determine a and b in terms of moments.
18. Let (X, Y) have correlation ρ and let $a(X)$ be the best linear predictor of Y. Show that $E(Y - a(X))^2 = \sigma_Y^2(1 - \rho^2)$.
19.† Let $\mathbf{X}$ be an n-dimensional random vector with covariance matrix $\mathbf{C}(\mathbf{X})$ of rank $k < n$. Show that there exists an orthogonal matrix, $\mathbf{B}$, such that $\mathbf{BX} = \binom{\mathbf{Y}}{\mathbf{w}}$, where $\mathbf{Y}$ is k-dimensional and $\mathbf{w}$ is a constant vector.
20. Show that $\mathbf{C}(\mathbf{X})$ is nonsingular if, and only if, the essential dimension of $\mathbf{X}$ is n.

36. SAMPLE MEAN AND SAMPLE VARIANCE

In this section we apply the results of the previous one to certain linear and quadratic functions of rv's. These quantities have a variety of uses in summarizing sample information and in statistical inference.

Suppose that a population consists of M individuals and t_i is some measurement (say, height) on the ith individual. Set $\bar{t}_M = \Sigma_{i=1}^{M} t_i/M$ and $v_M = \Sigma_{i=1}^{M} (t_i - \bar{t}_M)^2/M$. If X is the measured value for an individual chosen at random, then $\mu_X = \bar{t}_M$ and $\sigma_X^2 = v_M$.

Let $X_1, \ldots, X_n$ be rv's and consider $\overline{X}_n = \Sigma_{i=1}^{n} X_i/n$ and $V_n = D_n^2/n$, where $D_n^2 = \Sigma_{i=1}^{n} (X_i - \overline{X}_n)^2$. Then $\overline{X}_n$ is the sample analog of the population mean and V_n is the sample analog of the population variance.

Sample Mean and Variance

We call $\overline{X}_n$ the *sample mean* and V_n the *sample variance*.

Algebraic manipulations (or application of the computing formula for the variance) yield

$$D_n^2 = nV_n = \sum_{i=1}^{n} X_i^2 - n\overline{X}_n^2 = \sum_{i=1}^{n} X_i^2 - \left(\sum_{i=1}^{n} X_i\right)^2 \Big/ n. \qquad (1)$$

Below we display $E(\overline{X}_n)$ and $\text{Var}(\overline{X}_n)$ for the case in which $X_1, \ldots, X_n$ are iid. This includes the case of sampling with replacement. The formulas given are valid under more general conditions.

Suppose that $E(X_i) = \mu$ for $i = 1, \ldots, n$. Then

$$E(\overline{X}_n) = \mu. \qquad (2)$$

If, in addition, $X_1, \ldots, X_n$ are independent and $\text{Var}(X_i) = \sigma^2$ for $i = 1, \ldots, n$, then

$$\text{Var}(\overline{X}_n) = \frac{\sigma^2}{n}. \qquad (3)$$

Equations (2) and (3) follow from elementary properties of expectation and variance and were the results, respectively, of Problems 31.15 and 35.9.

Recall that $\text{SD}(Y) = \sqrt{\text{Var}(Y)}$ is a measure of dispersion for the distribution of Y about its mean. From (2) and (3) we see that $\overline{X}_n$ becomes "closer" to μ as n increases. This reinforces the natural idea that $\overline{X}_n$ is a better estimate of μ for larger n.

Under the conditions for (3),

$$E(V_n) = \frac{n-1}{n}\sigma^2. \qquad (4)$$

The proof of (4) is left as a problem. The calculation of $\text{Var}(V_n)$ requires additional assumptions.

Let $\mathbf{X}_1, \ldots, \mathbf{X}_n$ be k-dimensional random vectors. Let X_{ij} denote the jth component of $\mathbf{X}_i$. Again set $\overline{\mathbf{X}}_n = \sum_{i=1}^{n} \mathbf{X}_i/n$, a random vector whose jth component is $\overline{X}_{nj} = \sum_{i=1}^{n} X_{ij}/n$ ($j = 1, \ldots, k$).

Assume that $E(\mathbf{X}_i) = \boldsymbol{\mu}$ ($i = 1, \ldots, n$). Then $E(\overline{\mathbf{X}}_n) = \boldsymbol{\mu}$. If the covariance matrix of each $\mathbf{X}_i$ is $\mathbf{C}(\mathbf{X}_i) = \mathbf{A}$ and $\mathbf{X}_1, \ldots, \mathbf{X}_n$ are independent, then $\mathbf{C}(\overline{\mathbf{X}}_n) = \mathbf{A}/n$. The computations are matrix analogs of those for rv's.

We turn now to a determination of $E(\overline{X}_n)$ and $\text{Var}(\overline{X}_n)$ for the case of sampling without replacement.

Consider a random sample of size n, without replacement, from a population of size M where, as before, the ith member has value t_i. Let X_i be the ith number selected and μ and σ^2 be the population mean and variance, respectively. That is, $\mu = \bar{t}_M$ and $\sigma^2 = v_M$.

By Corollary 22.2:

(i) $X_1, \ldots, X_n$ are identically distributed so that $E(X_i) = \mu$ and $\mathrm{Var}(X_i) = \sigma^2$ [$\sum_{i=1}^{n} \mathrm{Var}(X_i) = n\sigma^2$].

(ii)   The pairs (X_i, X_j) $(i \neq j)$ are identically distributed, so that all have the same covariance, say c, independent of n. Hence $\sum_{i<j} \mathrm{Cov}(X_i, X_j) = n(n - 1)c$.

Example 35.1 gives a direct evaluation of c. What follows is a simpler, but indirect evaluation.

By (i), formula (2) is valid in this case. By (i) and (ii) together with the formula (35.9) for $\mathrm{Var}(\sum_{i=1}^{n} X_i)$,

$$\mathrm{Var}\left(\sum_{i=1}^{n} X_i \right) = n\sigma^2 + n(n - 1)c. \tag{5}$$

When $n = M$, the sample consists of the entire population, so that $\sum_{i=1}^{M} X_i$ is degenerate at $M\mu$ and $\mathrm{Var}(\sum_{i=1}^{M} X_i) = 0$. Setting $n = M$ in (5) yields $c = -\sigma^2/(M - 1)$. Hence

$$\mathrm{Var}\left(\sum_{i=1}^{n} X_i \right) = n\left(1 - \frac{n - 1}{M - 1} \right)\sigma^2 = \frac{n(M - n)}{M - 1}\sigma^2 \tag{6}$$

and

$$\mathrm{Var}(\bar{X}_n) = \frac{M - n}{M - 1} \cdot \frac{\sigma^2}{n}. \tag{7}$$

Compare (7) with (3) and observe that they differ only in the factor $(M - n)/(M - 1)$. Many statistical applications require the calculation of $\mathrm{SD}(\bar{X}_n)$ and the independence assumption is often justified by considering the population to be infinite. Motivated by these considerations, $\sqrt{(M - n)/(M - 1)}$ is called the *finite population correction factor*. This factor is (1) less than 1 and (2) close to 1 when M is large relative to n.

Summary

Sample mean: $\bar{X}_n = \sum_{i=1}^{n} X_i/n$.
Sample variance: $V_n = D_n^2/n$, where $D_n^2 = \sum_{i=1}^{n} (X_i - \bar{X}_n)^2$.

If $X_1, \ldots, X_n$ are independent, each with mean μ and variance σ^2, then $E(\bar{X}_n) = \mu$, $\mathrm{Var}(\bar{X}_n) = \sigma^2/n$, and $E(V_n) = (n - 1)\sigma^2/n$.

If $X_1, \ldots, X_n$ is a sample without replacement from a population of size M with mean μ and variance σ^2, then $E(\overline{X}_n) = \mu$ and $\text{Var}(\overline{X}_n) = [(M - n)/(M - 1)](\sigma^2/n)$.

Problems

1. Specialize (2), (3), and (7) to the case in which the X_i are indicators.
2. (a) A digit is chosen at random. Represent the result by a 10-dimensional vector, the jth component being 1 if the digit is $j - 1$ and 0 otherwise. Let $\mathbf{X}$ be this random vector. Find $E(\mathbf{X})$ and $C(\mathbf{X})$.
   (b) The experiment in part (a) is repeated n times and $\mathbf{X}_i$ is the random vector in (a) for ith trial. Find $E(\overline{\mathbf{X}}_n)$ and $\mathbf{C}(\overline{\mathbf{X}}_n)$.
3. Let $X_1, \ldots, X_n, X_{n+1}$ be independent rv's with $E(X_i) = \mu$ and $\text{Var}(X_i) = \sigma^2$. Suppose that you observe $X_1, \ldots, X_n$ in order to predict the value of X_{n+1}. You will use $\overline{X}_n$ to predict X_{n+1}, suffering loss $(\overline{X}_n - X_{n+1})^2$. Additionally, each of the observed X_i costs c units. How should n be chosen to minimize expected cost?
4. In verifying (3), it was assumed that $X_1, \ldots, X_n$ are independent. Can you replace this by a weaker condition?
5. (Stratified sampling) A population of size M is divided into two subpopulations (strata). Stratum 1 has M_1 individuals with mean μ_1 and variance σ_1^2, while stratum 2 has $M_2 = M - M_1$ individuals with mean μ_2 and variance σ_2^2.
   (a) Show that the population mean is $\mu = (M_1\mu_1 + M_2\mu_2)/M$ and the variance is

$$\sigma^2 = \frac{M_1\sigma_1^2 + M_2\sigma_2^2}{M} + \frac{M_1M_2(\mu_1 - \mu_2)^2}{M^2}.$$

   (b) A sample of size n is to be taken from the population according to the following scheme:
   (i) A random sample $X_1, \ldots, X_{n_1}$ is taken from stratum 1.
   (ii) A random sample $Y_1, \ldots, Y_{n_2}$ $(n_2 = n - n_1)$ is taken from stratum 2. Set $Z = (M_1\overline{X}_{n_1} + M_2\overline{Y}_{n_2})/M$ and show that $E(Z) = \mu$.
   (c) Assume that M_1 and M_2 are large compared with n, so that the finite population correction factor can be ignored. Also ignore the fact that n_1 and n_2 are integers. Assume that n is fixed and show that $\text{Var}(Z)$ is minimized by setting $n_1 = M_1\sigma_1 n/(M_1\sigma_1 + M_2\sigma_2)$.
   (d) Let $W_1, \ldots, W_n$ be a random sample from the entire population. Consider also Z with the allocation of (c). Show that $\text{Var}(\overline{W}_n) \geq \text{Var}(Z)$. When is this inequality strict?

6. (a) Verify (4). *Hint*: $E(X^2) = \mu_X^2 + \sigma_X^2$.

   (b) Obtain the analog of (4) for sampling without replacement from a finite population.

37. MARKOV AND CHEBYSHEV INEQUALITIES; WEAK LAW OF LARGE NUMBERS

Moments of rv's can be used to obtain bounds on probabilities of certain intervals. These bounds are used primarily for proving important limit theorems and for other analytic purposes. The numerical results obtained from applying these bounds to specific models have only limited practical interest. The required moments are all assumed to exist.

We begin with an upper bound on $P[X \geq a]$.

Theorem 1. (Markov inequality) *Let X be a nonnegative rv. For all $a > 0$,*

$$P[X \geq a] \leq \frac{E(X)}{a}. \tag{1}$$

Proof. Since X is nonnegative,

$$X = XI_{[X<a]} + XI_{[X\geq a]} \geq XI_{[X\geq a]} \geq aI_{[X\geq a]}.$$

Hence $E(X) \geq aE(I_{[X\geq a]}) = aP[X \geq a]$.

Example 1. The bound given by (1) is very crude, but universal. Let $E(X) = E(Y) = 1$ and let $a = 2$. From (1) we have $P[X \geq 2] \leq 1/2$ and $P[Y \geq 2] \leq 1/2$. Let X and Y be as follows:

$$P[X = 0] = \frac{1}{2} = P[X = 2]$$

$$P[Y = 0] = \frac{n-1}{2n}, \quad P\left[Y = \frac{2n}{n+1}\right] = \frac{n+1}{2n},$$

so that $E(X) = E(Y) = 1$. For X, the bound in (1) with $a = 2$ is attained. However, $P[Y \geq 2] = 0$ for all $n > 1$.

Example 2. The mean income in a certain population is \$20,000. Thus (1) guarantees that no more than 33% of the population have incomes of \$60,000 or more. Without additional information about the population, we cannot tell whether this bound is achieved or if the inequality is strict (see Problem 6).

Corollary 1. *For any rv X and a, $k > 0$,*

$$P[|X| \geq a] \leq \frac{E(|X|^k)}{a^k}. \tag{2}$$

Proof. Apply (1) to the event $[|X|^k \geq a^k]$. The remaining details are left as a problem.

Corollary 1 provides an elementary proof of the very useful Chebyshev inequality, which gives a bound on probabilities in terms of the standard deviation.

Corollary 2. (Chebyshev inequality) *If $E(X) = \mu$ and $\mathrm{Var}(X) = \sigma^2$, then for any $a > 0$,*

$$P[|X - \mu| \geq a] \leq \frac{\sigma^2}{a^2}. \tag{3}$$

Proof. In (2), replace X by $X - \mu$ and set $k = 2$.

An alternative form of (3), obtained by setting $a = \sigma t$, is

$$P[|X^*| \geq t] = P[|X - \mu| \geq t\sigma] \leq \frac{1}{t^2}, \tag{4}$$

where X^* is the standardization of X. If $t = a/\sigma < 1$, the bound is greater than 1 and is useless.

Example 3. In the population of Example 2, suppose that the standard deviation of incomes is \$5000. Then from (4) with $t = 4$, we see that at most 6.25% of the population can have incomes of \$40,000 or more. This is an improvement over the bound in Example 2. What can be said if the standard deviation is \$1,000?

A simple application of the Chebyshev inequality yields an important result concerning the long-run stability of arithmetic means and relative frequencies in random samples. This result forms the basis of the frequency interpretation of probability and is a confirmation of inferences suggested by empirical observation of repeatable experiments.

Theorem 2. (Weak law of large numbers) *Let $X_1, \ldots, X_n$ be iid rv's for which $E(X_i) = \mu$ and $\mathrm{Var}(X_i) = \sigma^2$. Then for any $a > 0$,*

$$\lim_{n \to \infty} P[|\overline{X}_n - \mu| \geq a] = 0. \tag{5}$$

Proof.   By (36.2) and (36.3), $E(\overline{X}_n) = \mu$ and $\text{Var}(\overline{X}_n) = \sigma^2/n$. Applying the Chebyshev inequality yields

$$0 \le P[|\overline{X}_n - \mu| \ge a] \le \frac{\sigma^2/n}{a^2} \longrightarrow 0$$

as $n \to \infty$.

Historically, the original version of Theorem 2 was the following corollary.

Corollary 3.   *In a sequence of n Bernoulli trials, let Y_n be the relative frequency of occurrences of the event A. Then for all $a > 0$,*

$$\lim_{n \to \infty} P[|Y_n - P(A)| \ge a] = 0. \tag{6}$$

Proof.   Let B_i be the event that A occurs on the ith trial. Then I_{B_1}, I_{B_2}, . . . satisfy the conditions of Theorem 2 with $E(I_{B_1}) = P(A)$. Furthermore, $Y_n = \sum_{i=1}^{n} I_{B_i}/n$. Hence (6) is a special case of (5).

Corollary 3 can be used in choosing between competing probabilities for an event A. Suppose that a choice must be made from among several specified theoretical probability models. Corollary 3 suggests choosing a model that comes close to the observed relative frequencies.

In Theorem 2 we assumed existence of $\text{Var}(X_i)$. Weaker conditions can be formulated that yield (5). In fact, it would suffice that $E(X_i)$ exist, but the proof in that more general case requires more advanced methods.

As a final illustration of the application of the Chebyshev inequality, consider the following example.

Example 4.†   Prior to elections the TV networks present polling results and, typically, announce that the accuracy is to within 3%. In the notation of Corollary 3, this means that $P[|Y_n - p| \le .03] \ge .95$, where $p = P(A)$, the proportion of those voters who favor candidate A. Since $\text{Var}(Y_n) = pq/n$, where $q = 1 - p$ (see GP 1), it follows that $P[|Y_n - p| \le .03] \ge 1 - pq/[(.03)^2 n]$, so that n is required to satisfy $pq/[(.03)^2 n] \le .05$ and hence $n \ge 200{,}000\, pq/9$. Since pq is at most $1/4$, the requirement is that $n \ge 5556$. In fact, preelection pools are rarely based on sample sizes greater than 1500. The matter will be discussed further in a later section.

Summary

Markov inequality: For a nonnegative rv X, $P[X \geq a] \leq \mu_X/a$.
Chebyshev inequality: $P[|X - \mu_X| \geq a] \leq \sigma_X^2/a$.
Weak law of large numbers: $\lim_{n \to \infty} P[|\overline{X}_n - \mu| \geq a] = 0$, where $\mu = E(X_i)$.

Problems

1. Consider an experiment and some event A in that experiment.
   (a) Give at least two bounds for the probability of k or more occurrences of A in n trials.
   (b) For which bounds in part (a) do you need to assume independence?
   (c) For independent trials, compare your bound in part (a) with the exact probability when $k = n$.
2. Under the setup of Problem 1, assume that the trials are independent and that $P(A) = 1/2$.
   (a) Obtain a bound for the probability that the number of occurrences of A differs from $n/2$ by at least a.
   (b) Use your bound in part (a) to obtain a bound on the probability of at least k occurrences for A for integer $k > n/2$.
   (c) Compare the bound in part (b) with the bounds in Problem 1(a).
   (d) Compare the bound in part (b) with the exact probability when $k = n$.
3. Let $b > 0$, $P[X = b] = p$ and $P[X = 0] = q = 1 - p$. For what values for $a > 0$ is (1) attained?
4. Let X and Y be rv's for which $\text{Var}(X) < \text{Var}(Y)$, but $E(X) = E(Y) = \mu$. Does this imply that $P[|X - \mu| \geq a] \leq P[|Y - \mu| \geq a]$ for all $a > 0$? Prove or give a counterexample.
5. Suppose that X has mean 20 and variance 4. Let $Y = X/4 + b$. Determine the Chebyshev bound for $P[b \leq Y \leq b + 10]$.
6. Under the conditions in Example 2, exhibit a population model such that the bound would be achieved.
7. Complete the proof of Corollary 1.
8. In Examples 2 and 3, suppose that a sample of size n is to be chosen to estimate the mean income. How large must n be if the mean income is to be estimated to within \$500 with probability at least .99%
9. Let $B_1, B_2, \ldots$ be pairwise independent events and set $P(B_i) = p_i$. Let Y_n be the proportion of the events $B_1, \ldots, B_n$ that occur. Show that $\lim_{n \to \infty} p_n = p_0$ implies that $\lim_{n \to \infty} P[|Y_n - p_0| \geq a] = 0$. See also GP 7.

GENERAL PROBLEMS

1.† Let $A_1, \ldots, A_n$ be independent events, each with probability $p = 1 - q$. Let $S_n = \sum_{i=1}^{n} I_{A_i}$. Show that $E(S_n) = np$ and $\text{Var}(S_n) = npq$.

2.† A population consists of M individuals and M_1 have property π. A random sample of size n is taken without replacement. Let A_i be the event that the ith individual in the sample has property π. Set $S_n = \sum_{i=1}^{n} I_{A_i}$ and $M_2 = M - M_1$. Show that $E(S_n) = nM_1/M$ and $\text{Var}(S_n) = [(M - n)/(M - 1)] \cdot (nM_1M_2/M^2)$.

3. Let $X_1, \ldots, X_n$ be iid rv's and assume that $\text{Var}(X_i) = 18$. We wish to estimate $\mu = E(X_i)$.
   (a) Suggest an estimate $a(X_1, \ldots, X_n)$.
   (b) How large must n be according to the Chebyshev inequality so that $P[|a(X_1, \ldots, X_n) - \mu| \geq 3] \leq 1/10$?

4. Let $I_{A_1}, \ldots, I_{A_n}$ be iid indicators and suppose that we wish to estimate $p = P(A_i)$.
   (a) Suggest an estimate $a(I_{A_1}, \ldots, I_{A_n})$.
   (b) How large must n be so that $P[|a(I_{A_1}, \ldots, I_{A_n}) - p| \geq .05] \leq .01$?

5. Consider the rat of Section 19. This rat runs a T-maze. On any given run, the left arm contains food with probability p_1 and the right arm contains food with probability p_2. For simplification assume that $P_1 = p$ and $p_2 = q = 1 - p$. The placement of food on successive runs is done independently. Suppose that the rat chooses an arm at random for the first run and follows the play the winner strategy thereafter. That is, it stays with an arm if it contained food on the previous run and switches arms otherwise.
   (a) Let L_i be the event the rat chooses the left arm on the ith run. Show, by induction, that $P(L_i \mid L_1) = p$ for $i = 2, 3, \ldots$.
   (b) Use part (a) to show that $P(L_i) = p$ for $i = 2, 3, \ldots$ and that the L_i are independent.
   (c) Let A_i be the event that the rat obtains food on the ith run. Show that $P(A_1) = 1/2$ and $P(A_i) = 1 - 2pq$ for $i = 2, 3, \ldots$.
   (d) Show that

$$P(A_i A_j) = \frac{1 - 2pq}{2} \qquad \text{if } i = 1; \ j = 2, 3, \ldots$$

$$= 1 - 3pq \qquad \text{if } i = 2, 3, \ldots; \ j = i + 1$$

$$= (1 - 2pq)^2 \qquad \text{if } i = 2, 3, \ldots;$$

$$j = i + 2, i + 3, \ldots.$$

For which i, j are A_i and A_j independent?

    (e) Let X be the number of times the rat obtains food in the first n runs. Write an expression for X in terms of the indicators of the A_i and compute $E(X)$ and $\text{Var}(X)$.

6. Let $X_1, \ldots, X_n$ be iid, positive rv's. Set $S_n = \Sigma_{i=1}^n X_i$ and $Y_i = X_i/S_n$. Determine $E(Y_i)$.

7. Let $B_1, B_2, \ldots$ be events such that B_i is unfavorable to B_j for every $i \neq j$ and $P(B_i) = p$ for all $i = 1, 2, \ldots$. Generalize Corollary 37.3 by showing that $\lim_{n \to \infty} P[|Y_n - p| \geq a] = 0$ for all $a > 0$ where $Y_n = \Sigma_{i=1}^n I_{B_i}/n$.

8. Generalize Theorem 37.2 to the case of negatively correlated X_i's.

9. Let $X_1, X_2, \ldots$ be iid continuous rv's. We say that X_i is a *record value* if $i = 1$ or if $X_i > \max(X_1, \ldots, X_{i-1})$.

    (a) Let Y_n be the number of record values among $X_1, \ldots, X_n$ and find $E(Y_n)$.

    (b) Let T_k be the waiting time (number of X_i's) for the kth record value and let $\Delta_k = T_k - T_{k-1}$ ($k \geq 1$). Examine $E(\Delta_k)$ and $E(T_k)$.

REFERENCES

DeGroot, Morris H. (1986). *Probability and Statitics*, 2nd ed. Reading, Mass.: Addison-Wesley Publishing Co., Inc.

Hoel, Paul G., Port, Sidney C., and Stone, Charles J. (1971). *Introduction to Probability Theory*. Boston: Houghton Mifflin Company.

Woodroofe, Michael(1975). *Probability with Applications*. New York: McGraw-Hill Book Company.

7.

Special Models on $\mathcal{R}^n$ and Their Applications

38. INTRODUCTION

The models in this chapter are special in that they are the most widely used and hence the most extensively studied among the models encountered in applications. They also serve to delineate some of the basic issues and techniques involved in the construction of a probability model for a specific circumstance.

In some situations a theoretical model for a random variable (or vector) is used because it emerges by deduction from a basic set of assumptions about certain events (e.g., independence, equal likelihood). In others, a model may be chosen merely because it appears to fit an historical record of observed quantities, or because it reflects the users' opinions about the quantity(ies) in question.

Ultimately, the decision as to which model to use for a particular "real" problem will depend on a combination of mathematical and nonmathematical considerations. Previous chapters dealt with issues which, although motivated by empirical phenomena, are primarily mathematical (such as probability calculus and representation of models). The nonmathematical ones involve such fuzzy notions as "convenience," "usefulness," "analytical simplicity," and "approximation," among others. As a consequence, the user's judgment and purpose, not to mention experience, are going to play important roles in the choice of model.

For instance, convenience and simplicity argue strongly in favor of using continuous models in some cases where a discrete one would seem to be appropriate. Suppose, for example, that we want a model for the reported incomes of U.S. taxpayers for the year 1980. Because money is a discrete quantity (everyone's reported income is some integer number of cents), we could conceivably use a discrete model to describe it. To do so might require a scale from 0 to, say, 10^{10} cents, along with the proportions of individuals with the various incomes. Needless to say, this would be a most unwieldy, cumbersome, and unnecessarily detailed representation. Such a situation is better treated by approximating the rv in question with a continuous one. As another example, imagine that the monthly demand for a certain commodity is an rv with integer values from 0 to 50 and that we want a model for next year's average monthly demand (or total demand). Even if we are willing to treat monthly demand as a discrete rv, the average of 12 of its values will generally be much too complicated mathematically to treat as a discrete quantity and we would look for some continuous model as an approximation. Finally, it should be noted that continuous models are used for familiar physical quantities such as length, distance, weight, temperature, velocity, and time, even though these may be reported on a discrete scale. In those cases the discreteness is attributable to the limited precision of the measuring device and not to the quantity itself.

In our view, it is best to think of all probability models as conditional models—conditioned on assumptions (and/or parameters) that may be (and generally are) subject to varying degrees of uncertainty when applied to real events or quantities. Methods of coping with these uncertainties are matters of extensive debate among users, researchers, and philosophers of probability. The question of which specific model (i.e., specific enough to produce numbers) to use for a given situation is a subject, among others, which properly belongs to the field of "statistics," to be explored in a later volume.

What we present here are, in fact, parametric families of probability models on $\mathcal{R}^n$. This means that to obtain numerical values for the probabilities of events related to a particular family it will generally be necessary to specify the value(s) of its parameter(s).

Each family will be given a name. The corresponding random variable (or vector), distribution function, density, or probability function will be given the same name. A table at the end of this chapter (pp. 333–334) summarizes these families of models.

Because of their prevalence in applications and the analytic complexity of some of the formulas, many of the models we shall discuss have a set of tables associated with them. Some of these are to be found in the

Appendix and we shall explain their use as we go along. Commercial software exists for computing these probabilities, although in some cases simple programs can be written based on formulas to be presented. Some of these families or specific members of them have been introduced in examples and problems in earlier chapters.

39. MODELS ASSOCIATED WITH FREQUENCIES

Frequency of occurrence or, simply, the number of times something happens is a primitive and widespread form of measurement. When appropriately scaled it yields relative frequency which, as we have stated, is one of the empirical touchstones of probability theory. The first families of theoretical models that will be developed are for integer-valued (hence discrete) rv's and are often applied to counts such as traffic accidents, team victories, defective parts, births of a given sex, votes for a candidate, jurors from a specified group, floods, and many others.

We begin by considering a collection of events, $\alpha = \{A_1, \ldots, A_n\}$. In Section 4 we introduced the idea of the partition generated by α. This partition can be represented by the set of all possible values of the random indicator vector $(I_{A_1}, \ldots, I_{A_n})$. Every member of the partition is identified with a sequence of 1's (where A_i occurs) and 0's (where A_i^c occurs) and any rv defined in terms of the A_i's can be expressed as a function of the random vector $(I_{A_1}, \ldots, I_{A_n})$. Currently, our interest is in the rv corresponding to the number of occurrences of the A_i's. We denote this by S_n and, in terms of indicators,

$$S_n = \sum_{i=1}^{n} I_{A_i}$$

(i.e., the number of 1's in the vector of indicators).

There is a general, rather forbidding expression for the probability function of S_n in terms of all possible intersections of the events in α. Our aim, however, is to develop formulas based on additional simple assumptions which greatly reduce the complexity of the resulting model. Before doing so, we give three important relations involving S_n which hold in complete generality. These are expressions for its mean and its variance and a recursion property. Each of them exploits the fact that S_n is a sum of random variables (see Sections 31, 35, and 36 and GP 2.5).

$$E(S_n) = \sum_{i=1}^{n} E(I_{A_i}) = \sum_{i=1}^{n} P(A_i) \tag{1}$$

$$\text{Var}(S_n) = \sum_{i=1}^{n} \text{Var}(I_{A_i}) + 2 \sum_{i<j} \sum \text{Cov}(I_{A_i}, I_{A_j})$$

$$\tag{2}$$

$$= \sum_{i=1}^{n} P(A_i)P(A_i^c) + 2 \sum_{i<j} \sum [P(A_iA_j) - P(A_i)P(A_j)]$$

$$P[S_n = k] = P[S_{n-1} = k - 1, I_{A_n} = 1] + P[S_{n-1} = k, I_{A_n} = 0]$$

$$= P(A_n \mid S_{n-1} = k - 1)P[S_{n-1} = k - 1] \tag{3}$$

$$+ P(A_n^c \mid S_{n-1} = k)P[S_{n-1} = k].$$

The mean involves only the marginal probabilities of the A_i's and the only additional quantities required for computing the variance are the probabilities of pairwise intersections. Equation (3) was the subject of GP 2.5 and will be used below to prove an important limit theorem. It can, in fact, be used to produce the probability function of S_n under certain conditions.

We now turn to some special models for S_n.

A. Binomial Models

We have shown that when rv's are iid with common probability function g, their joint probability function f assumes a simple form:

$$f(x_1, x_2, \ldots, x_n) = \prod_{i=1}^{n} g(x_i). \tag{4}$$

Binomial Model

The model for S_n which results from the assumption that the I_{A_i}'s are iid is called a *binomial model*. This assumption on the I_{A_i}'s is equivalent to the assumption that the events A_i, $i = 1, \ldots, n$, are independent and all have the same probability, $P(A_i) = p$, $i = 1, \ldots, n$. A binomial model is completely determined by, and varies with, the specific values of n and p. The set of all binomial models is therefore a two-parameter family of models (one model for each value of n and p).

If we call the occurrence of A_i "success on the ith trial," then S_n is the number of successes in n Bernoulli trials. In most applications, the trials are repetitions of the same experiment, B is some event in its sample space, and A_i is the event "B occurs on the ith trial." The labeling of a given event or its complement as "success" is a matter of convenience.

In connection with binomial models we use the following notation:

$\mathscr{B}(n, p)$ denotes the binomial model given n, p.

$\mathscr{B} = \{\mathscr{B}(n, p): n$ is a positive integer, $p \in [0, 1]\}$ denotes the family of all binomial models.

$b(k \mid n, p)$ denotes the binomial probability function evaluated at the integer k given n, p.

b denotes the family of binomial probability functions.

$B(t \mid n, p)$ denotes the binomial distribution function evaluated at t given n, p.

B denotes the family of binomial distribution functions.

This pattern of script, lowercase, and uppercase letters will be followed for most other models. To avoid cumbersome notation, in these parametric families we will not make a notational distinction between a function and its values. Context will make clear what is meant.

The probability functions can be derived from (3) as suggested in GP 2.5(b). Another derivation follows.

From (4) we get the following expression for the joint probability function of $I_{A_1}, \ldots, I_{A_n}$ when they are iid:

$$f(x_1, \ldots, x_n) = p^{\sum_{i=1}^{n} x_i} q^{n - \sum_{i=1}^{n} x_i} \qquad \text{if } x_i = 0, 1; \quad i = 1, \ldots, n, \quad (5)$$

where $0 \le p = 1 - q \le 1$.

Inspection of (5) reveals the important fact that for given n and p, f depends only on $\sum x_i$ [i.e., the number of 1's in the vector $\mathbf{x} = (x_1, x_2, \ldots, x_n)$]. The event $[S_n = k]$ refers to the collection of all vectors $\mathbf{x}$ for which $\sum x_i = k$ and, as can be seen from (5), each such vector has probability $p^k q^{n-k}$. The probability function of S_n is therefore

$$P(S_n = k \mid p) = b(k \mid n, p) = cp^k q^{n-k} \qquad \text{if } k = 0, 1, \ldots, n,$$

where c is the number of vectors $\mathbf{x}$ with the property that $\sum x_i = k$. The number of such vectors is the number of subsets of size k (positions for the 1's) from a set of size n (the number of available positions). Thus $c = \binom{n}{k}$ and hence

$$b(k \mid n, p) = \binom{n}{k} p^k q^{n-k} \qquad \text{if } k = 0, 1, 2, \ldots, n. \quad (6)$$

The mean and variance for this model are

$$E(S_n \mid p) = np \quad (7)$$

and

$$\text{Var}(S_n \mid p) = npq \quad (8)$$

as obtained in GP 6.1. Alternatively, (7) and (8) can be obtained from (6) (Problem 15).

Example 1. A fair six-sided die is rolled (independently) six times. The event A_i occurs if the ith roll produces the value i. For example, in the sequence 3 1 3 6 5 2 the events A_3, A_5 occurred while A_1, A_2, A_4, A_6 did not. Since the rolls are independent and $P(A_i) = 1/6$, $i = 1, 2, \ldots, 6$, the number, S_6, of occurrences of a given A_i's obeys a $\mathscr{B}(6, 1/6)$ model, so that

$$P\left(S_6 = k \mid p = \frac{1}{6}\right) = b\left(k \mid 6, \frac{1}{6}\right) = \binom{6}{k}\left(\frac{1}{6}\right)^k\left(\frac{5}{6}\right)^{6-k}$$

$$\text{if } k = 1, \ldots, 6.$$

Table 1 exhibits $b(k \mid 6, 1/6)$ to three decimal places. Furthermore,

$$E\left(S_6 \mid p = \frac{1}{6}\right) = 6 \cdot \frac{1}{6} = 1$$

and

$$\text{Var}\left(S_6 \mid p = \frac{1}{6}\right) = 6 \cdot \frac{1}{6} \cdot \frac{5}{6} = \frac{5}{6}.$$

Example 2. A device consists of five independent components, each with probability .9 of functioning. Let S_5 denote the number of functioning components and let U_5 denote the number of nonfunctioning components. The model for S_5 is $\mathscr{B}(5, .9)$; the model for U_5 is $\mathscr{B}(5, .1)$. Since $S_5 + U_5 = 5$, the value of S_5 determines that of U_5, and vice versa. Either model can be used to study the probability that any number of components function. The reliability of the device is obtained in three special cases below.

(a) *Series hookup:* Device functions if, and only if, every component functions.

$$P(\text{device functions}) = P(S_5 = 5 \mid p = .9) = P(U_5 = 0 \mid p = .1)$$

$$= b(5 \mid 4, .9) = b(0 \mid 5, .1) = (.9)^5 = .59049$$

Table 1

k	0	1	2	3	4	5	6
$b\left(k \mid p = \frac{1}{6}\right)$	.335	.402	.201	.054	.008	.001	.000

(b) *Parallel hookup*:   Device functions if, and only if, at least one component functions.

$$P(\text{device functions}) = P(S_5 \geq 1 \mid p = .9) = 1 - B(0 \mid 5, .9)$$
$$= P(U_5 \leq 4 \mid p = .1) = B(4 \mid 5, .1) = 1 - (.1)^5$$
$$= .99999.$$

(c) *A k out of n hookup with k = 3 and n = 5*:   Device functions if, and only if, at least three components function.

$$P(\text{device functions}) = P(S_5 \geq 3 \mid p = .9) = 1 - B(2 \mid 5, .9)$$
$$= P(U_5 \leq 2 \mid p = .1) = B(2 \mid 5, .1)$$
$$= b(0 \mid 5, .1) + b(1 \mid 5, .1) + b(2 \mid 5, .1)$$
$$= (.9)^5 + \binom{5}{1}(.9)^4(.1) + \binom{5}{2}(.9)^3(.1)^2 = .99144.$$

In each of (a), (b), and (c) exact answers are given to five decimal places. In practice we would settle for two-, perhaps three-decimal-place accuracy unless extreme precision were warranted by the context.

Direct calculations of binomial probabilities can become quite laborious. (See the note following Theorem 2.) If, for instance, in Example 2 we had 20 components and, say, a 12 out of 20 hookup, unaided computation of the reliability of the device would take a considerable amount of time and patience. To reduce much of the work, and because of its frequent applications, there exist extensive tables and computer software for $\mathcal{B}(n, p)$ models. In Table A of the Appendix the values of the distribution functions B for selected n and p are given. For each pair (n, p) that appears in the table, the values of $B(k \mid n, p)$ are listed for all k for which $b(k \mid n, p)$ is different from zero to three-decimal-place accuracy. For the missing k values, $B(k \mid n, p) = 0$, or 1 to three places. Thus part (c) of Example 2 requires $B(2 \mid 5, .1)$, given in the first binomial table to three decimals. The tabled value is .991.

These tables may also be used to obtain individual probabilities, $b(k \mid n, p)$, by differencing. Thus $b(2 \mid 5, .1) \doteq .0729$ (as calculated above), and from the table,

$$b(2 \mid 5, .1) = B(2 \mid 5, .1) - B(1 \mid 5, .1) = .991 - .919 = .072.$$

The discrepancy is due to rounding errors.

In Table A, values of B for $p > .5$ are not given, nor are they required. As illustrated in Example 2, to each $\mathcal{B}(n, p)$ model there corresponds the

$\mathcal{B}(n, q)$ model related by the fact that if $S_n = \sum_{i=1}^n I_{A_i}$ and $U_n = \sum_{i=1}^n I_{A_i^c}$, then $S_n + U_n = n$. The iid conditions on the I_{A_i}'s apply equivalently to the $I_{A_i^c}$'s, so that if S_n is $\mathcal{B}(n, p)$, then U_n is $\mathcal{B}(n, q)$. Thus

$$P(S_n \le r \mid p) = B(r \mid n, p) = P(U_n \ge n - r \mid q)$$
$$= 1 - B(n - r - 1 \mid n, q). \tag{9}$$

This correspondence between $\mathcal{B}(n, p)$ and $\mathcal{B}(n, q)$ eliminates the necessity of tabling B (or b) for values of $p > .5$.

Another type of calculation for which tables are useful is the determination of values of the rv's with specified tail probabilities (quantiles).

Example 3. A certain examination consists of 50 true–false questions. The objective is to establish a reasonable standard for passing in light of the fact that a student taking the exam is at liberty to use a guessing strategy, completely unrelated to his knowledge of the material being tested. One such strategy is to guess, independently and at random, the answer to each question. The score, S_{50}, that a student would achieve under such "pure guessing" would have a $\mathcal{B}(50, 1/2)$ model and would be expected to be near 25. To counter this, a passing score, k_0, is chosen such that a student adopting this strategy will have only a minute probability, say .01, of attaining it. Thus we want the smallest k_0 such that

$$P\left(S_{50} \ge k_0 \mid p = \frac{1}{2}\right) = 1 - B\left(k_0 - 1 \mid 50, \frac{1}{2}\right) \le .01$$

or, equivalently,

$$B\left(k_0 - 1 \mid 50, \frac{1}{2}\right) \ge .99.$$

Hence $k_0 - 1$ is the 99th percentile of $\mathcal{B}(50, 1/2)$. The solution of this inequality is obtained easily through the use of Table A with $n = 50$ and $p = .5$. Go down the column headed $p = .5$ until reaching the first r value for which $B(r \mid 50, .5) \ge .99$. This yields $r = 33 = k_0 - 1$ and establishes that the required passing grade should be set at 34. Using the less stringent probability requirement $B(k_0 - 1 \mid 50, .5) \ge .9$ yields $r = 30$, which, in turn, establishes $k_0 = 31$. Just over 60% correct would be required for passing.

What about other guessing strategies? It may, for example, occur to some students to try to guess the pattern of T's and F's rather than guessing each answer independently. Still others may write down all T's, in hope that there are likely to be more correct statements than false ones. In fact, the examination can be so designed that whatever guessing scheme a student uses, his score, S_{50}, will be $\mathcal{B}(50, 1/2)$. This is accomplished by first choosing a pattern of T's and F's independently and at random for each of the 50 questions, then choosing questions so that the correct answers are those that were selected by this process. Any pure guessing scheme will then produce a score S_{50} that obeys the $\mathcal{B}(50, 1/2)$ model.

Example 4. A machine produces items of variable quality. Suppose that each item produced is classified as either "defective" or "acceptable" and that p is the probability that the ith run (trial) produces a defective (A_i). The machine is considered to be operating properly if $p \leq p_0$, where p_0 is a small number representing the threshold of acceptable quality. Consider a sequence of items produced by the machine and the associated indicators $I_{A_1}, \ldots, I_{A_n}$. This is a sequence of Bernoulli trials if the events $A_1, \ldots, A_n$ can be assumed to be independent. These assumptions may or may not be realistic. If the machinery is subject to significant wear during the production run, this would argue against constancy of p from item to item. If, say, the wear on the machine depends on whether it produced a defective item or an acceptable one, this would argue against the independence assumption.

In using a Bernoulli sequence here, we are therefore assuming no wearout and no memory for the system. Then if the machine is operating properly, $\sum_{i=1}^{n} I_{A_i}$ will have a $\mathcal{B}(n, p)$ model with $p \leq p_0$.

The following theorems concerning binomial models and rv's are stated here for future reference. The proofs are left to the reader.

Theorem 1. *Each $\mathcal{B}(n, 1/2)$ model is symmetric about the value $n/2$. That is,*

$$b\left(\frac{n}{2} + t \mid n, \frac{1}{2}\right) = b\left(\frac{n}{2} - t \mid n, \frac{1}{2}\right) \tag{10}$$

for every t or, equivalently,

$$P\left(S_n \leq \frac{n}{2} - t \mid p = \frac{1}{2}\right) = P\left(S_n \geq \frac{n}{2} + t \mid p = \frac{1}{2}\right). \tag{11}$$

We call the $\mathcal{B}(n, 1/2)$ model the *symmetric binomial* model.

Theorem 2. *The probability function of a $\mathfrak{B}(n, p)$ model satisfies the recursion property*

$$b(k + 1 \mid n, p) = \frac{(n - k)p}{(k + 1)q} b(k \mid n, p), \qquad k = 0, 1, \ldots, n. \quad (12)$$

Note: Theorem 2 gives an algorithm that can be used in computing. Begin with $b(0 \mid n, p) = q^n$ and, then, recursively apply (12). This was the method used to compute Table A. Software is available for both computers and programmable calculators.

Theorem 3. *Let S_n, S'_m denote independent rv's such that S_n is $\mathfrak{B}(n, p)$ and S'_m is $\mathfrak{B}(m, p)$. Then the rv $S_n + S'_m$ has a $\mathfrak{B}(n + m, p)$ model.*

B. Hypergeometric Models

Typically, the object of sampling is to gain information concerning the population from which the sample is drawn. Since sampling *with* replacement (equivalent to sampling from an infinite population) entails possible redundancies (some individuals may be drawn more than once), the usual mode of sampling is *without* replacement. If the objects being sampled belong to one of two possible categories, the appropriate model for the sample composition belongs to the three-parameter family of hypergeometric models which we proceed to exhibit.

Suppose that a population consists of M individuals, of which M_1 have the property π and $M_2 = M - M_1$ do not. Consider a random sample of size n. Let A_i denote the event that the ith individual drawn has property π and $S_n = \Sigma_{i=1}^n I_{A_i}$. If sampling is with replacement, S_n is a $\mathfrak{B}(n, M_1/M)$ rv.

For the remainder of this subsection assume that sampling is *without* replacement. The fact that $P(A_i) = p = M_1/M$ has been noted earlier (see Theorem 13.1), but the events $A_1, \ldots, A_n$ are not independent.

Hypergeometric Model

The model for S_n in this case is denoted by $\mathfrak{H}(n, M_1, M_2)$ and is called a *hypergeometric model*. The family of hypergeometric models, the corresponding family of probability functions, and the corresponding family of distribution functions will be denoted by $\mathfrak{H}$, h, and H, respectively.

The joint probability function, f, of the indicators $I_{A_1}, \ldots, I_{A_n}$ can be derived either by counting the number of ordered samples with specified composition or by using the multiplication rule for events (12.10). Either method yields

$$f(x_1, \ldots, x_n) = \frac{(M_1)_k (M_2)_{n-k}}{(M)_n} = \frac{\binom{M-n}{M_1-k}}{\binom{M}{M_1}}, \tag{13}$$

where $k = \Sigma_{i=1}^n x_i$; $x_i = 0$ or 1. [Recall that $(m)_r = m!/(m-r)!$.] The numerator is the number of selections for which k specified draws yield an individual with property π, while the remaining $n - k$ do not. The denominator is the total number of ordered selections of length n from a set of size M.

The event $[S_n = k]$ corresponds to the set of all $\mathbf{x} = (x_1, \ldots, x_n)$ for which $\Sigma_i x_i = k$ and, as was shown earlier in the binomial case, there are $\binom{n}{k}$ such vectors, each with probability given by (13). Thus

$$P(S_n = k \mid M_1, M_2) = h(k \mid n, M_1, M_2)$$

$$= \binom{n}{k} \frac{(M_1)_k (M_2)_{n-k}}{(M)_n}$$

$$= \frac{\binom{n}{k}\binom{M-n}{M_1-k}}{\binom{M}{M_1}} \qquad \text{if } k = 0, \ldots, n \tag{14}$$

where $M = M_1 + M_2$.

The usual form of the hypergeometric probability function is obtained by counting the number of subsets of size n and the number of such subsets containing k with property π (Problem 16). This form is

$$h(k \mid n, M_1, M_2) = \frac{\binom{M_1}{k}\binom{M_2}{n-k}}{\binom{M}{n}} \qquad \text{if } k = 0, 1, \ldots, n. \tag{15}$$

The mean and variance for this model with $p = M_1/M = 1 - q$ are

$$E(S_n \mid M_1, M_2) = \frac{nM_1}{M} = np \tag{16}$$

and

$$\text{Var}(S_n \mid M_1, M_2) = \frac{M - n}{M - 1} \cdot \frac{nM_1M_2}{M^2} = \left(\frac{N - n}{M - 1}\right) npq \tag{17}$$

as obtained in GP 6.2. [See also (36.6).]

Comparing (16) with (7), observe that the expected value of S_n is the same for sampling with or without replacement. Comparing (17) with (8), observe that sampling without replacement decreases $\text{SD}(S_n)$ by the factor $\sqrt{(M - n)/(M - 1)}$, which, as defined in Section 36, is the finite population correction factor.

Example 5. In a group of 15 equally qualified applicants for jobs, 10 are female and 5 are male. Of the 15, six are selected at random to staff the available positions. If S_6 is the number of selected persons who are female, then

$$P(S_6 = k \mid M_1 = 10, M_2 = 5) = h(k \mid 6, 10, 5)$$

$$= \frac{\binom{10}{k}\binom{5}{6 - k}}{\binom{15}{6}} \qquad \text{if } k = 1, 2, \ldots, 6.$$

Note that $P(S_6 = 0 \mid M_1 = 10, M_2 = 5) = 0$. Table 2 gives the other probabilities:

Table 2

k	1	2	3	4	5	6
$h(k \mid 6, 10, 5)$	$\dfrac{2}{1001}$	$\dfrac{45}{1001}$	$\dfrac{240}{1001}$	$\dfrac{420}{1001}$	$\dfrac{252}{1001}$	$\dfrac{42}{1001}$

Example 6. Of 20 students enrolled in Astrology 101, 10 are randomly selected to receive a new self-taught, computerized training program and the remaining 10 will be taught by the usual lecture format. At the end of the course, the 20 students are given a common examination. The result of the examination is that 12 pass and 8 fail. Under the assumption that the two methods are equally effective, the number of passing students (or failing students) who were taught by the usual method (or by the new method) is a rv with a hypergeometric probability function (see, e.g., Problem 20). Let S_{10} be the number of passing students taught by the new method. Then S_{10} has an $\mathcal{H}(10, 12, 8)$ model and

$$h(9 \mid 10, 12, 8) = \frac{\binom{12}{9}\binom{8}{1}}{\binom{20}{10}} = .010$$

$$h(8 \mid 10, 12, 8) = \frac{\binom{12}{8}\binom{8}{2}}{\binom{20}{10}} = .075.$$

The value of $h(10 \mid 10, 12, 8)$ is negligibly small. Thus $P(S_{10} \geq 9 \mid M_1 = 12, M_2 = 8)$ is approximately .010. If, in fact, the observed value of S_{10} is 9, the assumption that the two methods are equally effective would be dubious.

Example 7. Suppose that 60% of all students in a certain group are men. A pollster interviews 25 students selected at random from the group without replacement. What is the probability that a majority of the sampled students are women (i.e., that the pollster is misled by the sample)? Letting S_{25} be the number of men in the sample, this probability is

$$P(S_{25} \leq 12 \mid M_1 = .6M, M_2 = .4M) = H(12 \mid 25, .6M, .4M)$$

$$= \sum_{k=0}^{12} \frac{\binom{.6M}{k}\binom{.4M}{25-k}}{\binom{M}{25}}, \quad (18)$$

where it has been assumed that $M_1 = .6M$ is an integer.

For a group of size $M = 100$, we can use (18) to compute this probability. A fast, eight-step BASIC program [using a recursion formula analogous to (12)] for $h(k \mid 25, 60, 40)/h(k - 1 \mid 25, 60, 40)$ (see Problem 22) gives the value 6.47×10^{-3}. However, if the group consists of all students in some state and $M = 1,000,000$, tables do not exist and computation by (18) is too time consuming.

If M is large relative to the sample size, sampling without replacement is very close to sampling with replacement. The theorem that follows demonstrates that for M large, (18) can be approximated by $B(12 \mid 25, .6)$ provided that the $M_i \to \infty$ appropriately.

Theorem 4. *For the hypergeometric probability function $h(k \mid n, M_1, M_2)$,*

$$\lim_{\substack{M \to \infty \\ M_1/M \to p}} h(k \mid n, M_1, M_2) = b(k \mid n, p). \tag{19}$$

Proof. The proof of (19) is by induction on n. For $n = 1$,

$$h(0 \mid 1, M_1, M_2) = \frac{M_2}{M} \longrightarrow 1 - p = q = b(0 \mid 1, p)$$

and

$$h(1 \mid 1, M_1, M_2) = \frac{M_1}{M} \longrightarrow p = b(1 \mid 1, p),$$

so that (19) holds.

Suppose that (19) is true for $n = m$. We prove that (19) is true for $n = m + 1$. Using recursion formula (3) yields

$$h(k \mid m + 1, M_1, M_2) = h(k \mid m, M_1, M_2) \frac{M_2 - (m - k)}{M - m}$$

$$+ h(k - 1 \mid m, M_1 \, M_2) \frac{M_1 - (k - 1)}{M - m}. \tag{20}$$

The inductive hypothesis and (20) yield

$$\lim_{\substack{M \to \infty \\ M_1/M \to p}} h(k \mid m + 1, N, M) = qb(k \mid m, p) + pb(k - 1 \mid m, p)$$

$$= b(k \mid m + 1, p). \tag{21}$$

The last equality follows by applying (3) to Bernoulli trials.

Operationally, Theorem 4 is interpreted to mean that the hypergeometric distribution with M large relative to n is approximated by the binomial distribution with $p = M_1/M$. Since the hypergeometric distribution involves four entries and the binomial distribution involves three entries,

Table 3

		$H(k \mid 20, M_1, M_2)$		
k	$B(k \mid 20, 1/4)$	$M_1 = 25, M_2 = 75$	$M_1 = 75, M_2 = 225$	$M_1 = 300, M_2 = 900$
3	.225	.196	.206	.223
4	.415	.398	.410	.414
5	.617	.624	.619	.618
6	.786	.809	.793	.787
		$H(k \mid 60, M_1, M_2)$		
k	$B(k \mid 60, 1/4)$	$M_1 = 25, M_2 = 75$	$M_1 = 75, M_2 = 225$	$M_1 = 300, M_2 = 900$
3	.232	.120	.204	.225
4	.451	.404	.440	.448
5	.569	.590	.573	.570
6	.775	.881	.799	.781

the latter is easier to table. In fact, far more extensive tables are available for the binomial distribution than for the hypergeometric distribution. Even with a computer, binomial probabilities can be obtained considerably faster than hypergeometric probabilities.

Theorem 4 applies to fixed n. If we let $n \to \infty$ as well, then $h(k \mid n, M_1, M_2)$ and $b(k \mid n, p)$ both converge to zero. We must, then, consider conditions under which the relative error of the approximation converges to zero. This occurs if in addition to the conditions in (18), we have $n^2/M \to 0$.

Table 3 is presented for the purpose of comparing $B(k \mid n, 1/4)$ with $H(k \mid n, M/4, 3M/4)$ for several values of k, $n = 20$ and 60, and $M = 100$, 300, and 1200.

C. Poisson Models

Poisson Family

The *Poisson family*, $\mathscr{P}_o$, is a one-parameter family of probability models, each of which assigns positive probability to every nonnegative integer. We denote a member of the family by $\mathscr{P}_o(\mu)$, where $\mu > 0$ is the value of the parameter. The Poisson probability function is[1]

$$\pi(k \mid \mu) = \frac{\mu^k}{k!} e^{-\mu} \quad \text{if } k = 0, 1, \ldots . \tag{22}$$

---

[1]Since p and P are used in so many contexts, we use the Greek equivalents here.

Selected values of $\Pi(r \mid \mu)$ are given in Table B, from which values of $\pi(k \mid \mu)$ may also be obtained.

Calculations can be facilitated by using the recursion formula

$$\pi(k + 1 \mid \mu) = \frac{\mu}{k + 1} \pi(k \mid \mu). \tag{23}$$

If X is a $\mathfrak{Po}(\mu)$ rv, then its rth factorial moment is

$$E((X)_r) = \sum_{k=0}^{\infty} k(k - 1) \cdots (k - r + 1) \frac{\mu^k}{k!} e^{-\mu}$$

$$= \mu^r \sum_{k=r}^{\infty} \frac{\mu^{k-r}}{(k - r)!} e^{-\mu} = \mu^r.$$

Elementary calculations yield

$$E(X) = \text{Var}(X) = \mu. \tag{24}$$

The first application of Poisson models is to frequency of occurrence of rare events in large numbers of trials. In particular, consider the $\mathfrak{B}(n, p)$ model with small p and large n. The following theorem shows that $b(k \mid n, p) \approx \pi(k \mid np)$. Of course, if p is close to 1, then

$$b(k \mid n, p) = b(n - k \mid n, q) \approx \pi(n - k \mid nq).$$

Theorem 5. *If $\lim_{n \to \infty} np_n = \mu$, then*

$$\lim_{n \to \infty} b(k \mid n, p_n) = \frac{\mu^k}{k!} e^{-\mu} = \pi(k \mid \mu) \tag{25}$$

for $k = 0, 1, \ldots$. That is, for n large and p small, $b(k \mid n, p)$ can be approximated by $\pi(k \mid np)$.

Proof. For $k = 0$,

$$b(0 \mid n, p_n) = (1 - p_n)^n = \left(1 - \frac{np_n}{n}\right)^n \longrightarrow e^{-\mu} = \pi(0 \mid \mu)$$

as $n \to \infty$, proving (25) for $k = 0$.

From (12)

$$\frac{b(m + 1 \mid n, p_n)}{b(m \mid n, p_n)} = \frac{(n - m)p_n}{(m + 1)(1 - p_n)},$$

which, under the conditions, converges to $\mu/(m + 1)$. Hence

$$\lim_{n \to \infty} b(m + 1 \mid n, p_n) = \frac{\mu}{m + 1} \lim_{n \to \infty} b(m \mid n, p_n)$$

and the conclusion follows from (23) and induction.

Table 4

| k | $p = .5$ | | $p = .1$ | |
	Binomial	Poisson approx.	Binomial	Poisson approx.
0	7.89×10^{-31}	1.93×10^{-22}	2.66×10^{-5}	4.54×10^{-5}
1	7.97×10^{-29}	9.84×10^{-21}	3.22×10^{-4}	4.99×10^{-4}
2	3.98×10^{-27}	2.51×10^{-19}	1.94×10^{-3}	2.77×10^{-3}
3	1.32×10^{-25}	4.27×10^{-18}	7.84×10^{-2}	1.03×10^{-2}
4	3.22×10^{-24}	5.45×10^{-17}	.0237	.0292
5	6.26×10^{-23}	5.57×10^{-16}	.0576	.0671
10	1.53×10^{-17}	6.45×10^{-12}	.583	.583
15	2.41×10^{-13}	6.36×10^{-9}	.960	.951
20	5.58×10^{-10}	1.24×10^{-6}	.999	.998

| k | $p = .05$ | | $p = .01$ | |
	Binomial	Poisson approx.	Binomial	Poisson approx.
0	5.92×10^{-3}	6.74×10^{-3}	.366	.368
1	.0371	.0404	.736	.736
2	.118	.125	.921	.920
3	.258	.265	.982	.981
4	.436	.440	.997	.996
5	.616	.616	1.000	.999
10	.988	.986	1.000	1.000
15	1.000	1.000	1.000	1.000
20	1.000	1.000	1.000	1.000

In Table 4 the distribution functions of the binomial and approximating Poisson distributions are given for selected values of p and k with $n = 100$. It should be noted that the approximation is best when k is near np. In fact, the relative error is large only when $(k - np)^2/n$ is substantial. We cannot discuss here the details of the rate of convergence.

In the next application the Poisson model is derived from physical assumptions concerning the rate at which certain events occur. Suppose that a source of radioactive emissions is observed for a time period of length t. For convenience set the time at which observation begins equal to 0. Let X_t be the number of emissions observed. What is the distribution of X_t?

The same problem may be posed for customers arriving at a store or for cars passing a given point. Certain assumptions will be made that are reasonable for radioactive emissions. Are they reasonable for the other examples?

Poisson Process

For any time interval J, let Z_J be the number of occurrences (emissions, arrivals, cars, etc.) during J. Assume that

(i) If $J_1, \ldots, J_n$ are disjoint time interval, then $Z_{J_1}, \ldots, Z_{J_n}$ are independent rv's.

(ii) The distribution of Z_J depends only on the length of J and, furthermore, if the length of J is t, then

$$P[Z_J = 1] = \lambda t + \delta_1(t);$$

$$P[Z_J = 0] = 1 - \lambda t + \delta_2(t);$$

$$P[Z_J \geq 2] = \delta_3(t),$$

where $\lim_{t \to 0} \delta_i(t)/t = 0$ for $i = 1, 2, 3$.

Let $X_t = Z_{(0,t)}$. Then X_t is called a *Poisson process with rate* λ.

In assumption (ii), $\delta_1(t) + \delta_2(t) + \delta_3(t) = 0$ for all t.

The probability of an occurrence at any specified time is 0. Thus in the theorem that follows, it does not matter if either or both endpoints of the interval are included.

Theorem 6. *For a Poisson process X_t with rate λ,*

$$P[X_t = k] = \frac{(\lambda t)^k}{k!} e^{-\lambda t} = \pi(k \mid \lambda t) \qquad \text{if } k = 0, 1, \ldots. \qquad (26)$$

That is, X_t is a $\mathfrak{Po}(\lambda t)$ rv.

Proof. Partition the interval $(0, t)$ into n intervals of equal length. Each interval has length t/n. Let Y_n be the number of intervals with at least one occurrence. Then Y_n is a $\mathfrak{B}(n, p_n)$ rv where $p_n = \lambda t/n - \delta_2(t/n)$. Since $np_n = \lambda t - t\delta_2(t/n)(t/n)^{-1} \to \lambda t$ as $n \to \infty$, from Theorem 5,

$$\lim_{n \to \infty} P[Y_n = k] = \pi(k \mid \lambda t).$$

It suffices to show that $\lim_{n \to \infty} P[Y_n = k] = P[X_t = k]$. Let A_n be the event that at least one of the intervals has more than one occurrence. Then $[X_t \neq Y_n] \subset A_n$ and it suffices to show that $\lim_{n \to \infty} P(A_n) = 0$.

For $i = 1, \ldots, n$, let B_{ni} be the event that the ith interval has more than one occurrence. Then $A_n = \cup_{i=1}^{n} B_{ni}$. From Theorem 7.5 it follows that

$$0 \leq P(A_n) \leq \sum_{i=1}^{n} P(B_{ni}) = n\delta_3 \left(\frac{t}{n}\right) \longrightarrow 0$$

as $n \to \infty$.

For the binomial case, we counted the number of occurrences of a fixed event A_0 at discrete times $1, \ldots, n$. In the context of a Poisson process we are counting the occurrences of A_0, an event that can occur at any instant during the time interval $(0, t)$. This will lead to a Poisson process whenever the occurrences of A_0 satisfy conditions (i) and (ii). The index t need not represent time. It may be distance or, as in Problem 7, may be multidimensional.

Corollary 1. *If X is a $\mathcal{P}_0(\mu_1)$ rv, Y is a $\mathcal{P}_0(\mu_2)$ rv and X and Y are independent, then $X + Y$ is a $\mathcal{P}_0(\mu_1 + \mu_2)$ rv.*

Proof. Consider a Poisson process with rate 1. Then $Z_{[0,\mu_1)}$ and $Z_{[\mu_1,\mu_1+\mu_2)}$ have the same distribution as X and Y, respectively. Furthermore, $Z_{[0,\mu_1)}$ and $Z_{[\mu_1,\mu_1+\mu_2)}$ are independent. Hence $Z_{[0,\mu_1+\mu_2)} = Z_{[0,\mu_1)} + Z_{[\mu_1,\mu_1+\mu_2)}$ has the same distribution as $X + Y$. But this distribution is $\mathcal{P}_0(\mu_1 + \mu_2)$.

In the examples that follow, Appendix Table B, a table of $\Pi(r \mid \mu)$, has been used.

Example 8. Suppose that a book has 400 pages and the probability that a page is free of misprints is .98. It is reasonable to assume that the occurrences of misprints on different pages are independent so that the number S_{400} of pages free of misprints is a $\mathcal{B}(400, .98)$ rv. Let $U_{400} = 400 - S_{400}$, the number of pages needing corrections. Then

$$P[U_{400} > 10] = 1 - B(10 \mid 400, .02) \approx 1 - \Pi(10 \mid 8) = .184.$$

The exact value is .182.

Example 9. Assume that a flood high enough to cover a certain bridge occurs, on the average, once in 10 years. Such a flood is called a 10-year flood. Assume that the occurrences of 10-year floods form a Poisson process. (Does this assumption seem reasonable?)

If time is measured in years, the rate of this Poisson process is .1. What is the probability that over a 50-year period, the number X_{50} of 10-year floods is at most 8? Since $\lambda t = 5$, it follows that $P[X_{50} \leq 8] = \Pi(8 \mid 5) = .932$.

Example 10. A manufacturing process involves a certain machine that fails, on the average, three times every four months. Assume that the failures of this machine form a Poisson process which then has rate .75 if time is measured in months. Suppose that each machine repair will cost $100,000 and that the company has budgeted $1.5 million next year for such repairs. What is the probability that repairs of this machine will be over budget next year?

Let X_{12} be the number of repairs required next year. Then $P[X_{12} > 15] = 1 - \Pi(15 \mid 9) = 1 - .978 = .022$ is the required probability.

Summary

$\mathcal{B}(n, p)$ rv: $S_n = \Sigma_{i=1}^n I_{A_i}$, where $A_1, \ldots, A_n$ are independent events and each $P(A_i) = p$.

$\mathcal{H}(n, M_1, M_2)$ rv: S_n is the number of individuals with property π in a sample without replacement from a population of size $M = M_1 + M_2$ in which M_1 have property π.

$\mathcal{P}o(\mu)$ probability function: $\pi(k \mid \lambda) = (\mu^k/k!)e^{-\mu}$ for $k = 0, 1, 2, \ldots$.

The $\mathcal{B}(n, p)$ probability function is

$$b(k \mid n, p) = \binom{n}{k} p^k q^{n-k} \qquad \text{if } k = 0, 1, \ldots, n,$$

where $q = 1 - p$. Furthermore, $E(S_n) = np$ and $\text{Var}(S_n) = npq$.

The $\mathcal{H}(n, M_1, M_2)$ probability function is

$$h(k \mid n, M_1, M_2) = \binom{n}{k} \frac{(M_1)_k (M_2)_{n-k}}{(M)_n}$$

$$= \frac{\binom{M_1}{k}\binom{M_2}{n-k}}{\binom{M}{n}} \qquad \text{for } k \geq 0.$$

Furthermore,

$$E(S_n) = nM_1/M \quad \text{and} \quad \text{Var}(S_n) = \frac{M - n}{M - 1}\left(\frac{nM_1 M_2}{M^2}\right).$$

For the Poisson model, $\mathcal{P}o(\mu)$, both the mean and variance are μ. If arrivals at a certain point obey the assumptions of a Poisson process with rate λ, the number of arrivals in T time units is a $\mathcal{P}o(\lambda T)$ rv.

Problems

1. Let A, B be independent events and let $X = I_A + I_B$. Show that X is a $\mathcal{B}(2, p)$ rv if, and only if, $P(A) = P(B) = p$.

2. A box contains four red and six white balls. A sample of five balls is drawn. Let X be the number of red balls in the sample. Make a table comparing the probability function of X in the cases:
   (a) Sampling is with replacement.
   (b) Sampling is without replacement.

3. Let X be a $\mathcal{P}_0(3)$ rv. Find $P[X \geq 5]$.

4. Find the mode(s) of:
   (a) $b(k \mid n, p)$
   (b) $h(k \mid n, M_1, M_2)$
   (c) $\pi(k \mid \mu)$
   Hint: What values of k satisfy $b(k \mid n, p)/b(k - 1 \mid n, p) \geq 1$?

5. Let S_n be the number of successes in n Bernoulli trials with success probability p.
   (a) $n = 10$, $p = .25$. Determine $P[S_n = 2]$.
   (b) $n = 15$, $p = .7$. Determine $P[8 \leq S_n \leq 12]$.
   (c) $n = 15$, $p = .8$. Determine k such that $P[S_n \geq k] \leq .25$ but $P[S_n \geq k - 1] > .25$.

6. Let S_n be a $\mathcal{B}(n, p)$ rv. Find approximate values of:
   (a) $P(S_{100} \leq 5 \mid p = .01)$
   (b) $P(S_{100} \geq 3 \mid p = .008)$
   (c) $P(2 \leq S_{100} < 5 \mid p = .02)$

7. The Poisson process as defined is a one-dimensional (time) phenomenon. Give assumptions for a two-dimensional phenomenon analogous to assumptions (i) and (ii) and derive the desired Poisson distribution. Give an illustration of a phenomenon for which your assumptions appear to be valid.

8. Customers arrive at a service station at the rate 8 per hour and according to a Poisson process. Let X_t be the number of customers who arrive in time t. Determine:
   (a) The probability that three customers arrive in a 15-minute period.
   (b) The probability that in a half-hour period there will be between five and eight customers (inclusive).
   (c) The probability that the waiting time for the tenth customer to arrive is less than 75 minutes.

9. Prove Theorem 1.

10. Prove Theorem 2.

11. Prove Theorem 3.

12.† Consider a playoff series between teams A and B in which the series is won by the first team to win n games out of a possible $2n - 1$.

Suppose that the games are independent and that the probability that team A wins any game is p. Let P_n be the probability that team A wins this series.

(a) Show that $p > 1/2$ implies that $P_{n+1} > P_n$. Hint: $P_n = P[S_{2n-1} \geq n]$ where S_m is a $\mathfrak{B}(m, p)$ rv.

Discuss the validity of the assumptions in the context of the baseball World Series or any similar playoff series in other sports.

13. Suppose that a venire of jurors contains 50 women and 50 men. A panel of 25 jurors is chosen at random from the venire and a jury of 12 is chosen at random from the panel. Let S_{12} be the number of women on the jury.
    (a) Find $P[S_{12} = 6]$.
    (b) Show that $P[S_{12} = k] = P[S_{12} = 12 - k]$.
14. In a very large population, suppose that the proportions of men and of women who support proposition A are both p. Samples are taken of m men and n women. In the combined sample, a total of t support A. Given this fact, what are conditional distributions of the numbers of men and of women in the samples who support A?
15. Find the factorial moments of $\mathfrak{B}(n, p)$ and use the result to obtain its mean and variance.
16. Verify (15) by direct counting of unordered samples.
17. Acceptance sampling. Consider a large lot in which a proportion p of items are defective. A sample of size n is taken and if c (the *acceptance number*) or fewer are defective, the lot is accepted. The function $B(c \mid n, p)$ is called the *operating characteristic*. In this problem, use and justify appropriate approximations where necessary.
    (a) For $n = 25$, find c so that the probability of accepting a lot with 10% defectives is at most .08. For this c, evaluate the operating characteristic for 10%, 20%, . . . , 90% defectives.
    (b) For $n = 200$, find c so that the probability of accepting a lot with 2.5% defectives is at most .05. For this c, evaluate the operating characteristic for 2%, 3%, 3.5%, and 4% defectives.
18. Compare Table 1 with a table of $h(k \mid 6, 2, 10)$. Why is this comparison relevant?
19. Compare Table 2 with a table of $b(k \mid 6, 2/3)$. Why is this comparison relevant.
20. Let S_n be a $\mathfrak{B}(n, p)$ rv where $n = r + s$. Let S_r be the number of successes in the first r trials. Find the conditional distribution of S_r given $S_n = k$.
21. Let X_t be a Poisson process with rate λ and $t = u + r$. Find the conditional distribution of X_u given $X_t = k$.
22. Derive a recursion formula for the hypergeometric probability function analogous to (12).

23.† An alphabet consists of 2^m letters encoded in binary bits. Thus if the alphabet is A, B, C, D, encode A by 00, B by 01, C by 10, and D by 11. A message of one encoded letter is sent. Bits may be transmitted erroneously. This happens independently and with probability p for each bit. The message is sent $2n - 1$ times, and if he sees n identical copies of the message, the receiver concludes that it is the message. If no message appears n times, the receiver is undecided. For what values of p is the probability of correct decoding increasing in n? (See Problem 12.)

24. Determine $E(|S_n - n/2|)$ for S_n a symmetric binomial rv.

25.† Let N be a $\mathscr{P}_o(\mu)$ rv and assume that given $N = n$, S_N is a $\mathscr{B}(n, p)$ rv. Show that S_N is a $\mathscr{P}_o(p\mu)$ rv.

40. MODELS ASSOCIATED WITH WAITING TIMES

In addition to counting occurrences, we are often concerned with the time it takes for certain things to happen. The models presented in this section are for waiting times related to the occurrences of some fixed event, A_0. If the "clock" used for measuring time is integer valued (as is the case for repeated trials of an experiment), the resulting models will, naturally, be discrete. If, on the other hand, A_0 can occur at any instant (as is the case for a Poisson process), the appropriate model for waiting times will be continuous.

Two cases will be considered.

Case 1: Discrete time. $A_1, A_2, \ldots, A_n, A_{n+1}, \ldots$ is a sequence of events (the subscript indicates the time). Let T_k denote the waiting time for k occurrences of the A_i's and let S_n denote the number of occurrences of the A_i's in time n. Since $[T_k > n] = [S_n < k]$,

$$P[T_k > n] = P[S_n < k]$$

or $\qquad\qquad\qquad\qquad\qquad\qquad\qquad\qquad\qquad\qquad\qquad (1)$

$$P[T_k \le n] = P[S_n \ge k].$$

Case 2: Continuous time. $\{A_s, s > 0\}$ is a collection of events. These events may occur at any moment s, and in any specified time interval only countably many A_s's can occur. The rv, X_t, is the number of A_s's that occur in time $[0, t]$ and T_k is the waiting time for k occurrences of the A_s's. Since $[T_k > t] = [X_t < k]$,

$$P[T_k > t] = P[X_t < k]$$

or $\qquad\qquad\qquad\qquad\qquad\qquad\qquad\qquad\qquad\qquad\qquad (2)$

$$P[T_k \le t] = P[X_t \ge k].$$

The only difference between (1) and (2) is that in (1) the "clock" is discrete, while in (2) the "clock" is continuous. In either case, the equality expresses the fact that an upper (lower) tail for waiting time is equivalent to a lower (upper) tail for frequency. This relation between tails yields the model for T_k in the Bernoulli case, the Poisson case, and in fact, in any case for which the model for the number of occurrences in specified time is given.

A. Geometric Models

Geometric Model

The model for T_1, the waiting time for one success in a sequence of Bernoulli trials with $p > 0$, is called a *geometric model*. From (1),

$$F_{T_1}(n \mid p) = P(T_1 \leq n \mid p) = P(S_n \geq 1 \mid p)$$

$$= 1 - P(S_n = 0 \mid p) = 1 - q^n \qquad \text{if } n = 1, 2, \ldots \quad (3)$$

and

$$f_{T_1}(n \mid p) = P(T_1 = n \mid p) = F_{T_1}(n \mid p) - F_{T_1}(n - 1 \mid p)$$

$$= q^{n-1} - q^n = pq^{n-1} \qquad \text{if } n = 1, 2, \ldots . \quad (4)$$

Alternatively, the probability function of T_1 can be found by noting that $[T_1 = n]$ if and only if the first $n - 1$ trials are failures and the nth trial is a success. From the iid assumptions on Bernoulli trials, the result, (4), follows.

All moments of T_1 exist as the following argument demonstrates. Let $\alpha > 0$ be fixed and let $\lambda_\alpha = \Sigma_{n=1}^\infty n^\alpha q^{n-1}$. Consider the ratio r_n of successive terms:

$$r_n = \frac{(n + 1)^\alpha q^n}{n^\alpha q^{n-1}} = \left(1 + \frac{1}{n}\right)^\alpha q.$$

Clearly, $r_n \to q < 1$, which, by the usual ratio test for convergence of a series, guarantees that $\lambda_\alpha < \infty$.

It was shown in Example 30.3 that

$$E(T_1 \mid p) = \frac{1}{p}. \quad (5)$$

In Example 34.4 and Problem 34.10 a recursive method was exploited that yields (5), but uses the fact shown above that moments of T_1 exist. This recursion method will now be used to obtain $\text{Var}(T_1)$. Let A be the event "success on the first trial" $[P(A) = p]$. The conditional expectations are

$$E\left(\left(T_1 - \frac{1}{p}\right)^2 \,\Big|\, A\right) = \left(1 - \frac{1}{p}\right)^2 = \frac{q^2}{p^2}$$

and

$$E\left(\left(T_1 - \frac{1}{p}\right)^2 \,\Big|\, A^c\right) = E\left(1 + T_1 - \frac{1}{p}\right)^2 = 1 + E\left(T_1 - \frac{1}{p}\right)^2$$

$$= 1 + \text{Var}(T_1).$$

Therefore, by Theorem 34.3,

$$\text{Var}(T_1) = \frac{q^2}{p^2} \cdot p + [1 + \text{Var}(T_1)]q.$$

Hence

$$(1 - q)\,\text{Var}(T_1) = \frac{q^2}{p} + q = \frac{q}{p}$$

and

$$\text{Var}(T_1 \mid p) = \frac{q}{p^2}. \tag{6}$$

Example 1. A pair of fair dice is rolled until the sum of the dice shows 7. Each roll has probability $1/6$ of producing a 7, so that T_1, the waiting time for the occurrence of a 7, has probability function

$$P\left(T_1 = n \,\Big|\, p = \frac{1}{6}\right) = f_{T_1}\left(n \,\Big|\, \frac{1}{6}\right) = \frac{1}{6}\left(\frac{5}{6}\right)^{n-1} \quad \text{if } n = 1, 2 \dots.$$

The expected number of rolls until we get a 7 is $E(T_1 \mid p = 1/6) = 6$ and the variance is $\text{Var}(T_1 \mid p = 1/6) = 30$. The probability that we must roll 10 or more times before obtaining a 7 is

$$P\left(T_1 \geq 10 \,\Big|\, p = \frac{1}{6}\right) = 1 - F_{T_1}\left(9 \,\Big|\, \frac{1}{6}\right) = \left(\frac{5}{6}\right)^9.$$

Example 2. A machine produces items with defective rate p. How many good items, N_0, will the machine produce before it produces a defective item? Under the usual Bernoulli trials assumptions, T_1, the waiting time for a defective item, has a geometric model and, clearly, $N_0 = T_1 - 1$. Hence

$$P(N_0 = n \mid p) = f_{T_1}(n + 1 \mid p) = pq^n \quad \text{if } n = 0, 1, 2, \dots.$$

In addition,

$$E(N_0 \mid p) = E(T_1 \mid p) - 1 = \frac{q}{p}$$

and

$$\text{Var}(N_0 \mid p) = \text{Var}(T_1 \mid p) = \frac{q}{p^2}.$$

The model for N_0 is also frequently referred to as a geometric model. It is merely a translated version of the model for T_1.

Example 3.　Imagine that you and an opponent are about to play a sequence of (independent) games as follows: (i) The probability that you win game i is $1/2$ for each i. (ii) The loser of game i pays the winner $\$m_i$, where $m_i \geq 0$ is the amount bet by you before the play of the ith game.

Your sequence of choices $m_1, m_2, \ldots$ is called a *betting scheme*. Consider the consequences of the following betting scheme: You begin with $m_1 = 1$ (bet \$1 on the first game) and keep doubling the bet each time $(m_{i+1} = 2m_i = 2^i)$ until your first win. Beyond that, set $m_i = 0$ (i.e., quit gambling).

The expectation of W, your winnings from this scheme is computed as follows. The value of W is completely determined by T_1, the waiting time for you to win one game. If $T_1 = n$, you win $\$2^{n-1}$ for winning the nth game and have lost $\sum_{i=0}^{n-2} 2^i = 2^{n-1} - 1$ for losing each of the previous games. Hence $W = 1$, regardless of how long it takes to win one game, and $E(W) = 1$.

This computation of $E(W)$ is valid, but the betting scheme is applicable only if you have enough capital to play as long as may be required to win one game. Now make the more realistic assumption that you have a limited amount, say $\$(2^N - 1)$. This means that if you lose N games before winning, you must quit because you have gone broke. Then using the same betting scheme, adjusted for the possibility of going broke, $W = 1$ as before for all values of $T_1 \leq N$. But if $T_1 > N$, then $W = -(2^N - 1)$ (i.e., you have lost all you had). Hence

$$E(W) = 1 \cdot P[T_1 \leq N] - (2^N - 1)P[T_1 > N]$$

$$= 1 - \left(\frac{1}{2}\right)^N - (2^N - 1)\left(\frac{1}{2}\right)^N = 0.$$

In summary, if you have limited capital, *no matter how large*, the expected amount you win is \$0.

Example 4. As in Example 2, assume that the output of a machine is a sequence of Bernoulli trials where p is the probability of a defective. Suppose that the only way to determine whether an item is defective is to destroy it. In such circumstances economy suggests what is called a *sequential sampling plan* to assess the quality of items being produced. One such plan is to inspect items one at a time until either (1) a defective is found or (2) a total of n_0 items has been inspected (and thus destroyed). Let N denote the number of items inspected when this play is used. Then the value of N is determined by its relation to the waiting time, T_1, for one defective to occur. That is, $N = \min(T_1, n_0)$ and

$$P(N = n \mid p) = P(T_1 = n \mid p) \qquad \text{if } n = 1, 2, \ldots, n_0 - 1$$

$$= P(T_1 \geq n_0 \mid p) \qquad \text{if } n = n_0.$$

Hence

$$P(N = n \mid p) = pq^{n-1} \qquad \text{if } n = 1, 2, \ldots, n_0 - 1$$

$$= q^{n_0 - 1} \qquad \text{if } n = n_0.$$

The rv, N, is called a *truncated geometric rv*. Computation of its mean and variance are left to the reader.

B. Negative Binomial Models

Let T_k denote the waiting time for k successes in a sequence of Bernoulli trials.

Negative Binomial Model

For $k \geq 1$, the waiting time T_k has a *negative binomial model*, which we denote by $\overline{\mathscr{B}}(k, p)$. Thus the $\mathscr{B}(1, p)$ model is the geometric model discussed in Subsection A.

Using (1) and taking differences yields the negative binomial probability function $\overline{b}(r \mid k, p)$. Alternatively, argue as follows. The event $[T_k = r]$ occurs if and only if the first $r - 1$ trials produced $k - 1$ successes (in any order) and the rth trial is a success. Hence, for $r = k, k + 1, \ldots$,

$$\overline{b}(r \mid k, p) = P(T_k = r \mid p) = b(k - 1 \mid r - 1, p)p$$

$$= \binom{r - 1}{k - 1} p^{k-1} q^{r-k} p = \binom{r - 1}{k - 1} p^k q^{r-k}. \qquad (7)$$

Relation (1) gives the family $\overline{B}$ of distribution functions of negative binomials in terms of binomials:

$$\overline{B}(r \mid k, p) = P(T_k \le r \mid p) = P(S_r \ge k \mid p)$$

$$= 1 - B(k - 1 \mid r, p) \quad \text{if } k = r, r + 1, \ldots . \quad (8)$$

Some tail probabilities for the negative binomial can be obtained from Table A.

Example 5. Consider a sequence of Bernoulli trials with $p = .6$. Table A can be used to evaluate the probability that the tenth success occurs between the sixteenth and twentieth trials (inclusive). This is

$$P(16 \le T_{10} \le 20 \mid p = .6) = \overline{B}(20 \mid 10, .6) - \overline{B}(15 \mid 10, .6)$$

$$= B(9 \mid 15, .6) - B(9 \mid 20, .6) = .469.$$

Theorem 1. *Let X and Y be independent rv's with $\mathcal{B}(k_1, p)$ and $\mathcal{B}(k_2, p)$ models, respectively. Then $Z = X + Y$ has a $\mathcal{B}(k_1 + k_2, p)$ model.*

Proof. The result follows by letting X be the waiting time for k_1 successes in a sequence of Bernoulli trials and Y be the waiting time for the next k_2 successes in the same sequence. Then Z is the waiting time for $k_1 + k_2$ successes and therefore must have a $\mathcal{B}(k_1 + k_2, p)$ model.

The proof of the following corollary is left to the reader.

Corollary 1. *Let $X_1, X_2, \ldots , X_k$ be independent $\mathcal{B}(1, p)$ rv's. Then $T_k = \Sigma_{i=1}^{k} X_i$ is a $\mathcal{B}(k, p)$ rv. Thus a $\mathcal{B}(k, p)$ model is the model for the sum of k independent $\mathcal{B}(1, p)$ rv's.*

Corollary 1 enables us to get the mean and variance of a $\mathcal{B}(k, p)$ rv from the general results for sums of iid rv's. Thus if T_k has a $\mathcal{B}(k, p)$ model, then

$$E(T_k \mid p) = kE(T_1 \mid p) = \frac{k}{p} \quad (9)$$

$$\text{Var}(T_k \mid p) = k \, \text{Var}(T_1 \mid p) = \frac{kq}{p^2}. \quad (10)$$

Example 6. The way baseball is played, there is no bound on the number of players who will bat in a given nine-inning game. We do know, however, that 27 outs must be made in order that a team complete nine innings. We can get some idea of the expected number of players who will bat if we are willing to make certain (unrealistic?) assumptions. Assume, for example, that each batter represents a Bernoulli trial with $q = .260$ of getting to base safely. (The value, .260, might be an approximation based on the team batting average, etc.) Then T_{27} is the waiting time (number of batsmen) that will come to bat in a nine-inning game. Thus

$$E(T_{27} \mid p = .740) = \frac{27}{.740} \approx 36.5$$

and

$$\text{Var}(T_{27} \mid p = .740) = \frac{(27)(.260)}{(.740)^2} \approx 12.8.$$

Thus we should expect about 36 or 37 players to come to bat and 9 or 10 of them to reach base. How do these speculations compare with real games?

We expect each of the nine batters to bat approximately four times per game. Consider the event that each batter bats at least five times, that is, $[T_{27} \geq 45]$. Then from the Chebyshev inequality (Corollary 37.2),

$$P(T_{27} \geq 45 \mid p = .740) \leq \frac{12.8}{(45 - 36.5)^2} = .177,$$

a value much larger than the exact value.

As a final note on this family of models, consider a problem in sequential sampling from a finite population. Suppose that sampling continues as long as it takes to produce k individuals with property π $[P(\pi) = p]$ and T_k is the eventual sample size. If sampling is *with* replacement, the $\overline{\mathcal{B}}(k, p)$ model is appropriate for T_k. Problem 14 asks for the appropriate model for T_k when sampling is *without* replacement. It can be shown that when k/p (the expected, with-replacement sample size) is small relative to M (the population size), the $\overline{\mathcal{B}}(k, p)$ is a reasonable approximation to the without-replacement sample-size model.

C. Exponential Models

Exponential Model

Consider a one-parameter family of continuous models with distribution function and densities, given as follows:

$$F(t \mid \beta) = 1 - e^{-t/\beta} \qquad \text{if } t > 0 \tag{11}$$

$$f(t \mid \beta) = \frac{1}{\beta} e^{-t/\beta} \qquad \text{if } t > 0. \tag{12}$$

This is the family of *exponential* distributions, a subfamily of the gamma family discussed later.

The parameter β is a scale parameter of the family with $\beta = 1$ yielding its "standard" member. This means that if X has an exponential model with parameter β, then $X = \beta Y$, where Y is an exponential rv with parameter 1. Explicitly, from (11),

$$P\left[\frac{X}{\beta} \le t\right] = P[X \le t\beta] = 1 - e^{-t} = F(t \mid \beta = 1) = P[Y \le t].$$

Because of this relationship many properties of the exponential model with parameter β can be deduced from those of the exponential model with parameter 1. In particular, if $Y = X/\beta$ is an exponential rv with parameter 1, then

$$E(Y) = \int_0^\infty [1 - F(t \mid \beta = 1)] \, dt = \int_0^\infty e^{-t} \, dt = 1$$

and

$$\text{Var}(Y) = \int_0^\infty t^2 e^{-t} - 1 = 1.$$

Since $X = \beta X$ is an exponential rv with parameter β,

$$E(X) = \beta E(Y) = \beta \tag{13}$$

and

$$\text{Var}(X) = \beta^2 \, \text{Var}(Y) = \beta^2. \tag{14}$$

Note that β is both the mean and the standard deviation of the exponential model.

Now, let T_1 denote the waiting time for one occurrence in a Poisson process with rate $\lambda > 0$. From (2),

$$P(T_1 > t \mid \lambda) = P(X_t < 1 \mid \lambda) = P(X_t = 0 \mid \lambda) = e^{-\lambda t}$$

or

$$P(T_1 \le t \mid \lambda) = 1 - e^{-\lambda t} \qquad \text{if } t > 0$$

$$= F\left(t \mid \beta = \frac{1}{\lambda}\right). \tag{15}$$

Hence the waiting time for one occurrence in a Poisson process with rate λ is an exponential rv with $\beta = 1/\lambda$. From (13) and (14),

$$E(T_1 \mid \lambda) = \frac{1}{\lambda}, \tag{16}$$

$$\text{Var}(T_1 \mid \lambda) = \frac{1}{\lambda^2}. \tag{17}$$

See Example 30.2.

The Poisson table (Table B) can be used to obtain tail probabilities for certain exponential models.

Example 7. A garage maintains a "store" from which mechanics obtain replacement parts for the cars being repaired. Suppose that arrivals at the store form a Poisson process with rate $\lambda = 9$ per hour. The waiting time between arrivals is, then, an exponential rv, T_1, with $\beta = 1/9$ in hours, or with $\beta = 20/3$ in minutes. Using Table B, the probability that more than 20 minutes will elapse between arrivals is ($\lambda = 1/10$ per minute)

$$P\left(T_1 > 20 \mid \beta = \frac{20}{3}\right) = P(X_{20} = 0 \mid \lambda = .15) = \Pi(0 \mid \mu = 3) = .050.$$

Also, $E(T_1 \mid \beta = 20/3) = SD(T_1 \mid \beta = 20/3) = 20/3$ minutes.

As noted in Section 23, the exponential distribution is frequently used as a model for lifetimes or, equivalently, as a model for waiting times for failures.

D. Gamma Models

Extension of the previous family entails the *gamma function*. This is the function, Γ, whose value for any $\alpha > 0$ is

$$\Gamma(\alpha) = \int_0^\infty x^{\alpha-1} e^{-x} \, dx.$$

The function Γ has many interesting and useful properties, among which are the following, which we state without proof.

$$\Gamma(\alpha + 1) = \alpha\Gamma(\alpha) \qquad \text{for every } \alpha > 0, \tag{18}$$

$$\Gamma(n + 1) = n!, \qquad \text{where } n \text{ is an integer,} \tag{19}$$

$$\Gamma(1/2) = \sqrt{\pi}, \tag{20}$$

$$\Gamma\left(n + \frac{1}{2}\right) = \frac{(2n)!}{2^{2n}n!} \sqrt{\pi} \qquad \text{for } n \text{ a nonnegative integer,} \tag{21}$$

and

$$\int_0^\infty x^{\alpha-1}(1 - x)^{\beta-1} \, dx = \frac{\Gamma(\alpha)\Gamma(\beta)}{\Gamma(\alpha + \beta)} \qquad \text{for any } \alpha > 0, \beta > 0. \tag{22}$$

Formula (21) is a consequence of (18) and (20).

Gamma Models

The *gamma family* of probability models, which we denote by $\mathcal{G}$, is a two-parameter family with densities

$$g(t \mid \alpha, \beta) = \frac{1}{\beta^\alpha \Gamma(\alpha)} t^{\alpha-1} e^{-t/\beta} \qquad \text{if } t > 0, \tag{23}$$

where $\alpha > 0$ and $\beta > 0$ are the parameters of the family. The parameter α is called the *shape parameter* of the family, and β, as in the exponential case, is the scale parameter of the family. The exponential model with parameter β is $\mathcal{G}(1, \beta)$.

Figure 1 shows $g(x \mid \alpha, 1)$ for $\alpha = 1/2, 1, 2$ and indicates why α is called the shape parameter. If Y is a $\mathcal{G}(\alpha, 1)$ rv, then $X = \beta Y$ is a $\mathcal{G}(\alpha, \beta)$ rv and moments of X are easily determined from moments of Y. For any $r > 0$,

$$E(Y^r \mid \alpha) = \frac{1}{\Gamma(\alpha)} \int_0^\infty t^{\alpha+r-1} e^{-t} \, dt = \frac{\Gamma(\alpha + r)}{\Gamma(\alpha)}.$$

This calculation also verifies that all moments of Y exist. In particular, using (18),

$$E(Y \mid \alpha) = \frac{\Gamma(\alpha + 1)}{\Gamma(\alpha)} = \alpha$$

and

$$E(Y^2 \mid \alpha) = \frac{\Gamma(\alpha + 2)}{\Gamma(\alpha)} = \alpha(\alpha + 1).$$

Hence

$$\text{Var}(Y \mid \alpha) = \alpha(\alpha + 1) - \alpha^2 = \alpha.$$

Therefore, the mean and variance of X, a $\mathcal{G}(\alpha, \beta)$ rv, are

$$E(X \mid \alpha, \beta) = \beta E(Y \mid \alpha) = \alpha\beta \tag{24}$$

$$\text{Var}(X \mid \alpha, \beta) = \beta^2 \, \text{Var}(Y \mid \alpha) = \alpha\beta^2. \tag{25}$$

The exponential model, $\mathcal{G}(1, 1/\lambda)$, is the model for the waiting time T_1 for one occurrence in a Poisson process with rate λ. As usual, let T_k be the waiting time for k occurrences. Then, by (2), the distribution function F of T_k is

$$F(t \mid \lambda) = P(T_k \le t \mid \lambda) = P(X_t \ge k \mid \lambda)$$

$$= 1 - \sum_{m=0}^{k-1} \frac{(\lambda t)^m}{m!} e^{-\lambda t} \qquad \text{if } t > 0. \tag{26}$$

Differentiation of (26) yields the density,

$$f_k(t \mid \lambda) = \frac{d}{dt} G_k(t \mid \lambda) = -\lambda \sum_{m=0}^{k-2} \frac{(\lambda t)^m}{m!} e^{-\lambda t} + \lambda \sum_{m=0}^{k-1} \frac{(\lambda t)^m}{m!} e^{-\lambda t}$$

$$= \frac{\lambda^k t^{k-1}}{(k-1)!} e^{-\lambda t} \qquad \text{if } t > 0.$$

Now let $\beta = 1/\lambda$ and note that $(k-1)! = \Gamma(k)$. It follows that

$$f(t \mid \beta) = \frac{1}{\beta^k \Gamma(k)} t^{k-1} e^{-t/\beta} \qquad \text{if } t > 0. \tag{27}$$

Hence T_k is a $\mathcal{G}(k, 1/\lambda)$ rv. Tail probabilities for $\mathcal{G}(k, \beta)$ can be obtained from Table B using (2). In addition, (26) becomes

$$\Pi(k - 1 \mid \lambda t) = 1 - F(t \mid \lambda) = \int_t^\infty g\left(u \mid k, \frac{1}{\lambda}\right) du.$$

Some examples of gamma densities are exhibited in Figure 1.

Example 8. In Example 7, what is the probability that less than 40 minutes elapse before the eighth arrival [i.e., $P(T_8 < 40 \mid \lambda = 1/10)$]? This is

$$G(40 \mid 8, 10) = P\left(X_{40} \geq 8 \mid \frac{1}{10}\right) = 1 - P\left(X_{40} \leq 7 \mid \frac{1}{10}\right).$$

The entry in row $r = 7$ in Table B with $\mu = \lambda t = (1/10)40 = 4$ is

$$G(40 \mid 8, 10) = 1 - .949 = .051.$$

Using (24) and (25), the expected waiting time for eight arrivals is

$$E(T_8 \mid \beta = 10) = 80 \text{ minutes}$$

and the variance of T_8 is

$$\text{Var}(T_8 \mid \beta = 10) = 8(10)^2 = 800.$$

If α is an integer, then (24) and (25) can also be obtained from a theorem, analogous to Theorem 1, relating all waiting times to T_1. The proof of the following theorem follows from the basic Poisson process assumptions and is left to the reader.

Theorem 2. Let $W_1, \ldots, W_k$ be iid $\mathcal{G}(1, \beta)$ (i.e., exponential) rv's and let $T_k = \sum_{i=1}^k W_i$. Then T_k is a $\mathcal{G}(k, \beta)$ rv.

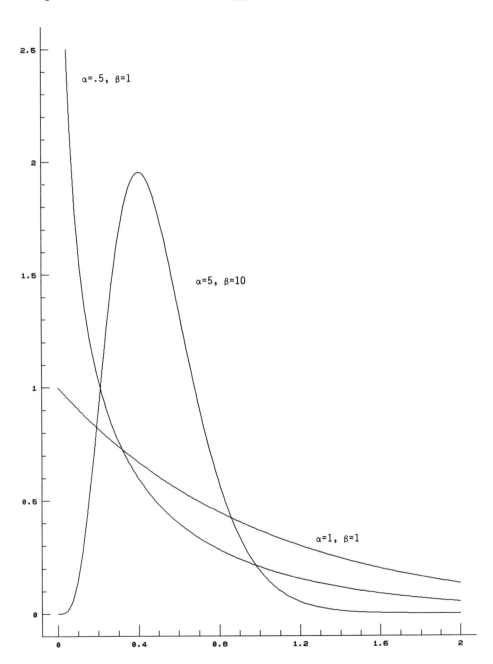

Figure 1   Gamma densities.

The entire subfamily $\mathcal{G}(\alpha, 1)$ [and hence the subfamily $\mathcal{G}(\alpha, \beta)$ for any fixed β] is closed under addition of independent rv's (see Example 49.1).

The following example examines the relationship between Poisson processes and exponential rv's.

Example 9. Imagine that a device (e.g., a lamp) consists of one component (e.g., a bulb) whose life L is an exponential [i.e., $\mathcal{G}(1, \beta)$] rv. When the component fails it is replaced immediately by a new one with the same life distribution. As the process develops, worn-out components are replaced by identical (independent) new ones. Such a process is called a *renewal process*. Each time a component fails and is replaced we say that a renewal has taken place. *Question*: In a given time interval, say $[0, t]$, what is the appropriate model for the number, X_t, of renewals?

Let $L_1, L_2, \ldots$ denote the sequence of lifetimes of replaced components. The rv L_1 is the time at which the first renewal takes place, and $L_1 + L_2$ is the time at which the second renewal takes place. Therefore, $[X_t < k] = [\Sigma_{i=1}^{k} L_i > t]$; that is, the number of renewals in time $[0, t]$ is less than k if, and only if, the kth renewal occurs beyond time t. This is precisely the same relation displayed in (2) with regard to waiting times. By Theorem 2, $\Sigma_{i=1}^{k} L_i$ is a $\mathcal{G}(k, \beta)$ rv. Hence

$$P(X_t < k \mid \beta) = P\left(\sum_{i=1}^{k} L_i > t \,\Big|\, \beta\right) = P(T_k > t \mid \beta),$$

where T_k is the waiting time for k occurrences in a Poisson process with rate $\lambda = 1/\beta$. Hence X_t, the number of renewals in $[0, t]$, is a Poisson process. The conclusion can be stated as follows: The number of occurrences of an event A_0 in time interval $[0, t]$ obeys a Poisson process with rate λ if, and only if, the waiting times between occurrences of A_0 are iid exponential rv's with scale parameter $\beta = 1/\lambda$.

Summary

Geometric rv: T_1 = waiting time for one success in a sequence of Bernoulli trials.

$\overline{\mathcal{B}}(k, p)$ *rv*: T_k = waiting time for k successes in a sequence of Bernoulli trials.

Exponential density: $f(t \mid \beta) = (1/\beta)e^{-t/\beta}$ for $t > 0$, where $\beta > 0$.

$\mathcal{G}(\alpha, \beta)$ *density*: $g(t \mid \alpha, \beta) = [1/(\beta^\alpha \Gamma(\alpha))]t^{\alpha-1}e^{-t/\beta}$ for $t > 0$, where α, $\beta > 0$.

T_1 is a $\overline{\mathcal{B}}(1, p)$ rv and $f(t \mid \beta) = g(t \mid 1, \beta)$.

$E(T_k \mid p) = k/p$ and $\text{Var}(T_k \mid p) = kq/p^2$.

The $\mathcal{G}(\alpha, \beta)$ mean and variance are, respectively, $\alpha\beta$ and $\alpha\beta^2$. The waiting time for k occurrences in a Poisson process with rate λ is a $\mathcal{G}(k, 1/\lambda)$ rv.

Problems

1. A fair coin is tossed repeatedly. Find the probability that the first head occurs between the third and seventh tosses inclusive.

2. Let T_k be the waiting time for the kth success in a sequence of Bernoulli trials.
   (a) $k = 7, p = .3$. Determine $P[T_k = 21]$.
   (b) $k = 15, p = .75$. Determine $P[16 \le T_k \le 21]$.
   (c) $k = 10, p = .5$. Find the median of T_k.

3. Let X be an exponential rv with parameter 3. Find the following probabilities.
   (a) $P[2 \le X \le 5]$
   (b) $P[X < 4]$
   (c) Find the interquartile range of X.

4. Let X be a $\mathscr{G}(2, 3)$ rv. Find the probabilities in Problem 3 and compare with your answers in Problem 3.

5. Let X be a positive integer-valued rv. Show that X is a geometric rv if, and only if, $P(X > k + 1 \mid X > 1) = P[X > k]$ for all $k = 1, 2, \ldots$.

6. Let X be a positive rv with continuous distribution function. Show that X is an exponential rv if, and only if, $P(X > t + s \mid X > s) = P[X > t]$ for all $s, t > 0$. *Hint*: The only continuous functions g on $[0, \infty)$ for which $g(t + s) = g(t) + g(s)$ are of the form $g(x) = cx$ for some c.

7. Prove Corollary 1.

8. Prove Corollary 2.

9. Let X be an exponential rv with parameter 1. Find the probability function of $[X]$. ($[a]$ is the largest integer less than or equal to a.) Can you identify the model?

10. Prove that β in the family $\mathscr{G}(\alpha, \beta)$ is a scale parameter. That is, show that if X is a $\mathscr{G}(\alpha, \beta)$ rv, then X/β is a $\mathscr{G}(\alpha, 1)$ rv.

11. Prove Theorem 2.

12. Find the mode of $g(t \mid \alpha, \beta)$.

13. Find the mode(s) of $\bar{b}(r \mid k, p)$.

14. A population consists of M_1 successes and M_2 failures. Sampling is without replacement.
   (a) Let T_k be the number of individuals drawn until the kth success is found. Find the probability function of T_k and relate it to the family $\mathscr{H}$.
   (b) Are the waiting times between consecutive successes independent? Are they identically distributed? Explain.

15. Of M deer in a forest, 10 are captured, tagged, and released. A new sample of deer is to be chosen successively without replacement until

either five tagged deer or a total of 20 deer are found. If $M = 200$, what is the probability function for the second sample size? *Note*: This sampling technique has application in estimating sizes of wildlife populations.

16. As in GP 1.7, assume that each year has 365 days and birthdays are assigned to k people at random with replacement. You found the probability that k people have distinct birthdays. Suppose that we admit people one at a time to a room, stopping as soon as two of those admitted have the same birthday. Let T be the number admitted to the room.
(a) Find the probability function of T.
(b) Find the mode of the probability function of T.
(c) Find the median of the probability function of T.

17. Let X be a $\mathcal{G}(\alpha, \beta)$ rv. Consider the Markov bound on $P[X > t]$ (Theorem 37.1) and the second moment bound (Corollary 37.1 with $k = 2$). Find conditions on t, α, and β making the former smaller than the latter.

18. Find the reverse factorial moments of $\overline{\mathcal{B}}(k, p)$ and use your results to obtain the mean and variance.

19. Prove the event relation $[T_k > n] = [S_n < k]$ used in Case 1 and the analogous relation used in Case 2.

20. Evaluate $\overline{b}(r + 1 \mid k, p)/\overline{b}(r \mid k, p)$ and find the mode(s) for $\overline{\mathcal{B}}(k, p)$.

41. MULTIDIMENSIONAL FREQUENCIES

Binomial and hypergeometric models have been applied to the number occurrences of an event A_0 and by implication, of its compliment, A_0^c. If we consider the partition $\{A_0, A_0^c\}$, each trial identifies which of the two members of the partition occurred. The models in this section are for frequencies associated with partitions consisting of (possibly) more than two members.

Let $\mathcal{a} = \{A_1, A_2, \ldots, A_r\}$ be a partition of the outcomes of an experiment $\mathcal{E}$. This means that on any trial of $\mathcal{E}$ one, and only one, member of $\mathcal{a}$ occurs.

Probability Vector

The probabilities of the events in $\mathcal{a}$ will be denoted by $p_i = P(A_i)$, $i = 1, 2, \ldots, r$. Since $\mathcal{a}$ is a partition, $0 < p_i < 1$ and $\sum_{i=1}^{r} p_i = 1$. Any vector, $\mathbf{p} = (p_1, p_2, \ldots, p_r)$ satisfying these conditions is called a *probability vector*.

We shall be concerned with n trials of $\mathscr{E}$ in two circumstances:

(A) Independent repetitions of $\mathscr{E}$.
(B) The trials of $\mathscr{E}$ consist of successive random sampling without replacement from a population. Each member of the population is assumed to have one, and only one, of r mutually exclusive properties $\pi_1, \ldots, \pi_r$. It is further assumed that among the M individuals comprising the population, M_i have property π_i ($i = 1, \ldots, r$), $\Sigma_{i=1}^r M_i = M$. The event A_i occurs for a given selection if the individual selected has property π_i.

In both cases we shall develop the model for the vector $\mathbf{S}_n = (S_{n1}, S_{n2}, \ldots, S_{nr})$, where S_{ni} is the number of times A_i occurs in the n trials of $\mathscr{E}$. The components of $\mathbf{S}_n$ are nonnegative integers and $\Sigma_{i=1}^r S_{ni} = n$. Due to this constraint, the vector $\mathbf{S}_n$ has essential dimension $r - 1$ provided that all $p_i > 0$. The case $r = 2$ was treated as one-dimensional in Section 39.

The following examples illustrate the distinctions between cases (A) and (B) and the use of the notation.

Example 1. A fair die is rolled n times. The event A_i occurs if a roll produces face i. The probability vector is $\mathbf{p} = (1/6, 1/6, \ldots, 1/6)$ and $\mathbf{S}_n = (S_{n1}, \ldots, S_{n6})$, where S_{ni} is the number of rolls that produce face i. Since the trials are independent, the model for $\mathbf{S}_n$ will depend only on $\mathbf{p}$.

Example 2. From a group consisting of 10 freshman (π_1), 15 sophomores (π_2), 15 juniors (π_3), and 10 seniors (π_4), 12 are selected at random without replacement. The events A_i for $i = 1, 2, 3, 4$ refer to drawing individuals with respective properties π_i for $i = 1, 2, 3, 4$. Here $M_1 = M_4 = 10$ and $M_2 = M_3 = 15$, so that $M = 50$. The vector of occurrences is $\mathbf{S}_5 = (S_{5,1}, S_{5,2}, S_{5,3}, S_{5,4})$, where $S_{5,i}$ is the number chosen that have property π_i. The probability vector in this case is $\mathbf{p} = (.2, .3, .3, .2)$ and represents the marginal probability model for each trial.

A. Multinomial Models

Multinomial Family

Trials of $\mathscr{E}$ are assumed to be independent. The family of models for $\mathbf{S}_n$ is called the *multinomial family* and is denoted by $\mathscr{B}_r$. An individual member of this family is denoted by $\mathscr{B}_r(n, \mathbf{p})$, where $\mathbf{p}$ is an r-dimensional probability vector.

Let S_{n1} be a $\mathscr{B}(n, p)$ rv and $S_{n2} = n - S_{n1}$. Then $\mathbf{S}_n = (S_{n1}, S_{n2})$ is a $\mathscr{B}_2(n, (p, q))$ random vector.

The probability function for $\mathbf{S}_n$ is obtained by generalizing the argument in Section 40 for the $\mathcal{B}(n, p)$ probability function. An outcome of a sequence of n trials of $\mathcal{E}$ can be thought of as a sequence of length n, each component being one of the integers $1, \ldots, r$. Let $\mathbf{k} \in \mathcal{R}^r$ have nonnegative integer components k_i with $\Sigma_{i=1}^r k_i = n$. The probability of any sequence containing k_i components equal to i is $\Pi_{i=1}^r p_i^{k_i}$. Thus $P[\mathbf{S}_n = \mathbf{k}] = c \, \Pi_{i=1}^r p_i^{k_i}$, where c is the number of sequences of length n with k_i components equal to i. Then c is the number of partitions of a set of size n into r subsets of sizes $k_1, k_2, \ldots, k_r$, respectively. By (10.3), $c = n!/(k_1! \, k_2! \cdots k_r!)$, so that the $\mathcal{B}_r(n, \mathbf{p})$ probability function is

$$b_r(\mathbf{k} \mid n, \mathbf{p}) = \frac{n!}{k_1! \cdots k_r!} \, p_1^{k_1} \cdots p_r^{k_r} \quad \text{for integers } k_i \geq 0, \ \sum_{i=1}^r k_i = n.$$

(1)

Example 3.   A machine produces items that are classified (and accordingly priced) in increasing order of their quality as A_1, A_2, A_3, A_4. Assume that the production process yields independent multinomial trials with $P(A_1) = .1$, $P(A_2) = .2$, $P(A_3) = .5$, and $P(A_4) = .2$. Let $\mathbf{S}_{10}$ denote the vector of occurrences of the A_i's among the 10 items. Then $\mathbf{S}_{10}$ has a $\mathcal{B}_4(10, \mathbf{p})$ model, where $\mathbf{p} = (.1, .2, .5, .2)$. For example, if $\mathbf{k} = (2, 3, 4, 1)$, then

$$P(\mathbf{S}_{20} = \mathbf{k} \mid \mathbf{p}) = b_4(\mathbf{k} \mid 10, \mathbf{p}) = \frac{10!}{2! \, 3! \, 4! \, 1!} (.1)^2 (.2)^3 (.5)^4 (.2)^1$$

$$= (12{,}600) \times 10^{-6} = .0126.$$

Suppose that the profit for each type of item is $0 for A_1, $1 for A_2, $2 for A_3, and $3.50 for A_4. The vector $\mathbf{k} = (2, 3, 4, 1)$ above represents a profit of $14.50 for the 10 items. Let W_{10} denote the total 10-item profit. Then the event $[W_{10} = \$14.50]$ consists of the vector $\mathbf{k}$ above and certain additional vectors. These vectors and their probabilities are displayed in Table 1. Adding the $b_4(\mathbf{k} \mid 10, p)$ over these $\mathbf{k}$, $P[W_{10} = \$14.50] = .023$.

Table 1

k	$b_4(\mathbf{k} \mid 10, \mathbf{p})$	k	$b_4(\mathbf{k} \mid 10, \mathbf{p})$
(0, 7, 2, 1)	.0002	(3, 4, 0, 3)	.00005
(1, 5, 3, 1)	.0040	(4, 2, 1, 3)	.0002
(2, 3, 4, 1)	.0126	(5, 0, 2, 3)	.00005
(3, 1, 5, 1)	.0063		

Example 4. Case (B), described earlier, concerns a finite population of individuals classified by properties $\pi_1, \ldots, \pi_r$. If sampling is *with* replacement, the vector $\mathbf{S}_n$, which gives the numbers of sampled individuals in each category, is then a $\mathcal{B}_r(n, \mathbf{p})$ random vector with $\mathbf{p} = (M_1/M, \ldots, M_r/M)$.

Example 5. Let X be a continuous rv with distribution function F. Suppose that the range of X is partitioned into a collection of r contiguous disjoint intervals $(-\infty, a_1], (a_1, a_2], \ldots, (a_{r-2}, a_{r-1}], (a_{r-1}, \infty)$. An iid collection of n observations on X generates n multinomial trials with $\mathbf{p} = (p_1, \ldots, p_r)$, where $p_i = F(a_i) - F(a_{i-1})$ $(a_0 = -\infty, a_r = \infty)$. This is similar to the construction of a histogram as an approximation to a density (Section 23). The random vector $\mathbf{S}_n$ gives the numbers of X's that fall in each interval and has model $\mathcal{B}_r(n, \mathbf{p})$. Such a scheme has application to problems in which measured values are grouped for one purpose or another. In statistics groupings are often made to compare an empirical model with a theoretical one.

The following theorems, whose proofs are left as problems, summarize some useful properties of the $\mathcal{B}_r$ family.

Theorem 1. *Let $\mathbf{S}_n$ be a $\mathcal{B}_r(n, \mathbf{p})$ random vector. Then, for any $m = 1, \ldots, r - 1$:*

(i) *$(S_{n1}, \ldots, S_{nm}, \Sigma_{i=m+1}^r S_{ni})$ is a $\mathcal{B}_{m+1}(n, \mathbf{p}_0)$ random vector, where* $\mathbf{p}_0 = (p_1, \ldots, p_m, \Sigma_{i=m+1}^r p_i)$.

(ii) *$\Sigma_{i=1}^m S_{ni}$ is a $\mathcal{B}(n, \varphi)$ rv, where $\varphi = \Sigma_{i=1}^m p_i$.*

(iii) *The conditional model for $(S_{n1}, \ldots, S_{nm})$ given $\Sigma_{i=m+1}^r S_{ni} = s$ is $\mathcal{B}_m(n - s, \mathbf{p}^*)$, where $p_i^* = p_i/\Sigma_{j=1}^m p_j$, $i = 1, \ldots, m$.*

Since the components of $\mathbf{S}_n$ can be exhibited in any order, Theorem 1 can be applied to any subcollection of components of $\mathbf{S}_n$.

Theorem 2. *Let $\mathbf{S}_n$, $\mathbf{S}'_m$ be independent $\mathcal{B}_r(n, \mathbf{p})$ and $\mathcal{B}_r(m, \mathbf{p})$ random vectors, respectively. Then $\mathbf{S}_n + \mathbf{S}'_m$ is a $\mathcal{B}_r(n + m, \mathbf{p})$ random vector.*

B. Multihypergeometric Models

Multihypergeometric Family

Case (B), described above, leads to a multidimensional version of the hypergeometric family, called the *multihypergeometric family*. Denote the family by $\mathcal{H}_r$ and individual members by $\mathcal{H}_r(n, M, \mathbf{p})$, where M is the population size, M_i ($i = 1, \ldots, r$) is the number of individuals in the population with property π_i, $p_i = M_i/M$, and $\mathbf{p} = (p_1, \ldots, p_r)$. The vector $\mathbf{S}_n$ gives the number of sampled individuals with the various properties. Elementary counting formulas give

$$P(\mathbf{S}_n = \mathbf{k} \mid M, \mathbf{p}) = h_r(\mathbf{k} \mid n, M, \mathbf{p})$$

$$= \frac{\binom{Mp_1}{k_1}\binom{Mp_2}{k_2} \cdots \binom{Mp_r}{k_r}}{\binom{M}{n}}$$

$$= \frac{\binom{M_1}{k_1}\binom{M_2}{k_2} \cdots \binom{M_r}{k_r}}{\binom{M}{n}}$$

$$\text{for integers } k_i \geq 0, \quad \sum_{i=1}^{r} k_i = n. \qquad (2)$$

The numerator gives the number of subsets of size n with k_i members with property π_i ($i = 1, \ldots, r$) and the denominator gives the total number of subsets of size n.

As defined above, $\mathcal{H}_r$ is an ($r + 2$) parameter family. However, one parameter can be eliminated by the linear constraint $\Sigma_{i=1}^r M_i = M$, that is, $\Sigma_{i=1}^r p_i = 1$.

The following theorem is analogous to Theorem 1.

Theorem 3. *Let $\mathbf{S}_n$ be an $\mathcal{H}_r(n, M, \mathbf{p})$ random vector. Then for any $m = 1, \ldots, r - 1$:*

(i) *$(S_{n1}, \ldots, S_{nm}, \Sigma_{i=m+1}^r S_{ni})$ is an $\mathcal{H}_{m+1}(n, M, \mathbf{p}_0)$ random vector, where $\mathbf{p}_0 = (p_1, \ldots, p_m, \Sigma_{i=m+1}^r p_i)$.*

(ii) *$\Sigma_{i=1}^m S_{ni}$ is an $\mathcal{H}(n, M\varphi, M(1 - \varphi))$ rv where $\varphi = \Sigma_{i=1}^m p_i$.*

(iii) *The conditional model for $(S_{n1}, \ldots, S_{nm})$ given $\Sigma_{i=m+1}^r S_{ni} = s$ is $\mathcal{H}_m(n - s, \Sigma_{i=1}^m M_i, \mathbf{p}^*)$, where $p_i^* = p_i/\Sigma_{j=1}^m p_j$, $i = 1, \ldots, m$.*

Proof. Left as a problem.

Theorem 3 can be applied to any subcollection of components of S_n.

Arguments similar to those used in Section 39 lead to the conclusion that when n/M is small the $\mathcal{H}_r(n, M, \mathbf{p})$ model is approximated by the $\mathcal{B}_r(n, \mathbf{p})$ model.

Example 6. In Example 2, the probability that among the 12 students selected, three are from each class, is obtained as follows. In (2), let $\mathbf{k} = (3, 3, 3, 3)$, $\mathbf{p} = (.2, .3, .3, .3)$, $M = 50$, and $n = 12$. Then

$$P(\mathbf{S}_{12} = \mathbf{k} \mid M, \mathbf{p}) = h_r(\mathbf{k} \mid M, \mathbf{p}) = \frac{\binom{10}{3}\binom{15}{3}\binom{15}{3}\binom{10}{3}}{\binom{50}{12}} = .025.$$

The moments of $\mathbf{S}_n$ can be given in a unified way for both cases. These moments are

$$E(S_{ni}) = np_i; \tag{3}$$

$$\mathrm{Var}(S_{ni}) = Knp_iq_i; \tag{4}$$

$$\mathrm{Cov}(S_{ni}, S_{nj}) = -Knp_ip_j \quad \text{for } i \neq j \text{ where } q_i = 1 - p_i;$$

$$K = 1 \text{ in case (A)}$$

$$= \frac{M - n}{M - 1} \text{ in case (B).} \tag{5}$$

Verification of (3), (4), and (5) are left as a problem.

Summary

Probability vector: $\mathbf{p} = (p_1, \ldots, p_r)$, where $p_i \geq 0$, $\Sigma_{i=1}^r p_i = 1$.

$\mathcal{B}_r(n, \mathbf{p})$ *random vector*: $\mathbf{S}_n = (S_{n1}, \ldots, S_{nr})$, where S_{ni} is the number of occurrences of A_i in n independent trials, $(A_1, \ldots, A_r)$ is a partition, and $p_i = P(A_i)$ $(i = 1, \ldots, r)$.

$\mathcal{H}_r(n, M, \mathbf{p})$ *random vector*: $\mathbf{S}_n = (S_{n1}, \ldots, S_{nr})$, where S_{ni} is the number of individuals with property π_i in a sample without replacement from a population of size M in which M_i have property π_i, $\Sigma_{i=1}^r M_i = M$, $p = (p_1, \ldots, p_r)$, and $p_i = M_i/M$.

Theorems 1 and 3 give marginal and conditional properties for $\mathcal{B}_r(n, \mathbf{p})$ and $\mathcal{H}_r(n, M, \mathbf{p})$ random vectors, respectively. The important moments of $\mathbf{S}_n$ are $E(S_{ni}) = np_i$, $\mathrm{Var}(S_{ni}) = Knp_iq_i$, and $\mathrm{Cov}(S_{ni}, S_{nj}) = -Knp_ip_j$ for $i \neq j$, where $q_i = 1 - p_i$, and

$$K = 1 \qquad \text{if } S_n \text{ is } \mathcal{B}_r(n, p)$$

$$= \frac{M - n}{M - 1} \qquad \text{if } S_n \text{ is } \mathcal{H}_r(n, M, p).$$

Problems

1. Let $\mathbf{S}_8$ be a $\mathcal{B}_3(8, \mathbf{p})$ random vector where $\mathbf{p} = (1/4, 1/2, 1/4)$.
   (a) Find $P[\mathbf{S}_8 = \mathbf{k}]$, where $\mathbf{k} = (3, 3, 2)$.
   (b) Show how this problem might be applied to a sequence of tosses of fair coins.
2. (a) Repeat Problem 1(a) with $\mathbf{S}_8$ a $\mathcal{H}_3(8, M, \mathbf{p})$ random vector with the same $\mathbf{p}$ and $M = 16, 24$.
   (b) Show how this problem might be applied to a population of two-child families.
3. Prove Theorem 1.
4. Prove Theorem 2.
5. Prove Theorem 3.
6. Let $\mathbf{S}_n$ be a $\mathcal{B}_r(n, \mathbf{p})$ random vector. What is the conditional model for S_{nj} given $(S_{n1}, \ldots, S_{n,j-1}) = (k_1, \ldots, k_{j-1})$?
7. Show that $b_r(\mathbf{k} \mid n, \mathbf{p}) = \prod_{j=1}^{r-1} b(k_j \mid n_j, \rho_j)$ for suitably chosen n_j and ρ_j. *Hint*: Use Problem 6 and the multiplication rule (14.3).
8. Repeat Problem 6 when $\mathbf{S}_n$ is an $\mathcal{H}_r(n, M, \mathbf{p})$ random vector.
9. Show that $h_r(k \mid n, M, \mathbf{p}) = \prod_{j=1}^{r-1} h(k_j \mid n_j, M_j, M_j^*)$ for suitably chosen n_j and M_j^*.
10. Let $\mathbf{S}_{15}$ be a $\mathcal{B}_3(15, \mathbf{p})$ random vector with $\mathbf{p} = (.4, .3, .3)$. Use Problem 7 and Table A to determine $b_3(\mathbf{k} \mid 15, \mathbf{p})$, where $\mathbf{k} = (5, 4, 6)$.
11. Verify (3), (4), and (5).
12. Let $\mathbf{S}_n$ be a $\mathcal{B}_r(n, \mathbf{p})$ random vector. Suppose that $np_i \to \mu_i > 0$ as $n \to \infty$ for $i = 1, \ldots, k < r$. Find the limit as $n \to \infty$ of the joint probability function of $X_1, \ldots, X_k$ and identify this limit.

42. NORMAL MODELS AND APPROXIMATIONS

The family of models that we are about to discuss occupies a unique and dominant position in the theory and application of probability. These models function primarily as approximations to other models. However, their capacity for providing useful, accurate approximations also make them quite suitable as assumed models for a considerable body of quantities, among which are measurement errors and population values. Each of the models discussed previously in this chapter can, under certain conditions, be approximated by a normal model.

There are two categories of normal models to be presented. First we present normal models on $\mathcal{R}$, then proceed to normal models on $\mathcal{R}^m$ for $m > 1$.

A. Normal Models on $\mathcal{R}$

Standard Normal

The continuous rv Z is called *standard normal* if it has the density

$$\varphi(x) = \frac{1}{\sqrt{2\pi}} e^{-x^2/2} \qquad \text{if } -\infty < x < \infty. \tag{1}$$

Its distribution function is denoted by Φ.

It is clear that $\varphi \geq 0$. What may not be obvious is that its integral has value 1. Let

$$J = \sqrt{2\pi}\ \Phi(\infty) = \int_{-\infty}^{\infty} e^{-x^2/2}\ dx.$$

It suffices to show that $J^2 = 2\pi$. But

$$J^2 = \left(\int_{-\infty}^{\infty} e^{-x^2/2}\ dx \right)\left(\int_{-\infty}^{\infty} e^{-y^2/2}\ dy \right) = \int_{-\infty}^{\infty}\int_{-\infty}^{\infty} e^{-(x^2+y^2)/2}\ dx\ dy.$$

Setting $x = r \cos \theta$ and $y = r \sin \theta$ gives

$$J^2 = \int_0^{\infty}\int_0^{2\pi} e^{r^2/2}\ r\ d\theta\ dr = 2\pi \int_0^{\infty} e^{-u}\ du = 2\pi.$$

The fact that $J = \sqrt{2\pi}$ provides a proof of (40.20), that is, that $\Gamma(1/2) = \sqrt{\pi}$ [i.e., $J = 2 \int_0^{\infty} e^{-x^2/2}\ dx$ and change the variable by setting $x = \sqrt{2y}$ to obtain $\Gamma(1/2) = J/\sqrt{2}$].

No simple expression exists for the evaluation of $\Phi(z)$. Table C provides values of $\Phi(z)$ for selected $z \geq 0$ and may be used whenever numerical evaluation of $\Phi(z)$ is required. The symmetry of φ implies that

$$\Phi(-z) = 1 - \Phi(z) \tag{2}$$

[see (33.4)], which is to be used for obtaining values of $\Phi(z)$ from Table C when $z < 0$. A standard normal rv, Z, has moments of all orders. That is, $E(Z^\alpha) < \infty$ for every $\alpha > 0$. This follows from the finiteness of the integrals

$$\int_0^{\infty} x^\alpha \varphi(x)\ dx = \int_0^{\infty} (\sqrt{2y})^\alpha \varphi(\sqrt{2y}) \frac{dy}{\sqrt{2y}}$$

$$= \frac{2^{\alpha/2-1}}{\sqrt{\pi}} \int_0^{\infty} y^{(\alpha+1)/2-1} e^{-y}\ dy = \frac{2^{\alpha/2-1}}{\sqrt{\pi}} \Gamma\left(\frac{\alpha+1}{2} \right).$$

Combining this with the symmetry of φ yields

$$E(Z^k) = \int_{-\infty}^{+\infty} x^k \varphi(x)\, dx = 0 \qquad \text{if } k \geq 0 \text{ is odd}$$

$$= \frac{2^{k/2}}{\sqrt{\pi}} \Gamma\left(\frac{k+1}{2}\right) \qquad \text{if } k \geq 0 \text{ is even.}$$

In particular, the mean and variance of Z are

$$E(Z) = 0$$

$$\text{Var}(Z) = E(Z^2) = \frac{2}{\sqrt{\pi}} \Gamma\left(\frac{3}{2}\right) = \frac{2}{\sqrt{\pi}} \cdot \frac{1}{2} \Gamma\left(\frac{1}{2}\right) = 1. \tag{3}$$

Let $X = \sigma Z + \mu$, where $\mu \in \mathcal{R}$ and $\sigma > 0$. Then X has the density

$$f(x \mid \mu, \sigma) = \frac{1}{\sigma}\, \varphi\left(\frac{x - \mu}{\sigma}\right) = \frac{1}{\sqrt{2\pi}\sigma}\, e^{-(x-\mu)^2/2\sigma^2}. \tag{4}$$

From (3),

$$E(X \mid \mu, \sigma) = \mu \quad \text{and} \quad \text{Var}(X \mid \mu, \sigma) = \sigma^2. \tag{5}$$

Normal

Any rv X with density (4) is called *normal with mean μ and variance σ^2*. The symbol $\mathcal{N}$ denotes the one-dimensional, two-parameter family of normal models. The fact that every normal random variable X can be expressed as $X = \sigma Z + \mu$ identifies $\mathcal{N}$ as the scale and location parameter family generated by the standard normal, which, accordingly, is denoted by $\mathcal{N}(0, 1)$.

For normal models we abandon the usual notational pattern for densities and distribution functions. Let F denote the distribution function of a $\mathcal{N}(\mu, \sigma^2)$ rv, X. Then

$$F(t \mid \mu, \sigma) = P\left(Z \leq \frac{t - \mu}{\sigma} \,\middle|\, \mu, \sigma\right) = \Phi\left(\frac{t - \mu}{\sigma}\right). \tag{6}$$

Relation (6) means that the value of $F(t \mid \mu, \sigma)$ can be obtained from Table C by looking up $\Phi(z)$, where $z = (t - \mu)/\sigma$, the standardized value of t (see Section 33). Thus if $\mu = 10$ and $\sigma = 2$, then $F(12 \mid 10, 2) = \Phi((12 - 10)/2) = \Phi(1) = .841$.

Example 1. Let X be a $\mathfrak{N}(3, 16)$ rv. Using (4), we obtain

$$P[2 < X < 5] = \Phi\left(\frac{5 - 3}{4}\right) - \Phi\left(\frac{2 - 3}{4}\right)$$

$$= \Phi(.5) - [1 - \Phi(.25)] = .692 + .599 - 1 = .291.$$

The table at the bottom of Table C simplifies calculations of the form in Example 2. We define $\Phi^{-1}(p)$ to be the value of z for which $\Phi(z) = p$ (i.e., the pth quantile of Φ).

Example 2. Let X be as in Example 1. Find c such that $P[X < c] = .95$. Then

$$.95 = F(c \mid 3, 16) = \Phi\left(\frac{c - 3}{4}\right).$$

Thus $(c - 3)/4 = \Phi^{-1}(.95) = 1.65$ and $c = 9.58$.

To find c such that $P[X < c] = .05$ or such that $P[|X - 3| \le c] = .95$, use the symmetry of φ. Thus in the first case $c = -3.58$ and in the second case $c = 7.84$.

B. Central Limit Theorem and Normal Approximations on $\mathcal{R}$

Let $X_1, \ldots, X_n$ be rv's and, as usual, let $S_n = \Sigma_{i=1}^n X_i = n\overline{X}_n$. For a variety of practical and theoretical reasons we require "good" approximations to the values of probabilities of events concerning S_n (or $\overline{X}_n$). Bounds (such as Markov and Chebyshev) and the weak law of large numbers provide crude or vague "guides" rather than approximations.

Theorem 1 below is a *limit* (not an approximation) theorem. It is quite remarkable in that its strong conclusion (normal limiting distribution) is valid under relatively mild conditions on the underlying model. Its proof, however, requires advanced mathematical tools. A partial proof is deferred to Chapter 9.

Reminder. (See Problem 33.3.) Consider independent X_i's each with mean μ and variance σ^2. If we denote the standardized sum by S_n^* and the standardized mean by $\overline{X}_n^*$, then

$$S_n^* = \frac{S_n - n\mu}{\sqrt{n}\sigma} = \frac{\overline{X}_n - \mu}{\sigma/\sqrt{n}} = \overline{X}_n^*.$$

Theorem 1. (Central limit theorem) *For each $n \geq 1$, let $X_1, \ldots, X_n$ be iid rv's with common distribution function F and assume that $E(X_i) = \mu$ and $\mathrm{Var}(X_i) = \sigma^2$. Let S_n^* and $\overline{X}_n^*$ denote, respectively, the standardized sum and the standardized sample mean of the X_i's. Then for every $\varepsilon > 0$ there is an n_0 (depending on ε and F) such that for $n \geq n_0$ and every z,*

$$\left| P[S_n^* \leq z] - \Phi(z) \right| = \left| P[\overline{X}_n^* \leq z] - \Phi(z) \right| < \varepsilon.$$

Here Φ denotes the standard normal distribution function.

Remark 1. An alternative statement of the conclusion of the theorem is that

$$\lim_{n \to \infty} P[S_n^* \leq z] = \Phi(z).$$

The version we chose emphasizes the fact that the rate of convergence depends on F.

Remark 2. Roughly speaking, when n is sufficiently large, the distribution of $\overline{X}_n$ is approximately $\mathfrak{N}(\mu, \sigma^2/n)$ and that of S_n is approximately $\mathfrak{N}(n\mu, n\sigma^2)$. This contrasts sharply with the Chebyshev bound. From the normal table, for example, we get $P[|\overline{X}_n - \mu| < 2\sigma/\sqrt{n}] \approx .95$. On the other hand, the bound gives only $P[(\overline{X}_n - \mu) < 2\sigma/\sqrt{n}] \geq 3/4 = .75$. However, the bound is exact and valid for all n, whereas the normal value is in error by some amount depending on n and F.

Remark 3. The distributions of standardized sums converge to a standard normal under much weaker conditions than those of Theorem 1. The identical distribution and independence of the X_i's may be replaced by conditions that would require substantial elaboration to state precisely. Roughly speaking, in the independence case, S_n will have an approximate normal distribution if none of the X_i's contributes (asymptotically) a disproportionate amount to either S_n or its variance. Moreover, the sum (or average) of a sample drawn without replacement (a dependent case) from a population whose size is large relative to the sample size (also assumed large) is approximately normal.

Theorem 1 and its various generalizations are often used to justify normality as an assumption or to explain the empirical fact that the normal distribution "fits" many observable quantities.

Let Y be a rv with mean μ and variance σ^2. By the *normal approximation* to Y we mean the approximation

$$F_Y(t) \approx \Phi\left(\frac{t - \mu}{\sigma}\right). \tag{7}$$

The following approximation theorem, also stated without proof, yields a useful variation of Theorem 1. It applies to nonidentically distributed rv's, but requires third moments. The conclusion provides a bound on the errors of normal approximations. Such theorems are called Berry–Esseen theorems.

Theorem 2. *Let* $X_1, \ldots, X_n$ *be independent rv's with* $E(X_i) = \mu_i$ *and* $\text{Var}(X_i) = \sigma_i^2$. *Let* $S_n = \Sigma_{i=1}^n X_i$, $m_n = \Sigma_{i=1}^n \mu_i$, $v_n^2 = \Sigma_{i=1}^n \sigma_i^2$, *and* $\gamma_n = \Sigma_{i=2}^n E|X_i - \mu_i|^3/v_n^3$. *Then*

$$|P[S_n^* \le t] - \Phi(t)| \le .8\gamma_n, \tag{8}$$

where, as before, S_n^* *is standardized* S_n *[i.e.,* $S_n^* = (S_n - m_n)/v_n$]. *In particular, in the iid case,* $\gamma_n = E|X_i^*|^3/\sqrt{n}$.

The bound given by (8) is too general to yield sharp results for specific cases, but adequate for proving convergence to the normal under the stated conditions if $\gamma_n \to 0$ (Lyapounov's central limit theorem) as in the case for iid rv's with third moments.

The following cases demonstrate various applications of Theorem 1. The bound in (8) is given for the binomial (see Table 1).

Case 1: Binomial Model

Recall that a binomial rv, S_n, is a sum of iid indicators, each with mean p and variance pq. Hence, by Theorem 1,

$$\lim_{n \to \infty} P\left[\frac{S_n - np}{\sqrt{npq}} \le z\right] = \Phi(z). \tag{9}$$

Table 1 shows the actual difference between the right- and left-hand sides of (9) for selected special cases of n and p.

Through more careful analysis it is possible to obtain an improved normal approximation for this model. Replace $b(k \mid n, p)$ by a histogram with intervals $[k - 1/2, k + 1/2)$. The area under the histogram equals 1, and in Figure 1 the superimposed density is that of a normal with $\mu = np = 6$ and variance $\sigma^2 = npq = 3.6$. This correspondence may be applied to the binomial for any n and p.

Table 1 Normal Approximations to Binomial ($n = 25$)

Rows: $B = B(k \mid 25, p)$; $\Phi = \Phi\left(\dfrac{k - 25p}{\sqrt{25pq}}\right)$; $\Phi^* = \Phi\left(\dfrac{k + \dfrac{1}{2} - 25p}{\sqrt{25pq}}\right)$

γ_n = error bound from Theorem 2 = $.8(1 - 2pq)/\sqrt{npq}$

	(a) $p = .1$, $\gamma_{25} = .437$			
k	2*	4	5	8
B	.537	.902	.967	1.000
Φ	.371	.841	.953	1.000
Φ^*	.500	.908	.977	1.000
$\lvert B - \Phi\rvert$	.166	.061	.014	.000
$\lvert B - \Phi^*\rvert$	.037	.006	.010	.000

	(b) $p = .2$, $\gamma_{25} = .272$					
k	2	4*	5*	8	10	12
B	.098	.421	.617	.953	.944	1.000
Φ	.067	.309	.500	.933	.994	1.000
Φ^*	.106	.401	.599	.960	.997	1.000
$\lvert B - \Phi\rvert$	.031	.113	.117	.020	.000	.000
$\lvert B - \Phi^*\rvert$	.008	.020	.018	.007	.003	.000

	(c) $p = .5$, $\gamma_{25} = .160$						
k	5	8	10	12	15	16	18
B	.002	.054	.212	.500	.885	.946	.993
Φ	.001	.036	.159	.421	.841	.919	.986
Φ^*	.003	.055	.212	.500	.885	.945	.992
$\lvert B - \Phi\rvert$	.001	.018	.053	.079	.044	.027	.007
$\lvert B - \Phi^*\rvert$	.001	.001	.000	.000	.000	.001	.001

*, Worst cases.

A modified normal approximation is obtained by substituting areas under the normal density for areas under the histogram. By construction, $P[m \leq S_n \leq k]$ (k, m integers) is the area under the histogram between $m - 1/2$ and $k + 1/2$. The corresponding area under the normal density is $\Phi((k + 1/2 - np)/\sqrt{npq}) - \Phi((k - 1/2 - np)/\sqrt{npq})$. The fact is that this approximation is generally superior to the one obtained by treating S_n as a $\mathfrak{N}(np, npq)$ rv. This is illustrated for special cases in Figure 1 and Table 1. The improved approximation is called the de Moivre–Laplace theorem and was the first case of the central limit theorem to be proved.

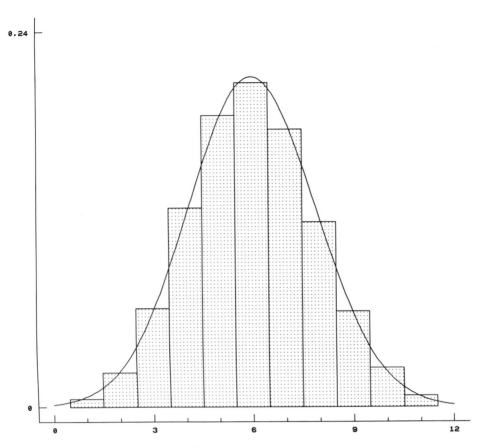

Figure 1 Normal fit to $b(15, .4)$.

The preferred normal approximation to $B(k \mid n, p)$ is

$$B(k \mid n, p) = P[S_n \le k] \approx \Phi\left(\frac{k + 1/2 - np}{\sqrt{npq}}\right). \qquad (10)$$

The addition of $1/2$ in (10) is called the *continuity correction*. Note that the continuity correction enables us to find a nonzero normal approximation to $P[S_n = k]$ for integer k. Thus

$$P[S_n = k] = P\left[k - \frac{1}{2} \le S_n \le k + \frac{1}{2}\right]$$
$$\approx \Phi\left(\frac{k + 1/2 - np}{\sqrt{npq}}\right) - \Phi\left(\frac{k - 1/2 - np}{\sqrt{npq}}\right) \qquad (11)$$

for n sufficiently large.

In general, a continuity correction will lead to improvement whenever the normal approximation is applied to discrete rv's with equally spaced values.

How large is sufficiently large in (10) and (11)? Of course, the answer depends on the accuracy desired in the approximation. A rule of thumb useful for most purposes is that the normal approximation to the binomial distribution will be reasonable if

$$n \cdot \min(p, q) \geq 6. \tag{12}$$

Table 1 gives error values for specific cases.

In Example 37.4 a sample size required to attain a specified accuracy in predicting an election result was determined. In particular, the requirement $P[|Y_n| \leq .03] \geq .95$ ($Y_n = S_n/n$) was met by taking $n \geq 5556$. This value of n uses the Chebyshev inequality. A better result can be found by using the normal approximation. Ignoring the continuity correction, the requirement becomes

$$\Phi\left(\frac{.03}{\sqrt{pq/n}}\right) - \Phi\left(\frac{.03}{\sqrt{pq/n}}\right) \geq .95$$

or, equivalently, that

$$\Phi\left(\frac{.03}{\sqrt{pq/n}}\right) \geq .975.$$

This is achieved if $.03\sqrt{n/(pq)} \geq 1.960$ or, using the fact that $pq \leq 1/4$, it is required that $n \geq 1098$. In fact, larger sample sizes are used by pollsters for two reasons.

1. The method of sampling is not random, so that the variance may, in fact, be slightly larger than pq/n.
2. Interviewing is by telephone, so that a larger sample is used to protect against unanswered calls and other sources of nonresponse.

Case 1(a): Hypergeometric Model

The standardized form of a $\mathcal{H}(n, M_1, M_2)$ rv is approximately $\mathcal{N}(0, 1)$ if n and $M - n$ are large and $p = M_1/M$ is moderate. This is clearly so when n is large and the conditions of Theorem 39.4 hold. In applying the normal approximation the continuity correction should again be used.

Example 3. Let S_{900} be a $\mathcal{H}(900, 1250, 1250)$ rv. Then $E(S_{900}) = 450$ and $\text{Var}(S_{900}) = 900(1/4)[(2500 - 900)/(2500 - 1)] \approx 144$. Thus

$$P[S_{900} \leq 475] \approx \Phi\left(\frac{475 + 1/2 - 450}{12}\right) = \Phi(2.12) = .983,$$

which agrees, to three places, with the exact value.

Case 2: Poisson Model

Let $X_1, X_2, \ldots$ be iid $\mathcal{P}_o(\lambda)$ rv's. For each $n = 1, 2, \ldots$ it follows from Corollary 40.1 that $\Sigma_{i=1}^n X_i$ is a $\mathcal{P}_o(n\lambda)$ rv. Applying Theorem 1 yields

$$\lim_{n \to \infty} P\left[\frac{\Sigma_{i=1}^n X_i - n\lambda}{\sqrt{n\lambda}} \leq z\right] = \Phi(z). \tag{13}$$

Hence if X_μ is a $\mathcal{P}_o(\mu)$ rv, then

$$\lim_{\mu \to \infty} P\left[\frac{X_\mu - \mu}{\sqrt{\mu}} \leq z\right] = \Phi(z). \tag{14}$$

That is, for μ large X_μ is approximately a $\mathcal{N}(0, 1)$ rv.

By Theorem 39.5, the $\mathcal{P}_o(np)$ and $\mathcal{B}(n, p)$ models are close when n is large and p is small. Since $np \geq 6$ is usually adequate for the approximation of $\mathcal{B}(n, p)$ by $\mathcal{N}(np, npq)$, it follows that $\mu \geq 6$ is usually adequate for the approximation of $\mathcal{P}_o(\mu)$ by $\mathcal{N}(\mu, \mu)$. Hence for the approximation of $\mathcal{P}_o(n\lambda)$ by $\mathcal{N}(n\lambda, n\lambda)$ to be adequate, we usually require that $n\lambda \geq 6$. Thus $\lambda = 1$ means that we require $n \geq 6$, and $\lambda = .1$ means that we require $n \geq 60$. In fact, there is no n simultaneously sufficiently large for all values of λ.

Example 4. Let X be a $\mathcal{P}_o(10)$ rv. Then using the continuity correction, we obtain

$$P[X \leq 14] = P[X \leq 14.5] = P\left[\frac{X - 10}{\sqrt{10}} \leq \frac{14.5 - 10}{\sqrt{10}}\right]$$

$$\approx \Phi(1.42) = .922.$$

The exact answer, from Table B, is $P[X \leq 14] = .917$.

By applying Theorem 2 we could give a bound on the error for the Poisson case, but it is particularly bad and we omit it.

Case 3: Gamma Model

Let $X_{\alpha,\beta}$ be a $\mathcal{G}(\alpha, \beta)$ rv. Then $X_{\alpha,\beta} = \sum_{i=1}^{n} U_i$, where the U_i are independent, $\mathcal{G}(\alpha/n, \beta)$ rv's. This was shown for α an integer by relating $X_{\alpha,\beta}$ to the waiting time for the αth occurrence in a Poisson process. The proof for noninteger α follows from the general closure property of the $\mathcal{G}$ family under addition of independent rv's (Example 48.6). From Theorem 1,

$$\lim_{\alpha \to \infty} P\left[\frac{X_{\alpha,\beta} - \alpha\beta}{\beta\sqrt{\alpha}} \leq z \right] = \Phi(z). \tag{15}$$

That is, for α large, the distribution function of the standardized version of $X_{\alpha,\beta}$ is approximately the $\mathcal{N}(0, 1)$ distribution function.

Example 5. In a Poisson process with rate 10 per hour, what is the probability that the waiting time for the fifteenth occurrence is at least 2 hours? The waiting time T_k for the kth occurrence is a $\mathcal{G}(k, 1/10)$ rv. Hence

$$P[T_{15} \geq 2] = P\left[\frac{T_{15} - 15/10}{\sqrt{15}/10} \right] \approx 1 - \Phi(1.29) = 1 - .900 = .100.$$

The exact answer is .105.

C. Normal Models on $\mathcal{R}^m$

This subsection requires knowledge of basic matrix algebra. Vectors will be assumed to be in column form. Various facts concerning multivariate normals will be stated, mostly without proof.

Let $Z_1, \ldots, Z_n$ be independent standard normal rv's so that the density of $\mathbf{Z} = (Z_1, \ldots, Z_n)'$ is

$$f(t) = \prod_{i=1}^{n} \varphi(t_i) = \left(\frac{1}{2\pi} \right)^{n/2} \exp\left(-\frac{1}{2} \sum_{i=1}^{n} t_i^2 \right) \tag{16}$$

The covariance matrix of $\mathbf{Z}$ is $\mathbf{I}_n$, the n-dimensional identity matrix. Let $\mathbf{X} = \mathbf{AZ} + \boldsymbol{\mu}$, where $\mathbf{A} \neq \mathbf{0}$ is an $m \times n$ matrix and $\boldsymbol{\mu}$ is a fixed m-dimensional vector. The distribution of $\mathbf{X}$ is determined by $\boldsymbol{\mu}$ and $\mathbf{AA}'$ (Theorem 4).

Normal Random Vector

The vector $\mathbf{X}$ given above is an *m-dimensional normally distributed random vector with mean $\boldsymbol{\mu}$ and covariance matrix* $\mathbf{C} = \mathbf{AA}'$ [see (35.14) and (35.15)]. The model is denoted by $\mathcal{N}_m(\boldsymbol{\mu}, \mathbf{C})$ and the components, $X_1, \ldots, X_m$, are said to be *jointly normal*. The distribution of $\mathbf{Z}$ is $\mathcal{N}_n(\mathbf{0}, \mathbf{I}_n)$, which is called the *standard normal model on $\mathcal{R}^n$*.

When **C** is nonsingular, **X** has a density given by

$$g(\mathbf{s} \mid \boldsymbol{\mu}, \mathbf{C}) = \frac{1}{(2\pi)^{m/2}(\det \mathbf{C})^{1/2}} e^{-(1/2)(\mathbf{s}-\boldsymbol{\mu})'\mathbf{C}^{-1}(\mathbf{s}-\boldsymbol{\mu})}. \tag{17}$$

The proof of (17) is in Chapter 8. For $m = 2$, (17) becomes

$$g(\mathbf{s} \mid \boldsymbol{\mu}, \mathbf{C}) = \frac{1}{2\pi\sigma_1\sigma_2\sqrt{1 - \rho^2}} e^{-Q/2}, \tag{18}$$

where

$$Q = \frac{(s_1 - \mu_1)^2/\sigma_1^2 - 2\rho(s_1 - \mu_1)(s_2 - \mu_2)/(\sigma_1\sigma_2) + (s_2 - \mu_2)^2/\sigma_2^2}{1 - \rho^2}$$

and ρ is the correlation between X_1 and X_2.

When **C** is singular (rank $\mathbf{C} = k < m$), **X** has essential dimension k (Theorem 35.7). There is an orthogonal transformation $\mathbf{BX} = \begin{pmatrix} \mathbf{Y} \\ \mathbf{w} \end{pmatrix}$, where

Y is k-dimensional and has a nonsingular covariance matrix and **w** is a constant (Problem 35.19).

Remark 4. Subvectors of normally distributed random vectors are normally distributed. Any linear combination of the components of a normally distributed random vector is a normally distributed rv. Furthermore, any linear transformation of a normally distributed random vector is normally distributed.

Remark 5. The conditional distribution of $\mathbf{Y}_1$ given $\mathbf{Y}_2 = \boldsymbol{\nu}$ is also normal. The case $m = 2$ is the subject of Problem 8.

Remark 6. The standardization process in $m > 1$ dimensions is analogous to standardization in one dimension. If **X** has mean vector $\boldsymbol{\mu}$ and nonsingular covariance matrix **C**, its standardized form is

$$\mathbf{X}^* = \mathbf{C}^{-1/2}(\mathbf{X} - \boldsymbol{\mu}),$$

where $\mathbf{C}^{1/2}$ is the symmetric "square root" of the covariance matrix **C** (known to exist in this case) and $\mathbf{C}^{-1/2}$ is the inverse of $\mathbf{C}^{1/2}$. Then $E(\mathbf{X}^*) = 0$ and

$$C(\mathbf{X}^*) = \mathbf{C}^{-1/2}E((\mathbf{X} - \boldsymbol{\mu})(\mathbf{X} - \boldsymbol{\mu})')\mathbf{C}^{-1/2} = \mathbf{C}^{-1/2}\mathbf{C}\mathbf{C}^{-1/2} = \mathbf{I}_m.$$

If **X** is $\mathfrak{N}_m(\boldsymbol{\mu}, \mathbf{C})$, then $\mathbf{X}^*$ is $\mathfrak{N}_m(0, \mathbf{I}_m)$, a standard normal random vector. Conversely, if **X** is $\mathfrak{N}_m(\boldsymbol{\mu}, \mathbf{C})$, then $\mathbf{X} = \mathbf{C}^{1/2}\mathbf{Z} + \boldsymbol{\mu}$, where **Z** is $\mathfrak{N}_m(0, \mathbf{I}_m)$.

Theorem 3. *Let* $\mathbf{X}$ *be a* $\mathfrak{N}_m(\boldsymbol{\mu}, \mathbf{C})$ *random vector and consider the partitions*

$$\mathbf{X} = \begin{pmatrix} \mathbf{Y}_1 \\ \mathbf{Y}_2 \end{pmatrix}, \ \boldsymbol{\mu} = \begin{pmatrix} \boldsymbol{\nu} \\ \boldsymbol{\xi} \end{pmatrix}, \ and \ \mathbf{C} = \begin{pmatrix} \mathbf{C}_{11} & \mathbf{C}_{12} \\ \mathbf{C}'_{12} & \mathbf{C}_{22} \end{pmatrix}, \ with \ \mathbf{Y}_1 \ and \ \boldsymbol{\nu} \ k\text{-}dimensional$$

and $\mathbf{C}_{11}$ *a* $k \times k$ *matrix. Then* $\mathbf{Y}_1$ *and* $\mathbf{Y}_2$ *are independent if, and only if,* $\mathbf{C}_{12} = \mathbf{0}$.

Proof. Assume that $\mathbf{C}$ (and hence $\mathbf{C}_{11}$ and $\mathbf{C}_{22}$) is nonsingular. The proof when $\mathbf{C}$ is singular is the subject of Problem 10.

Let $\mathbf{C}^{-1} = \begin{pmatrix} \mathbf{B}_{11} & \mathbf{B}_{12} \\ \mathbf{B}'_{12} & \mathbf{B}_{22} \end{pmatrix}$, where $\mathbf{B}_{11}$ is $k \times k$. The relation between $\mathbf{C}$

and $\mathbf{C}^{-1}$ is such that $C_{12} = 0$ is equivalent to $\mathbf{B}_{12} = 0$. Let $\mathbf{s} = \begin{pmatrix} \mathbf{t} \\ \mathbf{u} \end{pmatrix}$, where

$\mathbf{t} \in \mathfrak{R}^k$ is a possible value of $\mathbf{Y}_1$ and $\mathbf{u}$ is a possible value of $\mathbf{Y}_2$.

From (17),

$$-2 \log g(\mathbf{s} \mid \boldsymbol{\mu}, \mathbf{C}) = (\mathbf{s} - \boldsymbol{\mu})'\mathbf{C}^{-1}(\mathbf{s} - \boldsymbol{\mu})$$
$$= (\mathbf{t} - \boldsymbol{\nu})'\mathbf{B}_{11}(\mathbf{t} - \boldsymbol{\nu}) + (\mathbf{u} - \boldsymbol{\xi})'\mathbf{B}_{22}(\mathbf{u} - \boldsymbol{\xi})$$
$$+ 2(\mathbf{t} - \boldsymbol{\nu})'\mathbf{B}_{12}(\mathbf{u} - \boldsymbol{\xi}). \tag{19}$$

The last term on the right side of (19) vanishes for all $\mathbf{t}$ and $\mathbf{u}$, if, and only if, $\mathbf{B}_{12} = 0$. Hence $\mathbf{C}_{12} = 0$ is necessary and sufficient for

$$-2 \log g(\mathbf{s} \mid \boldsymbol{\mu}, \mathbf{C}) = h_1(\mathbf{t}) + h_2(\mathbf{u})$$

or, equivalently, for

$$g(\mathbf{s} \mid \boldsymbol{\mu}, \mathbf{C}) = f_1(\mathbf{t})f_2(\mathbf{u}). \tag{20}$$

Since (20) expresses the fact that $\mathbf{Y}_1$ and $\mathbf{Y}_2$ are independent, the proof is complete for the nonsingular case.

An immediate consequence of Theorem 3 is:

Corollary 1. *Jointly normal rv's are independent if, and only if, they are uncorrelated.*

Theorem 4. *If* $\mathbf{X}$ *is an m-dimensional, normally distributed random vector, its distribution is determined by its mean and its covariance matrix.*

Proof. It suffices to consider only mean $\mathbf{0}$ since the mean is a location vector. Let $Z_1, \ldots, Z_n$ be iid $\mathfrak{N}(0, 1)$ rv's, $\mathbf{Z} = (Z_1, \ldots, Z_n)'$, and $\mathbf{W} = (Z_1, \ldots, Z_k)'$. Let $\mathbf{X} = \mathbf{AZ}$ and $\mathbf{Y} = \mathbf{BW}$ and assume that $\mathbf{C} = \mathbf{AA'} = \mathbf{BB'}$, where $\mathbf{C}$ is $m \times m$. Then $\mathbf{X}$ and $\mathbf{Y}$ have the same covariance matrix, $\mathbf{C}$.

There exists an orthogonal matrix, **P**, such that **PCP'** is diagonal. Hence **PX** and **PY** have uncorrelated components. By Corollary 1, the components of **PX** are independent as are those of **PY**. Hence their distributions are determined by their respective marginal (normal) distributions, which are the same. Thus the distributions of **PX** and **PY** are the same. Finally, **X** = **P'PX** and **Y** = **P'PY** have the same distributions.

Let $\mathbf{X}_1, \ldots, \mathbf{X}_n$ be iid random vectors with mean $\boldsymbol{\mu}$ and nonsingular covariance matrix **C**. As in the case of rv's, the standardizations of $\mathbf{S}_n = \sum_{i=1}^n \mathbf{X}_i$ and $\overline{\mathbf{X}}_n = \mathbf{S}_n/n$ are the same,

$$\mathbf{C}^{-1/2} \frac{\mathbf{S}_n - n\boldsymbol{\mu}}{\sqrt{n}} = \sqrt{n}\mathbf{C}^{-1/2}(\overline{\mathbf{X}}_n - \boldsymbol{\mu}). \tag{21}$$

Even if **C** is singular,

$$\frac{\mathbf{S}_n - n\boldsymbol{\mu}}{\sqrt{n}} = \sqrt{n}(\overline{\mathbf{X}}_n - \boldsymbol{\mu}). \tag{22}$$

The following theorem, given without proof, is a generalization of Theorem 1 and suggests that under certain conditions, multivariate normal approximations are reasonable.

Theorem 5. (Central limit theorem) *Let* $\mathbf{X}_1, \mathbf{X}_2, \ldots$ *be a sequence of iid m-dimensional random vectors with mean* $\boldsymbol{\mu}$ *and covariance matrix* $\mathbf{C} \neq \mathbf{0}$. *Then the distribution of the random vector in* (22) *converges to* $\mathcal{N}_m(\mathbf{0}, \mathbf{C})$ *as* $n \to \infty$.

Example 6. Let $\mathbf{S}_n$ be a $\mathcal{B}_3(n, p)$ random vector so that $S_{ni} = \sum_{j=1}^n I_{B_{ij}}$, where B_{ij} is the event that A_i occurs on the jth trial. The random vector $(I_{B_{1j}}, I_{B_{2j}}, I_{B_{3j}})'$ has mean $\boldsymbol{\mu} = (p_1, p_2, p_3)'$ and covariance matrix

$$\mathbf{C} = \begin{pmatrix} p_1 q_1 & -p_1 p_2 & -p_1 p_3 \\ -p_1 p_2 & p_2 q_2 & -p_2 p_3 \\ -p_1 p_3 & -p_2 p_3 & p_3 q_3 \end{pmatrix}.$$

By Theorem 5, for large n, the distribution of $(\mathbf{S}_n - n\boldsymbol{\mu})/\sqrt{n}$ is approximately $\mathcal{N}_3(\mathbf{0}, \mathbf{C})$.

Summary

$\mathcal{N}(0, 1)$ *density:* $\varphi(x) = (1/\sqrt{2\pi})e^{-x^2/2}$.
$\mathcal{N}(\mu, \sigma^2)$ *rv X:* $X = \sigma Z + \mu$, where Z is a $\mathcal{N}(0, 1)$ rv.
$\mathcal{N}_m(\boldsymbol{\mu}, \mathbf{C})$ *random vector* **X**: **X** = **AZ** + $\boldsymbol{\mu}$, where **C** = **AA'** and **Z** = $(Z_1, \ldots, Z_n)'$ with $Z_1, \ldots, Z_n$ iid, $\mathcal{N}(0, 1)$.

Central limit theorem: If $X_1, \ldots, X_n$ are iid rv's with mean μ and variance σ^2, the standardization

$$\frac{\overline{X}_n - \mu}{\sigma/\sqrt{n}} = \frac{S_n - n\mu}{\sqrt{n}\sigma}$$

is approximately $\mathfrak{N}(0, 1)$ if n is large. Normal approximations to the binomial, hypergeometric, Poisson, and gamma distributions are discussed.

If $\mathbf{X}_1, \ldots, \mathbf{X}_n$ are iid random vectors with mean $\boldsymbol{\mu}$ and covariance matrix $\mathbf{C}$, $(\mathbf{S}_n - n\boldsymbol{\mu})/\sqrt{n} = \sqrt{n}(\overline{\mathbf{X}}_n - \boldsymbol{\mu})$ is approximately $\mathfrak{N}(\mathbf{0}, \mathbf{C})$ if n is large.

Problems

1. Let X be a $\mathfrak{N}(8, 25)$ rv. Find the following probabilities.
   (a) $P[9 < X < 20]$
   (b) $P[X > 10.2]$
   (c) $P[X \leq 8.3]$
2. Let X be as in Problem 1. Find c satisfying:
   (a) $P[X < c] = .01$
   (b) $P[|X - 8| < c] = .95$
3. Compare the $\mathfrak{B}(n, p)$ distribution function, with its Poisson and normal approximations when $n = 20$ and $p = .01, .1, .2, .3, .4, .5$.
4. In an election, 1,000,000 people are expected to vote. A pollster takes a sample of 1200 voters. Assume that 510,000 will vote for candidate A. Find the probability that the pollster does not correctly forecast the winner of the election. Will your answer change if 10,000,000 are expected to vote and 5,100,000 will vote for A?
5. Compare the $\mathfrak{Po}(\lambda)$ distribution function and its normal approximation when $\lambda = 1, 5, 10$.
6. Suppose that customers arrive in a store according to a Poisson process with rate two per minute. Let X be the number arriving during a particular 1-hour period. Find approximate values of:
   (a) $P[110 \leq X < 140]$
   (b) $P[X \geq 150]$
   (c) $P[X < 80]$
7. In Problem 6, let T_k be the waiting time in hours for the kth arrival. Find approximate values for:
   (a) $P[.5 \leq T_{100} \leq 1.5]$
   (b) $P[T_{120} > 3]$
   (c) $P[T_{20} \leq 1]$

8. Let X_i $(i = 1, \ldots, n)$ be independent indicators with $p_i = P[X_i = 1] = 1 - q_i$. Use Theorem 2 to show that $P[S_n^* \le t] \to \phi(t)$ if $\Sigma_{i=1}^{\infty} p_i q_i$ is divergent.
9. In Problem 8, suppose that $p_i = 1/2$ for each i and let $Y_i = 2X_i - 1$. Determine whether the distribution of $\Sigma_{i=1}^{n} iY_i$ (standardized) converges to a normal.
10. Let $\mathbf{X}$ be a $\mathcal{N}_2(\boldsymbol{\mu}, \mathbf{C})$ random vector. Show that the conditional distribution of X_2 given $X_1 = t$ is normal and identify its conditional mean and conditional variance.
11. Let $\mathbf{X}$ be a $\mathcal{N}_m(\boldsymbol{\mu}, \mathbf{C})$ random vector and $\mathbf{Y} = \mathbf{BX} + v$ $(\mathbf{B} \ne 0)$ be r-dimensional.
    (a) Verify that $\mathbf{Y}$ is a $\mathcal{N}_r(\boldsymbol{\eta}, \mathbf{D})$ random vector. Evaluate $\boldsymbol{\eta}$ and $\mathbf{D}$.
    (b) Use part (a) to verify that for any choice of $b_1, \ldots, b_m$, either $\Sigma_{i=1}^{m} b_i X_i$ is degenerate or it is a $\mathcal{N}(\xi, \tau^2)$ rv. Evaluate ξ and τ^2.
12. Complete the proof of Theorem 3. *Hint*: If $\mathbf{C}$ is singular, so is $\mathbf{C}_{11}$ or $\mathbf{C}_{22}$ (or both). Use Problem 35.19 on each of $\mathbf{Y}_1$ and $\mathbf{Y}_2$.
13.† Let $\mathbf{X}$ and $\mathbf{Y}$ be independent m-dimensional normally distributed random vectors. Show that $\mathbf{X} + \mathbf{Y}$ is normally distributed and find its mean and covariance matrix.
14. Verify (18), which gives the form of the bivariate normal density.
15. (a) Prove the statements in Remark 4.
    (b) Identify the means and covariance matrices of the random vectors referred to in Remark 4.
16. Show that for large n an approximation alternative to (11) is

$$\frac{1}{\sqrt{npq}} \, \Phi\left(\frac{k - np}{\sqrt{npq}}\right),$$

the $\mathcal{N}(np, npq)$ density evaluated at k.
17. Show that the formula for γ_n for the $\mathcal{B}(n, p)$ model is correct as in Table 1.

43. CONTINUOUS UNIFORM MODELS AND SIMULATION

A uniform model is a model for selecting a point at random from a set S in such a way that any subset of S has probability proportional to its "size." When $S \subset \mathcal{R}^n$ contains a nondegenerate, n-dimensional rectangle, the size of a subset is its n-dimensional volume, which for uniform models, must be finite.

A.   Uniform Models on $\mathcal{R}$

Let $a < b$ and

$$u(t \mid a, b) = \frac{1}{b - a} \qquad \text{if } a < t < b$$

$$= \frac{I_{(a,b)}(t)}{b - a}. \tag{1}$$

It is easy to verify that $u(t \mid a, b)$ is a density and that the corresponding distribution function is

$$U(t \mid a, b) = \frac{t - a}{b - a} \qquad \text{if } a \le t \le b$$

$$= 1 \qquad \text{if } t \ge b. \tag{2}$$

Let $a \le c \le d \le b$. For the model on $\mathcal{R}$ represented by the density $u(t \mid a, b)$, clearly $P(J) = (d - c)/(b - a)$, where J is an interval between c and d with or without the endpoints. Thus the probability of any interval $J \subset [a, b]$ is proportional to its length. If $J \subset [a, b]^c$, then $P(J) = 0$.

Uniform Model

The model on $\mathcal{R}$ represented by the density $u(t \mid a, b)$ is called the *uniform model* on (a, b) and is designated by $\mathcal{U}(a, b)$. The family is designated by $\mathcal{U}$.

Any member of $\mathcal{U}$ can be obtained from any other member by a change in location and/or scale. The usual choice of a standard member is the $\mathcal{U}(0, 1)$ model; it has mean $\mu = 1/2$ and variance $\sigma^2 = 1/12$. If X is a $\mathcal{U}(a, b)$ rv, then $X = a + (b - a)Y$ where Y is a $\mathcal{U}(0, 1)$ rv. It follows, therefore, that

$$E(X \mid a, b) = a + \frac{b - a}{2} = \frac{a + b}{2} \tag{3}$$

and

$$\text{Var}(X \mid a, b) = \frac{(b - a)^2}{12}. \tag{4}$$

In summary, a is a location parameter, $\theta = b - a$ is a scale parameter, and $(a + b)/2$ is a point of symmetry.

A $\mathcal{U}(0, 1)$ rv can be experimentally approximated by generating a sequence of independent random digits (i.e., independent, uniform selections from $\{0, 1, \ldots, 9\}$). A sequence of length n provides a rv, X_0, with the discrete uniform distribution on $\{i/10^n: i = 0, 1, \ldots, 10^n - 1\}$, which approximates the distribution of a $\mathfrak{U}(0, 1)$ rv.

Computers and many hand calculators make available pseudorandom numbers that simulate sequences of random digits. Use of pseudorandom numbers causes little difficulty if not too many are used together in the same simulation. A sequence of 5000 pseudorandom digits appears as Table D.

From a $\mathfrak{U}(0, 1)$ rv it is possible to generate a rv with any specified distribution. This process is called simulation.

The normal approximation in Section 42 yields a simple method of simulating normally distributed rv's starting from any sufficiently long sequence of iid rv's with two moments. The $\mathfrak{U}(0, 1)$, because of its availability, is a natural starting point.

Let $X_1, \ldots, X_n$ be $\mathfrak{U}(0, 1)$ rv's and $S_n = \Sigma_{i=1}^n X_i$. Then for n large,

$$S_n^* = \frac{S_n - n/2}{\sqrt{n/12}} \tag{5}$$

is approximately a $\mathfrak{N}(0, 1)$ rv and $\sigma S_n^* + \mu$ is approximately a $\mathfrak{N}(\mu, \sigma^2)$ rv. For most purposes, $n = 12$ will suffice and simplifies (5).

Example 1. Suppose that we wish to simulate a $\mathfrak{N}(10, 36)$ rv. Beginning in columns 16 to 20 of the seventh row of Table D and proceeding down those columns, we obtain the 12 values for iid $\mathfrak{U}(0, 1)$ rv's given in Table 1. Thus $S_n = 4.67531$ has been observed and hence $S_n^* = 4.67531 - 6 = -1.32469$ is a simulated observation of a $\mathfrak{N}(0, 1)$ rv. A simulated value of a $\mathfrak{N}(10, 36)$ rv is then $6(-1.32469) + 10 = 2.05186$.

The theorems that follow give the relationships between the $\mathfrak{U}(0, 1)$ distribution function and other distribution functions from which many simulation procedures are derived. We begin by defining the *inverse* of a distribution function G as

$$G^{-1}(p) = \min\{x: G(x) \geq p\} \tag{6}$$

for $0 < p < 1$. See Figure 1 for examples of $G^{-1}(p)$. We have defined $G^{-1}(p)$ to be the smallest pth quantile of G (see Problem 14).

Table 1

.91059	.14583	.19258	.48446
.95721	.38991	.02308	.09051
.04479	.55440	.23007	.65188

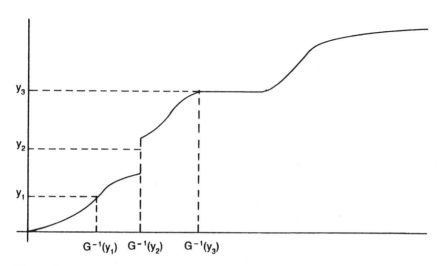

Figure 1

Theorem 1. *Let Y be a rv with continuous distribution function G. Then $G(Y)$ is a $\mathcal{U}(0, 1)$ rv.*

Proof. Let F be the distribution function of $G(Y)$. Clearly, $F(t) = 0$ if $t \leq 0$ and $F(t) = 1$ for $t \geq 1$. If $0 < t < 1$, then

$$F(t) = P[G(Y) \leq t] = P[Y \leq G^{-1}(t)] = G(G^{-1}(t)) = t. \qquad (7)$$

The validity of the second equality in (7) may be seen by examination of Figure 1.

The following theorem is a partial converse of Theorem 1; it does not require that G be continuous.

Theorem 2. *Let X be a $\mathcal{U}(0, 1)$ rv and let G be a univariate distribution function. Then $Y = G^{-1}(X)$ has distribution function G.*

Proof. By definition,

$$F_Y(t) = P[G^{-1}(X) \leq t] = P[X \leq G(t)] = G(t). \qquad (8)$$

See Figure 1 to verify the validity of the second equality in (8) in all cases.

Suppose that an iid sample from some distribution function G is to be simulated. By Theorem 2 we can do so if we are able to simulate a sample from the $\mathcal{U}(0, 1)$ distribution and G^{-1} is computationally manageable. In some cases, the rv Y may be a function $h(Z)$, where Z is some random vector with a known distribution function. We may simulate Z, then com-

pute $Y = h(Z)$. Performing n independent trials of this simulation experiment yields iid rv's $Y_1, \ldots, Y_n$ with the distribution function G of Y. We
can estimate $G(t)$ by the relative frequency of Y_i's, which are $\leq t$.

Other characteristics of Y or of G may be estimated in a similar fashion.
For instance, a natural estimate of $E(Y)$ is $\overline{Y}_n = \Sigma_{i=1}^n Y_i/n$, some of whose
properties were discussed in Sections 36 and 42.

Example 2.   One area in which simulation is often used is in the design
and analysis of queuing systems (i.e., waiting lines). For instance, consider
a facility (e.g., a repair shop) to which items arrive for some service (e.g.,
repair of the item), which can be performed by either of two servers (e.g.,
repair persons). Assume that the time X between arrivals has an exponential distribution and that the time Y that it takes to repair an item (by
either server) has a Weibull distribution. Specifically, for $t > 0$,

$$f_X(t) = \frac{5}{4} e^{-5t/4} \qquad \overline{F}_X(t) = e^{-5t/4}$$

$$f_Y(t) = \frac{8}{9} te^{-(2t/3)^2} \qquad \overline{F}_Y(t) = e^{-(2t/3)^2}.$$

(Assume that time is measured in hours.) Times between arrivals (X_i's)
are assumed independent with density f_X and service times (Y_i's) are assumed independent with density f_Y. Also, X_i's and Y_i's are assumed to be
mutually independent.

Quantities of general interest include W, the expected time an item
spends in the system; W_Q, the expected time waiting in line; and expected
times the servers are idle. Analytic expressions concerning such quantities
are easily obtained when f_Y is any exponential density. However, in the
case at hand f_Y is Weibull and no explicit formulas exist.

To simulate this process it is necessary to generate sequences with densities f_X and f_Y, respectively. Let U be a $\mathscr{U}(0, 1)$ rv. Then, from Theorem 2, $-(4/5)\log(1 - U)$ [or, equivalently, $-(4/5)\log U$] has a
$\mathscr{G}(1, 4/5)$ distribution and $(3/2)\sqrt{-\log U}$ has the appropriate Weibull distribution. By observing a sequence of iid $\mathscr{U}(0, 1)$ rv's we may, therefore,
simulate the queue and estimate quantities of interest.

Sixteen hours of the system were simulated (from time 0 to time 16)
with no items in the system initially. Table 2 displays the calculations up
to time 5.878. The evolution of the system is displayed completely in Figure 2.

In Table 2, the event A is an arrival, while S_i is a completion of service
by server i. The number in the system is N, so that the number in the
queue is $\max(0, N - 2)$.

Table 2

			Next service completion				
u_1	u_2	Arrival	First server	Second server	Event	Time	N
.00367	—	4.486	—	—	A	4.486	1
.97101	.42260	4.509	5.878	—	A	4.509	2
.64804	.91104	4.856	5.878	4.967	A	4.856	3
.87462	—	4.963	5.878	4.967	A	4.963	4
.05223	—	7.325	5.878	4.967	S_2	4.967	3
—	.44764	7.325	5.878	6.312	S_1	5.878	2
—	.11992	7.325	8.063	6.312	S_2	6.312	1

The uniforms for interarrival times were obtained from Table D by starting down columns 11 to 15 from row 6, continuing to the top of columns 16 to 20. These are denoted by u_1. The uniforms for service times (u_2) were similarly obtained starting in row 6, columns 21 to 25.

The first item arrived at time $T_1 = -(4/5)\log(.00367) = 4.486$, and since the system was previously empty, immediately entered service, finishing at time $V_1 = T_1 + (3/2)\sqrt{-\log(.42260)} = 5.878$. The second item arrived at time $T_2 = T_1 - (4/5)\log(.97101) = 4.509$, and since only one item was previously in the system, immediately entered service, finishing at time $V_2 = T_2 + (3/2)\sqrt{-\log(.91104)} = 4.967$. The third item arrived at time $T_3 = T_2 - (4/5)\log(.64804) = 4.856$ and entered the queue; and so on.

During the 16-hour period, 18 items arrived and the total of interarrival times was 15.994 hours, so that $\overline{X} = .889$ hour. Thirteen services were completed requiring a total service time of 19.733 hours, so that $\overline{Y} = 1.518$. Compare these with $E(X) = .8$ hour and $E(Y) = 3\sqrt{\pi}/4 = 1.33$ hours.

The average time in the system for the 13 that completed service was 2.133 hours. They spent an average of .615 hour in the queue. Out of the total of 32 hours of available server time, 10.274 hours or 32.1% were unused.

The simulation methods in this section are straightforward but not necessarily efficient, either with respect to computation time or the number of random digits required. More sophisticated and efficient procedures are available for special models and there is continuing activity and interest in developing new ones.

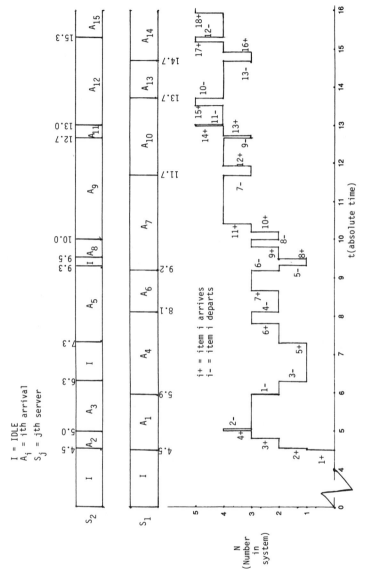

Figure 2 Simulated queue summary.

B.   Uniform Models on $\mathcal{R}^n$

For any $C \subset \mathcal{R}^n$, let $M(C)$ be the n-dimensional volume of C.

Uniform Model

Let $\mathcal{S} \subset \mathcal{R}^n$ and $0 < M(\mathcal{S}) < \infty$. The model on $\mathcal{R}^n$ that assigns probability $M(\mathcal{S}B)/M(\mathcal{S})$ to $B \subset \mathcal{R}^n$ is called the uniform model on $\mathcal{S}$ and is denoted by $\mathcal{U}_n(\mathcal{S})$. The $\mathcal{U}_n(\mathcal{S})$ density is

$$u_n(\mathbf{t} \mid \mathcal{S}) = \frac{I_\mathcal{S}(\mathbf{t})}{M(\mathcal{S})}. \tag{9}$$

Example 3.†   Let $X_1, \ldots, X_n$ be iid $\mathcal{U}(a, b)$ rv's. Then $\mathbf{X} = (X_1, \ldots, X_n)$ is a $\mathcal{U}_n(\times_{i=1}^n (a, b))$ random vector.

If $Y_1 < Y_2 < \cdots < Y_n$ are the X_i arranged in order, then Y_i is called the *ith order statistic from* $X_1, \ldots, X_n$. Under our assumptions, $Y = (Y_1, \ldots, Y_n)$ is a $\mathcal{U}_n(\mathcal{S})$ random vector where $\mathcal{S} = \{t: t \in \mathcal{R}^n, a < t_1 < t_2 < \cdots < t_n < b\}$. Note that $M(\mathcal{S}) = (b - a)^n/n!$. We will consider order statistics in greater generality in Section 44 and in Chapter 8.

Probabilities connected with $\mathcal{U}_n(\mathcal{S})$ random vectors are often easily computed by geometric considerations. This is particularly true when $n = 2$.

Example 4.   Let $\mathbf{X}$ be a $\mathcal{U}_2(\mathcal{S})$ random vector where $\mathcal{S}$ is the triangle bounded by $t_1 = 0$, $t_2 = 0$, and $t_1 + t_2 = 1$, so that $M(\mathcal{S}) = 1/2$. From Figure 3 we compute $P[X_1 + X_2 < 1/2]$. Now $P[X_1 + X_2 < 1/2] = P[X \in B]$, where B is the triangle bounded by $t_1 = 0$, $t_2 = 0$, and $t_1 + t_2 = 1/2$. Thus $M(B) = 1/8$ and

$$P\left[X_1 + X_2 < \frac{1}{2}\right] = \frac{M(B)}{M(\mathcal{S})} = \left(\frac{1}{8}\right)\bigg/\left(\frac{1}{2}\right) = \frac{1}{4}.$$

Example 5.   Let f be a density on R^n and let $\mathbf{X}$ be a $\mathcal{U}_{n+1}(\mathcal{S})$ random vector, where

$$\mathcal{S} = \{(x_1, \ldots, x_{n+1}): x_{n+1} < f(x_1, \ldots, x_n)\}.$$

The random vector $(X_1, \ldots, X_n)$ obtained by deleting X_{n+1} from X has marginal density f. This is a generalization of the result in Problem 23.9.

Summary

$\mathcal{U}(a, b)$ *density:* $u(t \mid a, b) = I_{(a,b)}(t)/(b - a)$.
$\mathcal{U}_n(\mathcal{S})$ *density:* $u_n(t \mid \mathcal{S}) = I_\mathcal{S}(t)/M(\mathcal{S})$, where $M(\mathcal{S})$ is the n-dimensional volume of $\mathcal{S}$.

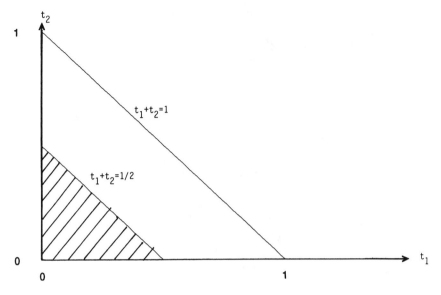

Figure 3

The $\mathcal{U}(a, b)$ distribution function is

$$U(t \mid a, b) = \frac{t - a}{b - a} \qquad \text{if } a < t < b$$

$$= 1 \qquad \text{if } t \geq b$$

and its mean and variance are $(a + b)/2$ and $(b - a)^2/12$, respectively. The family is a location and scale family; the usual choices are a for the location parameter and $b - a$ for the scale parameter. Then $\mathcal{U}(0, 1)$ is the standard member.

$\mathcal{U}(0, 1)$ rv's are useful in simulation of rv's. See Examples 1 and 2.

Problems

1. Let X be a $\mathcal{U}(3, 5)$ rv.
   (a) Find $P[3.5 < X < 4]$.
   (b) Find the .75 quantile of X.
   (c) Find the median of X^2.
2. In Example 4, find:
   (a) $P[X_1 < 1/2]$
   (b) $P[X_2 > 3/4]$
   (c) $P[|X_1 - X_2| < 1/3]$

3. Let X_1 and X_2 be independent, $\mathcal{U}(0, 1)$ rv's. Find:
   (a) $P[X_1 + X_2 < 3/4]$
   (b) $P[|X_1 - X_2| > 1/2]$
   (c) $P[Y_1 < 1/2, Y_2 > 3/4]$, where $Y_1 = \min(X_1, X_2)$ and $Y_2 = \max(X_1, X_2)$
   (d) The distribution function and density of $X_1 + X_2$

In Problems 4 and 5, let **X** be a $\mathcal{U}_2(\mathcal{S})$ random vector. Solve these problems geometrically where possible.

4. Let $\mathcal{S}$ be the interior of the circle of radius 1 centered at the origin. Let $Y_1 = (X_1^2 + X_2^2)^{1/2}$ and Y_2 be the angle, in radians, between the t_1 axis and the line through the origin and X.
   (a) Find the joint distribution function of Y_1 and Y_2.
   (b) Find the joint density of Y_1 and Y_2.
   (c) Are Y_1 and Y_2 independent?
   (d) Find the marginal densities of Y_1 and of Y_2.
5. Let $\mathcal{S}$ be the parallelogram bounded by $t_2 = 0$, $t_2 = 2$, $t_1 = t_2$, and $t_1 = t_2 + 2$.
   (a) Find the joint distribution function of $X_1 + X_2$ and $X_1 - X_2$.
   (b) Find the joint density of $X_1 + X_2$ and $X_1 - X_2$.
   (c) Are $X_1 + X_2$ and $X_1 - X_2$ independent?
   (d) Find the marginal densities of $X_1 + X_2$ and $X_1 - X_2$.
6. Let $Z_1, Z_2, \ldots$ be a sequence of random digits. Observe Z_1. If $Z_1 = 9$, observe Z_2. Define X by

$$X = 0 \quad \text{if } Z_1 = 0, 1 \text{ or } (Z_1 = 9 \text{ and } Z_2 = 0, 1, 2, 3, 4)$$

$$= 1 \quad \text{if } Z_1 = 2, 3, 4, 5, 6$$

$$= 2 \quad \text{if } Z_1 = 7, 8 \text{ or } (Z_1 = 9 \text{ and } Z_2 = 5, 6, 7, 8, 9).$$

   (a) Show that X is a $B(2, .5)$ rv.
   (b) Let T be the number of Z_i observed and find $E(T)$.
   (c) Is this method of generating a $B(2, .5)$ rv better than direct application of Theorem 2? Explain.
7. How would you use a sequence of random digits to generate a sequence of n independent tosses of a fair coin? Of a fair die?
8. Let F and G be distribution functions and assume that F is continuous. Suppose that X has distribution function F. Find a function h so that $h(X)$ has distribution function G.
9. Suppose that X is a $\mathcal{G}(1, 1)$ rv. Apply the result of Problem 8 to obtain a rv $h(X)$ with the distribution function G of Example 25.1.
10. Let X be a $\mathcal{G}(1, 1)$ rv.
    (a) Compute $E(\sqrt{X})$.
    (b) Simulate 25 independent $\mathcal{G}(1, 1)$ rv's.

(c) Use the 25 rv's of part (b) to estimate $E(\sqrt{X})$ and compare your estimate with the value in part (a).

11. Give a complete discussion of the existence and value of the αth moment of the $\mathcal{U}(a, b)$ distribution allowing $-\infty < \alpha < \infty$.

12. Let θ be a $\mathcal{U}(0, 2\pi)$ rv and let $Y = \tan \theta$. Simulate rv's Y_{ij} ($i = 1, 2$; $j = 1, \ldots, 25$), each with the same distribution as Y. Let G be the distribution function of Y.
    (a) Use the Y_{1j} to estimate $G(-2)$, $G(-1)$, $G(0)$, $G(1)$, and $G(2)$.
    (b) Repeat part (a) using the Y_{2j}.
    (c) Let $\overline{Y}_j = (Y_{1j} + Y_{2j})/2$ and use the $\overline{Y}_j$ to estimate their common distribution function at the same points as in part (a).
    (d) Compare the estimates in parts (a), (b), and (c).

13. Let h be a nonnegative function on the finite interval $[a, b]$ and assume that h is bounded by M finite. Consider the experiment consisting of selecting a point (U, V) at random in the rectangle $[a, b] \times [0, M]$ and let X be the indicator of the event $[V \leq h(U)]$.
    (a) Determine the distribution of X.
    (b) Determine the conditional density of U given $X = 1$.
    (c) The above forms the basis of a simulation technique (called *Monte Carlo*) for estimating the integral of a bounded function on a finite interval. Explain how this might be done.

14. Let G be a distribution function.
    (a) Show that $G^{-1}(p)$ is the smallest pth quantile of G.
    (b) When is the pth quantile of G unique?

15. (a) Redo Problem 13 in the case in which h is a function on a bounded n-dimensional rectangle.
    (b) Use (a) and Example 5 to develop a method of simulating an n-dimensional random vector with a specified density satisfying appropriate conditions.

44. BETA MODELS

For $\alpha, \beta > 0$, the function

$$B(\alpha, \beta) = \frac{\Gamma(\alpha)\Gamma(\beta)}{\Gamma(\alpha + \beta)} = \int_0^1 x^{\alpha-1}(1 - x)^{\beta-1} \, dx,$$

given in (40.22), is called the *beta function*. For $\alpha, \beta > 0$, a family of densities related to the beta function is given by

$$be(x \mid \alpha, \beta) = \frac{\Gamma(\alpha + \beta)}{\Gamma(\alpha)\Gamma(\beta)} x^{\alpha-1}(1 - x)^{\beta-1} \qquad \text{if } 0 < x < 1. \qquad (1)$$

The model on $\mathcal{R}$ represented by this density is called the *beta model* with parameters α, β and is denoted by $\mathcal{B}e(\alpha, \beta)$. The $\mathcal{U}(0, 1)$ model is the $\mathcal{B}e(1, 1)$ model.

If X is a $\mathcal{B}e(\alpha, \beta)$ rv, then

$$E(X^\gamma \mid \alpha, \beta) = \frac{\Gamma(\alpha + \beta)\Gamma(\alpha + \gamma)}{\Gamma(\alpha)\Gamma(\alpha + \beta + \gamma)}$$

so that

$$E(X \mid \alpha, \beta) = \frac{\alpha}{\alpha + \beta} \tag{2}$$

and

$$\text{Var}(X \mid \alpha, \beta) = \frac{\alpha\beta}{(\alpha + \beta)^2(\alpha + \beta + 1)}. \tag{3}$$

Beta models quite often arise in connection with binomials, as will be illustrated in this section. Another application is to the distribution of order statistics of iid $\mathcal{U}(0, 1)$ rv's (see Example 1 and Problem 6).

A. Binomial and Beta Tails

The beta density be is clearly similar in form to the binomial probability function b. The following lemma provides an analytic relation between them.

Lemma 1. *The functions b and be satisfy*

$$\frac{d}{dp} b(r \mid n, p) = be(p \mid r, n - r + 1)$$

$$\begin{aligned}
&\quad - be(p \mid r + 1, n - r) &&\text{if } r = 1, \ldots, n - 1; \\
&= be(p \mid n, 1) &&\text{if } r = n; \\
&= -be(p \mid 1, n - 1) &&\text{if } r = 0. \tag{4}
\end{aligned}$$

Proof. First for $r = 0, 1, \ldots, n$ note that

$$\frac{d}{dp}(p^r q^{n-r}) = rp^{r-1}q^{n-r} - (n - r)p^r q^{n-r-1}. \tag{5}$$

By definition of b and Γ,

$$b(r \mid n, p) = P(X = r \mid n, p) = \frac{\Gamma(n + 1)}{\Gamma(r + 1)\Gamma(n - r + 1)} p^r q^{n-r}.$$

Now differentiate and use (5) to obtain

$$\frac{d}{dp} b(r \mid n, p) = \frac{\Gamma(n + 1)}{\Gamma(r)\Gamma(n - r + 1)} p^{r-1}q^{n-r}$$

$$- \frac{\Gamma(n + 1)}{\Gamma(r + 1)\Gamma(n - r)} p^{r}q^{n-r-1}$$

$$= be(p \mid r, n - r + 1) - be(p \mid r + 1, n - r)$$

$$\text{if } r = 1, \ldots, n - 1,$$

which was to be shown. Completion for the extreme values $r = 0$ and $r = n$ is left to the reader.

A simple application of Lemma 1 provides an important identity between binomial and beta tails.

Theorem 1.   *Let Y be a $\mathcal{B}(n, p)$ rv and let X be a $\mathcal{B}e(k, n - k + 1)$ rv. Then, $P[Y \geq k] = P[X \leq p]$ or, in terms of distribution functions,*

$$1 - B(k - 1 \mid n, p) = Be(p \mid k, n - k + 1). \tag{6}$$

Relation (6) is valid for $k = 1, 2, \ldots, n$.

Proof. $P[Y \geq k] = \sum_{r=k}^{n} b(r \mid n, p)$. Hence

$$\frac{d}{dp} P[Y \geq k] = \sum_{r=k}^{n} \frac{d}{dp} b(r \mid n, p)$$

$$= \sum_{r=k}^{n-1} [be(p \mid r, n - r + 1) - be(p \mid r + 1, n - r)]$$

$$+ be(p \mid n, 1) \tag{7}$$

$$= be(p \mid k, n - k + 1).$$

The last equality follows because the sum is a telescoping series. Equation (6) results from integration in (7).

Example 1.†   In Example 43.3 the joint density of the order statistics for a sample of $\mathcal{U}(a, b)$ rv's was given. Consider a sample $X_1, \ldots, X_n$ from $\mathcal{U}(0, 1)$ and let Y_k be the kth-order statistic. Then, calling $[X_i \leq p]$ a "success," $[Y_k \leq p]$ is the event that there are at least k successes in n trials. Hence since $P[X_i \leq p] = p$,

$$F_{Y_k}(p) = P[Y_k \leq p] = 1 - B(k - 1 \mid n, p) = Be(p \mid k, n - k + 1).$$

Thus Y_k is a $\mathcal{B}e(k, n - k + 1)$ rv.

B. Beta Models as Priors

In Problem 28.8, Y is a $\mathcal{B}(4, p)$ rv and the $\mathcal{U}(0, 1)$, that is, $\mathcal{B}e(1, 1)$, prior density was assumed for p. The posterior density of p given $Y = k$ ($k = 0, 1, 2, 3, 4$) is

$$f(t \mid k) = c(k)t^k(1 - t)^{4-k} \qquad \text{if } 0 < t < 1,$$

that is, the $\mathcal{B}e(1 + k, 1 + 4 - k)$ density.

This is an example of a more general property stated in the following theorem.

Theorem 2. *Let S_n be a $\mathcal{B}(n, p)$ rv and assume that p has the prior density $\mathcal{B}e(\alpha, \beta)$. Then the posterior density of p given $S_n = k$ is $\mathcal{B}e(\alpha + k, \beta + n - k)$.*

Proof. Since $P(S_n = k \mid p) = \binom{n}{k} p^k(1 - p)^{n-k}$, the posterior density of p is

$$f(t \mid k) = c(\alpha, \beta, k)t^{\alpha+k-1}(1 - t)^{\beta+n-k-1},$$

which is $be(t \mid \alpha + k, \beta + n - k)$.

The closure property of Theorem 2 illustrates the concept of one family ($\mathcal{B}e$) being "conjugate" with respect to another ($\mathcal{B}$) (see Problem GP11). This closure property makes beta priors convenient in connection with binomial experiments. In addition, varying α and β leads to a wide variety of shapes, one of which may be a good approximation to your prior opinion. Various shapes are demonstrated in Figure 1. Reversing the t scale, that is, interchanging t and $1 - t$, results in an interchange of α and β. Also $be(t \mid \alpha, \beta) \to \infty$ as $t \to 0$ if $\alpha < 1$ and as $t \to 1$ if $\beta < 1$.

Summary

$\mathcal{B}e(\alpha, \beta)$ density: $be(t \mid \alpha, \beta) = [\Gamma(\alpha + \beta)/(\Gamma(\alpha)\Gamma(\beta))]t^{\alpha-1}(1 - t)^{\beta-1}$ for $0 < t < 1$.

The mean and variance for the $\mathcal{B}e(\alpha, \beta)$ model are $\alpha/(\alpha + \beta)$ and $\alpha\beta/[(\alpha + \beta)^2(\alpha + \beta + 1)]$, respectively. Two useful relations for the $\mathcal{B}e$ family are:

1. $B(k \mid n, p) = Be(q \mid n - k, k + 1) = 1 - Be(p \mid k + 1, n - k)$
    (see Problem 5).
2. If S_n is a $\mathcal{B}(n, p)$ rv and the prior distribution of p is $\mathcal{B}e(\alpha, \beta)$, the posterior distribution of p given $S_n = k$ is $\mathcal{B}e(\alpha + k, \beta + n - k)$.

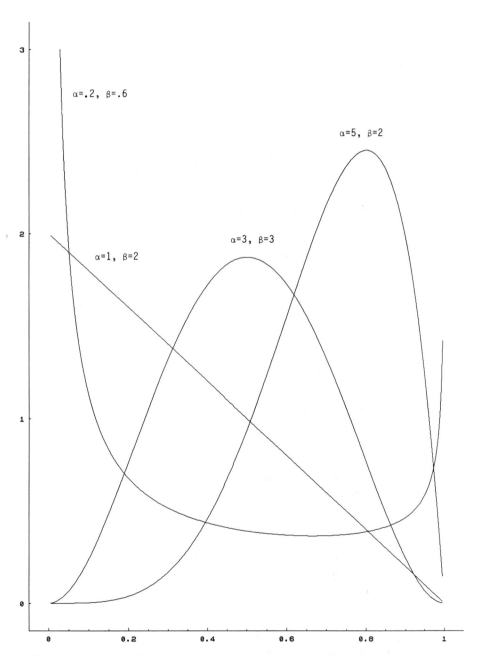

Figure 1 Beta densities.

Problems

1. Let X be a $\mathscr{B}e(3, 2)$ rv.
   (a) Find $P[.1 < X \le .5]$
   (b) Find the mode of X.
   (c) Use trial and error to find the median of X.
2. Verify (2) and (3).
3. Let S_n be a $\mathscr{B}(n, p)$ rv and assume the $\mathscr{B}e(\alpha, \beta)$ prior density for p.
   (a) Find $E(p \mid S_n)$.
   (b) Suppose that you have observed S_n and wish to estimate (guess a value of) p. If your estimate is a, you will lose $(p - a)^2$. Why should your estimate be $E(p \mid S_n)$?
   (c) Determine the marginal probability function of S_n. The result is called the *beta binomial* probability function.
   (d) Find $E(S_n)$.
4. Let T_k be a $\mathscr{B}(k, p)$ rv and assume the $\mathscr{B}e(\alpha, \beta)$ prior density for p. What is the posterior density of p given $T_k = r$?
5. Verify the expressions in the summary relating B to Be.
6. Use the result of Example 1 to determine the density of Y_k when the X_i are:
   (a) $\mathscr{U}(a, b)$
   (b) $\mathscr{G}(1, \beta)$
   Hint: $X_i < X_j$ if, and only if, $F(X_i) < F(X_j)$.

GENERAL PROBLEMS

1. (a) Show that for every $\lambda > 0$ and every integer $n > 0$ there exist iid rv's X_i ($i = 1, \ldots , n$) such that $\Sigma_{i=1}^n X_i$ is distributed $\mathscr{P}(\lambda)$.
   (b) What conclusion can you draw from part (a) about the normal approximation?
2. In a sequence of Bernoulli trials, let S_n be the number of successes in the first n trials and T_k be the waiting time for the kth success.
   (a) Show how normal approximations to the distribution of S_n and of T_k can be used to approximate the probability of the same event.
   (b) Do these approximations yield the same value?
3. In a Poisson process, let S_t be the number of occurrences in $(0, t)$ and T_k be the waiting time for the kth occurrence. Repeat parts (a) and (b) of Problem 2 in this case.
4. (a) In Problem 39.12, show that $p > 1/2$ implies that $\lim_{n \to \infty} P_n = 1$.
   (b) What implication does the result in part (a) have for the probability of correct decoding in Problem 39.23.

Families of Univariate Discrete Probability Functions

Name	Probability function	Mean	Variance	Application
Binomial, $\mathcal{B}(n, p)$ $n = 1, 2, \ldots$ $0 \le p \le 1$ $q = 1 - p$	$b(k \mid n, p) = \binom{n}{k} p^k q^{n-k},$ $k = 0, 1, \ldots, n$	np	npq	Number of successes in a sequence of n Bernoulli trials.
Hypergeometric, $\mathcal{H}(n, M_1 M_2)$ $n, M_1, M_2 \ge 0,$ integers, $n \le M_1 + M_2$	$h(k \mid n, M_1, M_2)$ $= \dfrac{\binom{M_1}{k}\binom{M_2}{n-k}}{\binom{M_1+M_2}{n}},$	$np,$ where $p = \dfrac{M_1}{M_1 + M_2}$	$npq\left(\dfrac{N-n}{N-1}\right),$ where $q = 1 - p$ and $N = M_1 + M_2$	Number of items with property π in sample of size n without replacement.
Poisson, $\mathcal{P}_o(\mu)$ $\mu > 0$	$p(k \mid \mu) = \dfrac{\mu^k}{k!}\, e^{-\mu},$ $k = 0, 1, \ldots.$	μ	μ	Number of occurrences during $(0, t)$ in a Poisson process with rate λ $\cdot$ $\mu = \lambda t.$
Negative binomial $\mathcal{B}^-(\lambda, p)$ $0 < p \le 1$ $q = 1 - p$ $k = 1, 2, \ldots.$ ($k = 1$ is geometric)	$b^-(r \mid k, p)$	$\dfrac{k}{p}$	$\dfrac{kq}{p^2}$	Number of Bernoulli trials to produce K successes.

Families of Univariate Densities

Name	Density	Mean	Variance	Application
Uniform on (a,b), $\mathfrak{U}(a,b)$ $a<b$	$u(t\mid a,b)=\dfrac{1}{b-a}$, $a<t<b$	$\dfrac{a+b}{2}$	$\dfrac{(b-a)^2}{12}$	Simulation
Normal, $\mathfrak{N}(\mu,\sigma^2)$ $\sigma>0$	$f(t\mid\mu,\sigma)=\dfrac{1}{\sqrt{2\pi}\,\sigma}e^{-(t-\mu)^2/2\sigma^2}$	μ	σ^2	Approximate distribution of standardized form of $\overline{X}_n$ for n large is $\mathfrak{N}(0,1)$ under certain conditions.
Gamma, $\mathfrak{G}(\alpha,\beta)$ $\alpha,\beta>0$ ($\alpha=1$ is exponential)	$g(t\mid\alpha,\beta)=\dfrac{t^{\alpha-1}}{\beta^\alpha\Gamma(\alpha)}\,e^{-t/\beta}$, $t>0$	$\alpha\beta$	$\alpha\beta^2$	For integer α, waiting time for αth occurrence in a Poisson process with rate $1/\beta$.
Beta, $\mathfrak{Be}(\alpha,\beta)$ $\alpha,\beta>0$ ($\alpha=\beta=1$ is $\mathfrak{U}(0,1)$)	$be(t\mid\alpha,\beta)=\dfrac{\Gamma(\alpha+\beta)}{\Gamma(\alpha)\Gamma(\beta)}\,t^{\alpha-1}(1-t)^{\beta-1}$, $0<t<1$	$\dfrac{\alpha}{\alpha+\beta}$	$\dfrac{\alpha\beta}{(\alpha+\beta)^2(\alpha+\beta+1)}$	Prior density for p in Bernoulli trials.

5. Let

$$f(x) = ke^{-(Ax^2 + Bx + C)} \qquad \text{if } -\infty < x < \infty.$$

   (a) For what values of A, B, C, and k is f a density?
   (b) Can you identify any families discussed in this chapter whose densities are of the form of f?
   (c) For the families found in part (b), find the expressions for the usual parameters in terms of A, B, C, and k.
   (d) Are there any densities of the form of f not belonging to any family found in part (b)?
6. Suppose that the density in Problem 5 is only positive when $x > 0$. Repeat Problem 5 in this case.
7. In a sequence of Bernoulli trials, let T be the number of trials until at least one success and at least one failure occur.
   (a) Find $E(T)$ and $\text{Var}(T)$ in more than one way.
   (b) Find the probability function of T.
8. (a) Let f be a bounded density and assume that $A = \{x: f(x) > 0\}$ is bounded. How can the result of Problem 23.9 be used to simulate a rv with density f?
   (b) If A is unbounded, how can the method in part (a) be used to simulate a rv whose density approximates f? How good is the approximation?
9. (a) Let f and g be densities on $\mathcal{R}$ such that $f/g \le c$ on $\{x: f(x) > 0\}$. Let $Y, Y_1, Y_2, \ldots$ be independent, each with density g and let $U, U_1, U_2, \ldots$ be independent, $\mathcal{U}(0, 1)$ rv's. Assume that each of the Y's and the corresponding U's are independent. Let N be the smallest index i for which $U_i \le f(Y_i)/[cg(Y_i)]$ and let $X = Y_N$. Show that X has density f. *Hint*: Evaluate

$$P[X \le x] = P\left(Y \le x \mid U \le \frac{f(Y)}{cg(Y)}\right).$$

   (b) Use the result in part (a) to obtain a method of simulating $|X|$ where X is $\mathcal{N}(0, 1)$. Let Y_i be $\mathscr{E}(1, 1)$ rv's. Use a small integer value for c.
   (c) Show that the result in part (a) generalizes the simulation method in Problem 8.
10. An *exponential family* of probability models for a rv X is one that is represented as follows:

$$f(x \mid \theta) = c(\theta)h(x)e^{v(\theta)t(x)} \qquad \text{for } \theta \in (a, b),$$

   where (i) f is either a probability function or a density, (ii) the interval (a, b) may be bounded or unbounded, (iii) the domain of f is $\mathcal{R}$, but the only relevant values of x are those for which $h(x) > 0$ (this does not depend on θ), and (iv) v is continuous and strictly monotone.

(a) Which families or subfamilies of models presented in this chapter are exponential families? Identify c, h, v, and t for each of them.

(b) Reparametrize the family so that $\omega = v(\theta)$. Verify that the family parametrized by ω is an exponential family. The parameter ω is called the *natural parameter* of the family.

(c) Find the mean and variance of $t(X)$ as a function of θ. *Hint*: In an exponential family of densities,

$$\frac{d}{d\theta} \int_{-\infty}^{\infty} g(x)f(x \mid \theta) \, dx = \int_{-\infty}^{\infty} g(x) \frac{d}{d\theta} f(x \mid \theta) \, dx$$

for any g such that $E(g(X) \mid \theta)$ exists. A similar conclusion holds for exponential families of probability functions when $\int$ is replaced by Σ.

11. Let $\mathcal{P} = \{P_\theta : \theta \in \Theta\}$ be a family of probability models and suppose that each P_θ is a conditional model for **X** given θ. The family $\mathcal{Q}$ of prior distributions over Θ is said to be *conjugate with respect to $\mathcal{P}$* if the posterior $q(\theta \mid x)$ belongs to $\mathcal{Q}$ for every **x**. In each of the following, identify a conjugate family.

(a) Poisson: **X** $= (X_1, \ldots, X_n)$, where the X_i are iid $\mathcal{P}_o(\lambda)$.

(b) Gamma: **X** $= (X_1, \ldots, X_n)$, where the X_i are iid $\mathcal{G}(\alpha_0, 1/\lambda)$, α_0 fixed.

(c) Normal: **X** $= (X_1, \ldots, X_n)$, where the X_i are iid $\mathcal{N}(\mu, \sigma_0^2)$ rv, σ_0 fixed

12. Let $X_1, \ldots, X_n$ be iid, $\mathcal{N}(\mu, 1/\theta)$ rv's. Consider the family $\mathcal{F}$ of priors for (μ, θ) given by:

(i) The conditional distribution of μ given θ is $\mathcal{N}(\varphi, 1/(\lambda\tau))$.

(ii) The marginal distribution of τ is $\mathcal{G}(\alpha, \beta)$.

Verify that $\mathcal{F}$ is conjugate with respect to the given family of normals.

13. Let X be a $\mathcal{B}e(\alpha, \beta)$ rv. Use Theorem 44.1 to show that for suitable α and β, the standardization X^* is approximately $\mathcal{N}(0, 1)$.

REFERENCES

Bhattacharya, R. N., and Rao, R. R. (1976). *Normal Approximation and Asymptotic Expansions*. New York: John Wiley and Sons, Inc. Includes a very complete discussion of bounds on errors in normal approximations and other Berry–Esseen theorems at an advanced level.

DeGroot, Morris H. (1986). *Probability and Statistics*, 2nd ed. Reading, Mass: Addison-Wesley Publishing Co., Inc. Contains a discussion of conjugate families as well as an orderly development of the special distributions.

Feller, William (1970). *An Introduction to Probability Theory and Its Applications*, Vol. 1, 3rd ed. New York: John Wiley and Sons, Inc. Proof of the deMoivre–Laplace theorem. Development of the binomial, hypergeometric, and Poisson distributions and their relationships.

Hillier, F. S., and Lieberman, G. J. (1990). *Introduction to Operations Research*, 5th ed. New York: McGraw-Hill Book Company. Good introduction to simulation, including a discussion of computer algorithms for generating pseudo-random numbers.

Ross, Sheldon M. (1989). *Introduction to Probability Models*, 4th ed. Orlando, Fla. Academic Press, Inc. Excellent chapter on simulation techniques. Poisson processes, related distributions, and applications.

Woodroofe, Michael (1975). *Probability with Applications*. New York: McGraw-Hill Book Company. Central limit theorem and its extension.

8.

Functions of Random Vectors

45. INTRODUCTION

Let $\mathbf{X}$ be an n-dimensional random vector with a specified distribution and suppose that $\mathbf{Y} = g(\mathbf{X})$, where g is a function taking values in $\mathcal{R}^k$. The object of this chapter is to describe and apply various techniques for obtaining the distribution of $\mathbf{Y}$. Section 46 (sums), 47 (Jacobians), and 48 (sample statistics) deal with special classes of problems for which reasonably complete solutions can be explicitly determined. The principles on which these methods are based are described briefly in this section.

Case I: $\mathbf{X}$ Is Discrete with Probability Function f

This is the easiest case in principle but often leads to difficult analytic problems. The direct approach is simply to evaluate $f_{\mathbf{Y}}$ as follows. Let $A_{\mathbf{y}} = \{\mathbf{x}: g(\mathbf{x}) = \mathbf{y}\}$ and then

$$f_{\mathbf{Y}}(\mathbf{y}) = P[\mathbf{Y} = \mathbf{y}] = P(A_{\mathbf{y}}) = \sum_{\mathbf{x} \in A_{\mathbf{y}}} f(\mathbf{x}). \tag{1}$$

Example 1. Let X_1 and X_2 be independent, each distributed uniformly on $J_N = \{1, 2, \ldots, N\}$. Then any event $A \subset J_N \times J_N$ has probability n_A/N^2, where n_A is the number of points in A. Suppose now that $\mathbf{Y} = (Y_1, Y_2)$, where $Y_1 = X_1 + X_2$ and $Y_2 = \max(X_1, X_2)$ Then for specified

338

$\mathbf{y} = (y_1, y_2)$, $A_\mathbf{y}$ contains either one point (if $y_1 = 2y_2$) or two points (if $y_2 + 1 \le y_1 < 2y_2$). Hence for positive integers $y_1 \le 2N$ and $y_2 \le N$,

$$f_\mathbf{Y}(\mathbf{y}) = \frac{1}{N^2} \quad \text{if } y_1 = 2y_2$$

$$= \frac{2}{N^2} \quad \text{if } y_2 + 1 \le y_1 < 2y_2.$$

Obviously, applying (1) can yield explicit results only when f and g cooperate to produce analytically simple $P(A_\mathbf{y})$.

Another approach that sometimes works involves distribution functions instead of probability functions. Suppose that g is real-valued ($k = 1$) and let $B_t = \{\mathbf{x}: g(x) \le t\}$. If $P(B_t) = F_Y(t)$ can be easily determined, then f_Y can be obtained from F_Y by differencing.

Example 2. Let $X_1, X_2, \ldots, X_r$ be iid rv's with probability function

$$f(k) = \frac{2k}{N(N + 1)} \quad \text{if } k = 1, 2, \ldots, N.$$

Let $Y = \max_{1 \le i \le r} X_i$. Then

$$F_Y(t) = P\left[\max_{1 \le i \le r} X_i \le t\right] = [F(t)]^r = \left[\frac{t(t + 1)}{N(N + 1)}\right]^r \quad \text{if } 1 \le t \le N$$

and

$$f_Y(t) = \frac{t^r[(t + 1)^r - (t - 1)^r]}{[N(N + 1)]^r} \quad \text{if } t = 1, 2, \ldots, N.$$

Case II: **X** Is Continuous with Density *f* and Distribution Function *F*

If g is sufficiently "smooth," $\mathbf{Y}$ will have a density $f_\mathbf{Y}$, and the object is to find it or to find $F_\mathbf{Y}$. If it can be shown that for every k-dimensional rectangle B,

$$P[\mathbf{Y} \in B] = \int_B h(\mathbf{y}) \, d\mathbf{y} \tag{2}$$

for some function h, then $f_\mathbf{Y} = h$. The same conclusion follows if (2) is valid for any set of the form $B = \{\mathbf{y}: y_i \le t_i, i = 1, \ldots, k\}$, in which case $P[\mathbf{Y} \in B] = F_\mathbf{Y}(\mathbf{t})$.

If $k = 1$, so that $g(\mathbf{X}) = Y$ is a rv, it is often best to proceed by obtaining F_Y first, then differentiating to obtain f_Y.

Example 3. Let $\mathbf{X} = (X_1, X_2)$, $Y = X_2/X_1$, and assume that f satisfies $f(x_1, x_2) > 0$ only if $0 \leq x_2 < x_1$. Then

$$P[Y \leq t] = P(X_2 \leq tX_1] = \int_0^\infty \left[\int_0^{tx_1} f(x_1, x_2) \, dx_2 \right] dx_1$$

$$= \int_0^\infty \left[\int_0^{tx_1} f(x_1, x_1 v) \, dv \right] dx_1$$

$$= \int_0^t \left[\int_0^\infty x_1 f(x_1, x_1 v) \, dx_1 \right] dv.$$

The last expression implies that

$$f_Y(t) = \int_0^\infty x_1 f(x_1, x_1 t) \, dx_1. \tag{3}$$

Consider the following choices for f:

(a) If $f(x_1, x_2) = e^{-x_1}$ for $0 \leq x_2 < x_1$, then

$$f_Y(t) = \int_0^\infty x_1 e^{-x_1} \, dx_1 = 1 \qquad \text{for } 0 \leq t < 1.$$

(b) If $f(x_1, x_2) = 2e^{-(x_1 + x_2)}$ for $0 \leq x_2 < x_1$, then

$$f_Y(t) = \int_0^\infty 2 x_1 e^{-(x_1 + x_1 t)} \, dx_1 = \frac{2}{(1 + t)^2} \qquad \text{for } 0 \leq t < 1.$$

(c) If $f(x_1, x_2) = c x_1 x_2^2$ for $0 \leq x_2 < x_1 \leq 1$, then $f_Y(t) = 3t^2$ for $0 \leq t < 1$.

The results of this example can be used to obtain distribution of $Y = \min(X_2/X_1, X_1/X_2)$ for any nonnegative iid X_1, X_2.

Example 4. Let $\mathbf{X} = (X_1, \ldots, X_n)$ and $\mathbf{Y} = (Y_1, Y_2)$, where $Y_1 = \max X_i$ and $Y_2 = \min X_i$. Let $t > s$. Then

$$P[Y_1 \leq t, Y_2 > s] = P\left(\bigcap_{i=1}^n B_i \right),$$

where $B_i = [s < X_i \leq t]$. Hence

$$F_\mathbf{Y}(t, s) = P[Y_1 \leq t] - P[Y_1 \leq t, Y_2 > s]$$

$$= F_\mathbf{X}(t, \ldots, t) - P\left(\bigcap_{i=1}^n B_i \right).$$

In the special case of independent X_i's with continuous distribution functions, F_i,

$$F_{\mathbf{Y}}(t, s) = \prod_{i=1}^{n} F_i(t) - \prod_{i=1}^{n} [F_i(t) - F_i(s)].$$

When the model for $\mathbf{X}$ is uniform, some problems are readily solved by appealing to geometric arguments, as in the following examples.

Example 5. Let (X_1, X_2) be uniform on the unit square and let $Y = X_1 + X_2$. Then (see Figure 1) $P[Y \leq t]$ is the area in the lower-triangular region bounded by the line $x_1 + x_2 = t$ if $t \leq 1$. For $t > 1$, $P[Y > t]$ is the comparable area in the upper-triangular region. Thus

$$F_Y(t) = \frac{1}{2} t^2 \qquad \text{if } 0 \leq t \leq 1$$

$$= 1 - \frac{1}{2}(2 - t)^2 \qquad \text{if } 1 < t \leq 2.$$

Finally, Y has the triangular density

$$f_y(t) = t \qquad \text{if } 0 \leq t \leq 1$$

$$= 2 - t \qquad \text{if } 1 < t \leq 2.$$

Example 6. Let $\mathbf{X} = (X_1, \ldots, X_n)$ be uniform in the unit sphere in $\mathcal{R}^n$ and let $Y = \sqrt{\sum_{i=1}^{n} X_i^2}$. Then $P[Y \leq t] = V_n(t)/V_n(1)$, where $V_n(t)$ is the volume of an n-dimensional sphere of radius t. Since $V_n(t)$ is proportional to t^n, it follows that $F_Y(t) = t^n$ and $f_Y(t) = nt^{n-1}$ for $0 \leq t < 1$.

The result readily extends to spheres of arbitrary radius r. The family thus generated has scale parameter r.

There are times where it becomes necessary, or simply convenient, to express points $\mathbf{x} \in \mathcal{R}^n$ in polar coordinates $\mathbf{y}$ rather than in Cartesian coordinates. This is accomplished by letting $\mathbf{y} = g(\mathbf{x})$, where g is a one-to-one function on $\mathcal{R}^n$ to $\mathcal{R}^n$ such that y_1 is the length of the vector $\mathbf{x}$ and $(y_2, \ldots, y_n)$ identifies the angles that $\mathbf{x}$ makes with the axes.

Suppose that $n = 2$ and consider the random column vector $\mathbf{X}$. Then, in polar coordinates,

$$Y_1 = \sqrt{X_1^2 + X_2^2}$$

$$Y_2 = \arctan \frac{X_1}{X_2}. \tag{4}$$

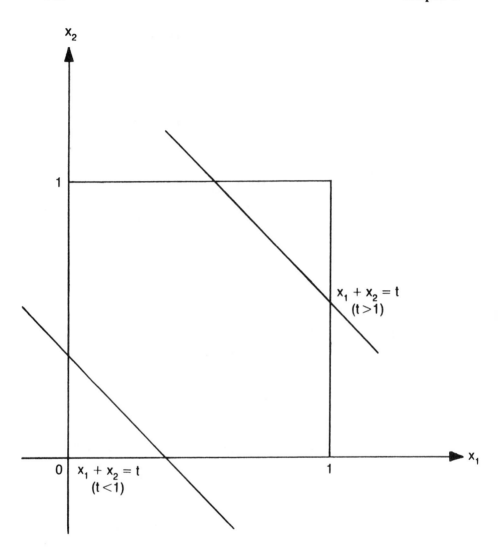

Figure 1

Conversely,

$$\mathbf{X} = \begin{pmatrix} X_1 \\ X_2 \end{pmatrix} = Y_1 \mathbf{U} = Y_1 \begin{pmatrix} \cos Y_2 \\ \sin Y_2 \end{pmatrix}. \tag{5}$$

Hence the random vector $\mathbf{X}$ can be generated by generating a length Y_1 and an angle Y_2 (or alternatively, a point on the unit circle) and applying (5).

Relation (5) can be generalized to $\mathcal{R}^n$ by

$$\mathbf{X} = Y_1 \mathbf{U}, \tag{6}$$

where Y_1 is the length of $\mathbf{X}$ and

$$\mathbf{U} = \begin{pmatrix} \cos Y_2 \\ \sin Y_2 \cos Y_3 \\ \vdots \\ \sin Y_2 \cdots \sin Y_{n-1} \cos Y_n \\ \sin Y_2 \cdots\cdots\cdots \sin Y_n \end{pmatrix}.$$

The random vector $\mathbf{U}$ is selected according to some specified conditional distribution (given Y_1) on the surface of the unit sphere in $\mathcal{R}^n$.

Example 7. Let (X_1, X_2) be a pair of independent standard normals. It can be shown that $Y_1 = \sqrt{X_1^2 + X_2^2}$ is the square root of a $\mathcal{G}(1, 2)$ rv (i.e., χ_2^2). This is done in Section 48, where the $\mathcal{G}(n/2, 2)$ subfamily is defined to be the χ_n^2 family. Futhermore, $Y_2 = \arctan X_1/X_2$ is independent of Y_1 and is uniform on $(0, 2\pi)$.

Conditioning is another method of deriving distributions. For simplicity, suppose that $Y = g(\mathbf{X})$ is a rv and that W is a rv for which $f_Y(y \mid W = w)$ can be readily identified. Then the result of Problem 34.8 is that

$$f_Y(y) = E[f_Y(y \mid W)], \tag{7}$$

where $f_Y(y \mid W)$ is the rv with values $f_Y(y \mid W = w)$.

Example 8. Let $(\mathbf{X}, W)$ be a random vector in $\mathcal{R}^{n+1}$ with density

$$f(x_1, \ldots, x_n, w) = w^n e^{-w(1 + \sum_{i=1}^n x_i)} = e^{-w} \prod_{i=1}^n [w e^{-w x_i}] \qquad \text{if } x_i, w > 0.$$

To find the density of $Y = \sum_{i=1}^n X_i$, note first that W is a $\mathcal{G}(1, 1)$ rv and that given $W = w$, the conditional density of $\mathbf{X}$ is that of n iid $\mathcal{G}(1, 1/w)$

(i.e., exponential) rv's. Hence the conditional density of Y given $W = w$ is $g(y \mid n, 1/w)$, that is,

$$f_Y(y \mid W = w) = \frac{w^n y^{n-1}}{\Gamma(n)} e^{-wy} \qquad \text{if } y > 0.$$

Hence

$$f_Y(y) = E(f_Y(y \mid W))$$

$$= \int_0^\infty f_Y(y \mid W = w) e^{-w} \, dw = \frac{n y^{n-1}}{(1 + y)^{n+1}} \qquad \text{if } y > 0 \quad (8)$$

and the corresponding distribution function is

$$F_Y(y) = \left(1 - \frac{1}{1 + y}\right)^n \qquad \text{if } y > 0.$$

Thus Y has the same distribution as $\max(Z_1, \ldots, Z_n)$, where the Z_i are iid with distribution function

$$F(y) = 1 - \frac{1}{1 + y} \qquad \text{if } y > 0.$$

See Problem 46.19 for another interpretation and derivation of the density in (8).

Summary

If $\mathbf{X}$ is discrete with probability function f and $\mathbf{Y} = g(\mathbf{X})$, then

$$f_{\mathbf{Y}}(\mathbf{y}) = \sum_{\mathbf{x} \in A_{\mathbf{y}}} f(\mathbf{x}),$$

where $A_{\mathbf{y}} = \{\mathbf{x} : g(\mathbf{x}) = \mathbf{y}\}$. If $Y = g(\mathbf{X})$ is a rv, then

$$F_Y(y) = \sum_{\mathbf{x} \in C_y} f(\mathbf{x}) \qquad \text{if } \mathbf{X} \text{ is discrete}$$

$$= \int_{C_y} f(\mathbf{x}) \, d\mathbf{x} \qquad \text{if } \mathbf{X} \text{ has density } f,$$

where $C_y = \{\mathbf{x} : g(\mathbf{x}) \le y\}$. If Y has a density, then f_Y can be found by differentiating F_Y. When $\mathbf{X}$ has a uniform distribution on some subset of $\mathcal{R}^2$, it may be possible to obtain F_Y by geometric considerations.

Problems

1. A box contains 10 balls numbered 1, . . . , 10. Three balls are selected at random from the box. For each part below, solve for the cases of sampling with replacement and sampling without replacement.
   (a) Find the probability function of the three-dimensional random vector $\mathbf{X}$, where X_i is the number on the ith ball drawn ($i = 1, 2, 3$).
   (b) Let g be the function on $\mathcal{R}^3$ taking values in $\mathcal{R}^2$ defined by $g_1(\mathbf{t}) = \min(t_1, t_2, t_3)$ and $g_2(\mathbf{t}) = \max(t_1, t_2, t_3)$. Find the probability function of $g(\mathbf{X})$.

2. A box contains 10 light bulbs, of which 2 are defective. Bulbs are drawn one at a time at random without replacement until the second defective bulb is found.
   (a) Find the probability function of the two-dimensional random vector $\mathbf{X}$, where X_i is the number of draws required to find the ith defective bulb ($i = 1, 2$).
   (b) Let g be the function on $\mathcal{R}^2$ taking values in $\mathcal{R}^2$ defined by $g_1(t) = t_1$ and $g_2(t) = t_2 - t_1$. Find the probability function of $g(\mathbf{X})$. Explain the components of $g(\mathbf{X})$ in terms of waiting times.

3. Let X be a $\mathcal{G}(1, 1)$ rv. Find the density of $(X + 1)^2$.

4. Let the rv X have density

$$f(t) = t^{-2} \qquad \text{if } t > 1$$
$$= 0 \qquad \text{otherwise.}$$

   Find the density of $Y = -1/X^3$.

5. A spinner of length 1 is spun around the origin. Let θ be the angle between the end of the spinner and a fixed line through the origin. Assume that θ is a $\mathcal{U}(-\pi, \pi)$ rv.
   (a) Find the distribution function and density of $\cos \theta$, the projection of the end of the spinner on the fixed line.
   (b) Find the distribution function and density of $\tan \theta$.
   (c) Compare your results in part (b) with those of Problem 44.12.

6. Let (X_1, X_2) be uniformly distributed on the unit square.
   (a) Find the density of $|X_1 - X_2|$.
   (b) Without calculations, show that the marginal distributions of $|X_1 - X_2|$ and $|X_1 + X_2 - 1|$ must be the same.

7. Let X be a $\mathcal{U}(0, 1)$ rv and $Y = 4X - [4X]$, that is, the fractional part of $4X$.
   (a) Find the density of Y.
   (b) Find the probability function of $[4X]$.

8. Each month the amount that a salesman sells in units of \$10,000 is distributed $\mathcal{G}(1, 10)$. He receives a salary of \$1000 per month. In ad-

dition, each month he receives a commission of 5% on all sales over $20,000 and a bonus of $1000 if he sells at least $150,000. Let Y be his monthly income in units of $1000 and find F_Y.

46. DISTRIBUTIONS OF SUMS AND RATIOS

A few properties of sums of rv's have been presented in earlier sections. Among these were formulas for means and variances of sums and, in special cases, their distributions (such as sums of independent Poissons, normals, and binomials). Various statistical applications require consideration of certain ratios of rv's. Since the techniques for determining the distributions of sums and ratios are similar, both are presented in this section.

Only sums of two rv's are considered here. Extensions to more than two may be accomplished by induction or by methods yet to be introduced (Section 47).

For the discrete case, the technique used here is the direct evaluation of the probability function by summing over the appropriate subset of $\mathcal{R}^2$. In the continuous case, the density is obtained from the distribution function.

Theorem 1.   *For any rv's, X and Y, set $Z = X + Y$.*

(i)   *In the discrete case,*

$$f_Z(t) = \sum_x f_{X,Y}(x, t - x) = \sum_y f_{X,Y}(t - y, y). \tag{1}$$

(ii)   *In the continuous case,*

$$f_Z(t) = \int_{-\infty}^{\infty} f_{X,Y}(x, t - x)\, dx = \int_{-\infty}^{\infty} f_{X,Y}(t - y, y)\, dy. \tag{2}$$

Proof.   In the discrete case,

$$[Z = t] = \bigcup_x [X = x, Y = t - x],$$

so that

$$f_Z(t) = \sum_x P[X = x, Y = t - x],$$

and the result follows.

In the continuous case,

$$F_Z(t) = P[X + Y \le t] = \int_{-\infty}^{\infty} \left[\int_{-\infty}^{t-x} f_{X,Y}(x, y)\, dy \right] dx.$$

In the inner integral, make the change of variable $u = y + x$ to obtain

$$F_Z(t) = \int_{-\infty}^{\infty} \left[\int_{-\infty}^{t} f_{X,Y}(x, u - x) \, du \right] dx$$

$$= \int_{-\infty}^{t} \left[\int_{-\infty}^{\infty} f_{X,Y}(x, u - x) \, dx \right] du. \tag{3}$$

Since (3) is valid for all t, the inner integral is the density of Z evaluated at u.

Corollary 1. *Let X and Y be independent rv's and $Z = X + Y$.*

(i) *In the discrete case,*

$$f_Z(t) = \sum_x f_X(x) f_Y(t - x) = \sum_y f_X(t - y) f_Y(y). \tag{4}$$

(ii) *In the continuous case,*

$$f_Z(t) = \int_{-\infty}^{\infty} f_X(x) f_Y(t - x) \, dx = \int_{-\infty}^{\infty} f_X(t - y) f_Y(y) \, dy. \tag{5}$$

Proof. Immediate from Theorem 1.

Convolution

The function f_Z in (4) or (5) is called the *convolution of f_X and f_Y*.

Example 1. For $i = 1, 2$, let X_i be a $\mathcal{P}_o(\lambda_i)$ rv. Assume that X_1 and X_2 are independent. For $t = 0, 1, \ldots$ apply (4) to obtain the probability function of $X_1 + X_2$, namely

$$f(t) = \sum_{k=0}^{\infty} \pi(k \mid \lambda_1) \pi(t - k \mid \lambda_2) = \sum_{k=0}^{t} \frac{\lambda_1^k}{k!} e^{-\lambda_1} \frac{\lambda_2^{t-k}}{(t - k)!} e^{-\lambda_2}$$

$$= \frac{1}{t!} e^{-(\lambda_1 + \lambda_2)} \sum_{k=0}^{t} \binom{t}{k} \lambda_1^k \lambda_2^{t-k} = \frac{1}{t!} (\lambda_1 + \lambda_2)^t e^{-(\lambda_1 + \lambda_2)} \tag{6}$$

$$= \pi(t \mid \lambda_1 + \lambda_2).$$

The next-to-last equality in (6) is a consequence of the binomial theorem (Theorem 10.1). Thus $X_1 + X_2$ is a $\mathcal{P}_o(\lambda_1 + \lambda_2)$ rv, as was shown in Corollary 39.1 in the context of a Poisson process.

The following example deals with a special case of Problem 42.13 without using matrix algebra.

Example 2. For $i = 1, 2$, let X_i be a $\mathfrak{N}(0, \tau_i^2)$ rv, where $\tau_1^2 + \tau_2^2 = 1$. Assume that X_1 and X_2 are independent. From (5), the density of $X_1 + X_2$ is

$$h(t) = \int_{-\infty}^{\infty} \frac{1}{\sqrt{2\pi}\,\tau_1} e^{-u^2/2\tau_1^2} \frac{1}{\sqrt{2\pi}\,\tau_2} e^{-(t-u)^2/2\tau_2^2} \, du. \tag{7}$$

By completing the square and using $\tau_1^2 + \tau_2^2 = 1$, the exponent in (7) becomes $-[(u - \xi)^2/\delta^2 + t^2]/2$, where $\xi = t\tau_1^2$ and $\delta = \tau_1\tau_2$. Then rearranging (7) yields

$$h(t) = \frac{1}{\sqrt{2\pi}} e^{-t^2/2} \int_{-\infty}^{\infty} \frac{1}{\sqrt{2\pi}\delta} e^{-(u-\xi)^2/(2\delta^2)} \, du. \tag{8}$$

The integrand in (8) is the $\mathfrak{N}(\xi, \delta^2)$ density and hence

$$h(t) = \frac{1}{\sqrt{2\pi}} e^{-t^2/2},$$

the $\mathfrak{N}(0, 1)$ density.

For $i = 1, 2$, let Y_i be a $\mathfrak{N}(\mu_i, \sigma_i^2)$ rv and let $\sigma^2 = \sigma_1^2 + \sigma_2^2$. Assume that Y_1 and Y_2 are independent. The previous result can be used to obtain the distribution of $Y_1 + Y_2$. For $i = 1, 2$, the rv's

$$X_i = \frac{Y_i - \mu_i}{\sigma}$$

are $\mathfrak{N}(0, \tau_i^2)$, where $\tau_i^2 = \sigma_i^2/\sigma^2$. Thus, X_1 and X_2 satisfy the previous conditions and $X_1 + X_2$ is a $\mathfrak{N}(0, 1)$ rv. Since $Y_i = \sigma X_i + \mu_i$, it follows that

$$Y_1 + Y_2 = \sigma(X_1 + X_2) + (\mu_1 + \mu_2).$$

Hence $Y_1 + Y_2$ is a $\mathfrak{N}(\mu_1 + \mu_2, \sigma_1^2 + \sigma_2^2)$ rv.

Example 3. Let X be a $\mathfrak{U}(0, 1)$ rv and Y be a $\mathfrak{G}(1, 1)$ rv. Assume that X and Y are independent and find the density of $X + Y$. That is, find the convolution of $u(s \mid 0, 1)$ and $g(t \mid 1, 1)$. Now $u(s \mid 0, 1)g(t - s \mid 1, 1) > 0$ if, and only if, $0 < s < \min(1, t)$ so that two cases are to be considered.

(i) $0 < t \le 1$:

$$f(t) = \int_0^t e^{-(t-u)} \, du = 1 - e^{-t}$$

(ii) $t > 1$:

$$f(t) = \int_0^1 e^{-(t-u)} \, du = e^{-t}(e - 1)$$

Examples 1 and 3 illustrate the need for careful handling of boundaries.

The results of Examples 1 and 2 are that sums of independent rv's from certain families are distributed by some other member of the family. These families are said to be *closed under convolution*.

We list here for reference some families that are closed under convolution.

(a)   Subfamilies of $\mathcal{B}$ with common p (Theorem 39.3)
(b)   Subfamilies of $\overline{\mathcal{B}}$ with common p (Theorem 40.1)
(c) $\mathcal{P}_a$ (Example 1)   (9)
(d) $\mathfrak{N}$ (Example 2)
(e)   Subfamilies of $\mathcal{G}$ with common scale parameter. (Problem 7)

The remainder of this section is devoted to the general problem of distributions of ratios. A special case was considered in Example 45.3.

Theorem 2.   For any continuous rv's X and Y, *set* $T = X/Y$. *Then*

$$f_T(t) = \int_0^\infty y[f_{X,Y}(ty, y) + f_{X,Y}(-ty, -y)] \, dy. \tag{10}$$

Proof

$$F_T(t) = P\left[\frac{X}{Y} \le t\right] = P[X \le tY, Y > 0] + P[X \ge tY, Y < 0]$$

$$= \int_0^\infty \left[\int_{-\infty}^{ty} f_{X,Y}(x, y) \, dx\right] dy + \int_{-\infty}^0 \left[\int_{ty}^\infty f_{X,Y}(x, y) \, dx\right] dy$$

$$= \int_{-\infty}^t \left[\int_0^\infty y f_{X,Y}(uy, y) \, dy\right] du + \int_t^\infty \left[\int_{-\infty}^0 y f_{X,Y}(uy, y) \, dy\right] du.$$

Differentiating yields (10) after an appropriate change of variable.

Corollary 2.   *Under the assumptions of Theorem 2, if X and Y are independent, then*

$$f_T(t) = \int_0^\infty y[f_X(ty)f_Y(y) + f_X(-ty)f_Y(-y)] \, dy. \tag{11}$$

Proof.   Immediate from Theorem 2.

Example 4. Let X and Y be iid, $\mathfrak{N}(0, 1)$ rv's and, as in Theorem 1, let $T = X/Y$. Then, using symmetry,

$$f_T(t) = \frac{1}{\pi} \int_0^\infty y e^{-(t^2+1)y^2/2} \, dy = \frac{1}{\pi} \int_0^\infty e^{-(t^2+1)u} \, du = \frac{1}{\pi(t^2 + 1)}.$$

This is the *standard Cauchy density* and the location-scale family it generates is the *Cauchy family* of densities.

Analogs of (10) and (11) exist in the discrete case, but are not very useful. Instead, direct computation can be used in some cases as in the following example.

Example 5. Let X_1 and X_2 be independent geometrics, each with $p = 1/2$ so that $P[X_1 = k] = P[X_2 = k] = (1/2)^k$ for $k = 1, 2, \ldots$. Consider the rv $Y = X_1/X_2$. The model for Y has very interesting mathematical properties. It assigns positive probability to every positive rational and 0 to every irrational. This produces a distribution function that has a jump at each positive rational but is continuous at every irrational. If $r = m/n$ is a rational in reduced form (m and n have no common divisors), then

$$P[Y = r] = \sum_{k=1}^\infty P[X_1 = km, X_2 = kn]$$

$$= \sum_{k=1}^\infty [(1/2)^{m+n}]^k = \frac{(1/2)^{m+n}}{1 - (1/2)^{m+n}} = \frac{1}{2^{m+n} - 1}.$$

Summary

Formulas (1) and (2) give the probability function and the density, respectively, of $Z = X + Y$. When X and Y are independent, (1) reduces to the convolution formula

$$f_Z(t) = \sum_x f_X(x) f_Y(t - x)$$

and (2) reduces to the convolution formula

$$f_Z(t) = \int_{-\infty}^\infty f_X(x) f_Y(t - x) \, dx.$$

Formula (10) gives the density of $T = X/Y$. When X and Y are independent, (10) reduces to

$$f_T(t) = \int_0^\infty y[f_X(ty) f_Y(y) + f_X(-ty) f_Y(-y)] \, dy.$$

Some families closed under convolution are given in (9).

Problems

1. Let X and Y be any rv's. Obtain the analog of Theorem 1 for $X - Y$ and specialize your result to the case of independence.

2. Let the probability function of (X, Y) be

$$f(k, m) = \frac{2}{n(n + 1)} \qquad \text{if } k = 1, \ldots, n; \quad m = 1, \ldots, k.$$

   Find the marginal probability functions of $X + Y$ and of $X - Y$.

3. Let the joint density of X and Y be

$$f(t, u) = 2 \qquad \text{if } t, u > 0; \quad t + u < 1.$$

   Find the marginal densities of $X + Y$ and of $X - Y$.

4. For $i = 1, 2$, let X_i be a $\overline{\mathcal{B}}(1, p_i)$ rv. Assume that X_1 and X_2 are independent. Find the probability function of $X_1 + X_2$. Simplify your answer when $p_1 = p_2$.

5. For $i = 1, 2$, let X_i be a $\mathcal{G}(1, \beta_i)$ rv. Assume that X_1 and X_2 are independent. Find the marginal densities of $X_1 + X_2$ and of $X_1 - X_2$. Simplify your answers when $\beta_1 = \beta_2$.

6. Show that in Example 2 the density of $X_1 + X_2$ and of $X_1 - X_2$ are the same, that is, that $X_1 + X_2$ and $X_1 - X_2$ are identically distributed. What is the distribution of $Y_1 - Y_2$?

7.† For $i = 1, 2$, let X_i be a $\mathcal{G}(\alpha_i, \beta)$ rv. Assume that X_1 and X_2 are independent. Show that $X_1 + X_2$ is a $\mathcal{G}(\alpha_1 + \alpha_2, \beta)$ rv. This problem has already been solved when the α_i are integers (Theorem 41.2) by considering waiting times in a Poisson process.

8. Let $X_1, \ldots, X_n$ be independent rv's and assume X_i is a $\mathcal{G}(\alpha_i, \beta)$ rv. Find the density of $\Sigma_{i=1}^n X_i$. *Hint:* Use Problem 7. Compare with Theorem 41.2.

9. Let $X_1, \ldots, X_n$ be independent rv's. Find the model for $\Sigma_{i=1}^n X_i$ if
   (a) X_i is a $\mathcal{P}o(\lambda_i)$ rv.
   (b) X_i is a $\mathcal{N}(\mu_i, \sigma_i^2)$ rv.

10. Let X and Y be iid $\overline{\mathcal{B}}(1, p)$ rv's and find the probability function of $X - Y$.

11. Let X_1 and X_2 be independent and assume that their densities are

$$f_{X_i}(x) = (i + 1)x^i \qquad \text{if } 0 < x < 1.$$

   Find the marginal densities of X_1/X_2 and of X_2/X_1.

12. Show that if T is a standard Cauchy rv, then $1/T$ has the same distribution.

13. For any rv's X and Y with $P[X + Y = 0] = 0$, let $W = X/(X + Y)$. Develop a general form of f_W based on Theorem 2. Specialize the result to the case of independence.

14. Let X_1 and X_2 be independent and assume that X_i is a $\mathcal{G}(\alpha_i, \beta)$ rv. Find the marginal densities of X_1/X_2 and of $X_1/(X_1 + X_2)$.

15. Let

$$f_{X,Y}(x, y) = cxy(1 - x - y) \qquad \text{if } x, y > 0, \quad x + y < 1.$$

Find the marginal densities of $X + Y$ and of X/Y.

16. Two fair dice are rolled. Let X_i be the result on the ith die and find the probability function of X_1/X_2.

17. In a sequence of Bernoulli trials with probability p of success, let T_i be the number of trials to the ith success. Find the distribution of T_1/T_2, the proportion of the waiting time for the first success relative to the total waiting time for two successes.

18. Show that the Cauchy family is closed under convolution.

19. Let $W = \Sigma_{i=1}^n X_i$ and $Y = X_{n+1}/W$, where $X_1, \ldots, X_{n+1}$ are iid $\mathcal{G}(1, \beta)$ rv's. Show that the denstiy of Y is $f_Y(y) = ny^{n-1}/(1 + y)^{n+1}$ for $y > 0$, the density given in (45.8).

47. DENSITIES OF DIFFERENTIABLE FUNCTIONS; JACOBIANS

Throughout this section $\mathbf{X}$ is a continuous n-dimensional random vector, $\mathbf{y} = g(\mathbf{x})$ takes values in $\mathcal{R}^k$, and g is assumed to possess whatever differentiability properties are called for in the expressions to be derived. (If $n = 1$, X replaces $\mathbf{X}$; if $k = 1$, Y replaces $\mathbf{Y}$.) The distribution function of $\mathbf{X}$ is F and its density is f; $F_{\mathbf{Y}}$ and $f_{\mathbf{Y}}$ denote the corresponding functions for $\mathbf{Y}$.

It will be necessary to consider the set $S^+ = S^+(F)$ defined by $S^+(F) = \left\{ \mathbf{x}; \dfrac{\partial^n}{\partial x_1 \cdots \partial x_n} F(\mathbf{x}) > 0 \right\}$ (i.e., the set where the derivative of F exists and is positive). According to (24.19), f can be chosen so that $S^+(F) = \{x; f(\mathbf{x}) > 0\}$ and we make this choice so that $f(\mathbf{x}) = 0$ if, and only if, $\mathbf{x} \notin S^+$.

Only the behavior of g on S^+ is relevant. In particular, consider $A_{\mathbf{y}} = \{\mathbf{x} \in S^+ : g(\mathbf{x}) = \mathbf{y}\}$, a set defined for every $\mathbf{y} \in \mathcal{R}^k$. Whenever $A_{\mathbf{y}}$ is empty, we set $f_{\mathbf{Y}}(\mathbf{y}) = 0$. If $A_{\mathbf{y}} = \phi$, it means either that $\mathbf{y}$ is not a possible value of g or that $f(\mathbf{x}) = 0$ when $g(\mathbf{x}) = \mathbf{y}$. The set of $\mathbf{y}$ values with which we need to be concerned is $g(S^+) = \{\mathbf{y}: A_{\mathbf{y}} \neq \phi\}$.

Most of the results to be described depend on the additional assumption that g is one-to-one on S^+. This means that for every $\mathbf{y} \in g(S^+)$, the set $A_{\mathbf{y}}$ contains exactly one point, which we denote by $g^{-1}(\mathbf{y})$ (read "g inverse of $\mathbf{y}$").

Presently we will examine various analytic expressions relating f_Y to f and g^{-1} and the derivatives of g^{-1}. Some of these results can be extended to many-to-one functions. In any case, the methods require that for every $y \in g(S^+)$, the equation $g(\mathbf{x}) = \mathbf{y}$ can be solved explicitly for $\mathbf{x}$.

The topic is presented in two subsections; Subsection A deals with the special case $n = k = 1$ and Subsection B deals with the other cases.

A. Functions of Random Variables

Let X be a continuous rv and assume that g is one-to-one on S^+. Then g is either increasing or decreasing (i.e., g is monotone) and, in either case, $\{x: g(x) \le t\}$ is an interval with one endpoint equal to $g^{-1}(t)$ (see Figure 1).

Theorem 1. *Assume that g on S^+ is monotone, that $Y = g(X)$, and let*
$$J(t) = \frac{d}{dt} g^{-1}(t). \text{ For every } t \in g(S^+):$$

(i) *If g is increasing, then*
$$F_Y(t) = F(g^{-1}(t)). \tag{1}$$

(ii) *If g is decreasing, then*
$$F_Y(t) = 1 - F(g^{-1}(t)). \tag{2}$$

(iii) *In either case, where $J(t)$ exists*
$$f_Y(t) = |J(t)| f(g^{-1}(t)). \tag{3}$$

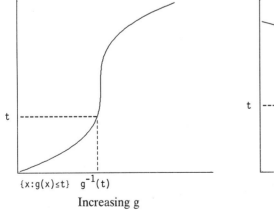
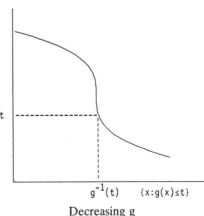

Increasing g   Decreasing g

Figure 1

Proof. Relations (1) and (2) follow from

$$[Y \le t] = [X \le g^{-1}(t)] \qquad \text{if } g \text{ is increasing}$$
$$= [X \ge g^{-1}(t)] \qquad \text{if } g \text{ is decreasing.}$$

Continuity of F is used only to prove (2). Relation (3) follows by differentiation in (1) and (2). Note that if g is increasing, then $J(t) > 0$; if g is decreasing, then $J(t) < 0$. The condition $g'(t) \neq 0$ guarantees that $J(t)$ is finite.

Example 1. Let $Y = a + bX$, where a, b are arbitrary constants (except that $b \neq 0$). Applying (3) with $g^{-1}(t) = (t - a)/b$ and $J(t) = 1/b$ yields

$$f_Y(t) = \frac{1}{|b|} f\left(\frac{t - a}{b}\right) \qquad \text{if } \frac{t - a}{b} \in S^+. \tag{4}$$

This is a slight extension of (34.3) which involved only $b > 0$.

Example 2. Consider $g(x) = x^r$, where $r \neq 0$, and let $v = 1/r$. There are two cases: (I) $S^+ \subset \mathcal{R}^+ = \{x: x > 0\}$ and (II) S^+ contains negative x's. In case (I) g is monotone on S^+ — increasing if $r > 0$, decreasing if $r < 0$. Then $g^{-1}(t) = t^v$ and $J(t) = vt^{v-1}$. In case (II) g is only defined on S^+ for certain values of r. The complete investigation of case (II) is left as a problem.

For the remainder of this example assume case (I). Applying (1), (2), and (3):

(i)   If $r > 0$, then $F_Y(t) = F(t^v)$;
(ii)   If $r < 0$, then $F_Y(t) = 1 - F(t^v)$; $\qquad\qquad$ (5)
(iii) If $r \neq 0$, then $f_Y(t) = |v| t^{v-1} f(t^v)$

for every t such that $t^v \in S^+$.

Now to illustrate what happens for specific models, consider (α) X is $\mathcal{U}(a, b)$ and (β) X is $\mathcal{G}(1, 1)$. In order that case (α) satisfy (I), it is necessary that $a > 0$.

For (α), let $r = -1/2$, so that $v = -2$, and applying (5)(iii),

$$f_Y(t) = \frac{2}{t^3} \cdot \frac{1}{b - a} \qquad \text{if } a < t^{-2} < b \qquad \left(\frac{1}{\sqrt{b}} < t < \frac{1}{\sqrt{a}}\right).$$

For (β), let $r > 0$. Apply 5(i) or (iii) to obtain

$$f_Y(t) = vt^{v-1} e^{-t^v} \qquad \text{if } t > 0.$$

This family of densities is a subfamily of the Weibull family (Section 23).

Remark 1. Theorem 1 can be derived from the change-of-variable rules for integration. Consider, for example,

$$G(t_1, t_2) = \int_{\lambda(t_1)}^{\lambda(t_2)} f(x)\, dx, \tag{6}$$

where $t_1 < t_2$ and γ is increasing on (t_1, t_2). Let $y = g(x)$ be such that $g^{-1} = \gamma$ on (t_1, t_2) and make the substitution $x = \gamma(y)$ in the integrand. Then

$$G(t_1, t_2) = \int_{t_1}^{t_2} \gamma'(y) f(\gamma(y))\, dy. \tag{7}$$

With some modifications (7) produces the result of Theorem 1. If in (6), $t_2 < t_1$ and γ is decreasing, then

$$G(t_1, t_2) = \int_{t_2}^{t_1} - \gamma'(y) f(\gamma(y))\, dy. \tag{8}$$

Relations (7) and (8) can be used to get results similar to Theorem 1 for more complicated g's. For simplicity, assume that g is defined on all of $\mathcal{R}$. Assume also that g alternates between increasing and decreasing segments as follows. There is a partition, $\mathcal{Q} = \{A_i: i = 1, 2, \ldots\}$, of $\mathcal{R}$ into adjoining intervals $A_i = (a_{i-1}, a_i]$ such that g satisfies

$$g(x) = \sum_i g_i(x) I_{A_i}(x), \tag{9}$$

where g_i is monotone on A_i for each i and is increasing on A_i if, and only if, g_{i+1} is decreasing on A_{i+1} (see Figure 2).

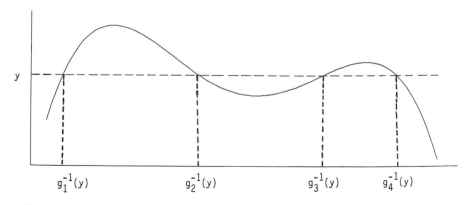

$$g_1^{-1}(y) \qquad\qquad g_2^{-1}(y) \qquad\qquad g_3^{-1}(y) \qquad\qquad g_4^{-1}(y)$$

Figure 2

Let g_i^{-1} denote the inverse of g_i on A_i and let

$$K_t = \{i: t \in g_i(A_i)\},$$

where $g_i(A_i) = \{g_i(t): t \in A_i\}$. Then for $Y = g(X)$,

$$f_Y(t) = \sum_{i \in K_t} |J_i(t)| f(g_i^{-1}(t)) \qquad \text{if } K_t \neq \phi, \tag{10}$$

where $J_i(t) = \dfrac{d}{dt} g_i^{-1}(t)$ and is finite for all $i \in K_t$.

The proof of (10) is based on a fairly obvious (but tedious) application of (7) and (8) and will be omitted.

Example 3. If $g(x) = x^2$, then $A_i = (-\infty, 0]$, $A_2 = (0, \infty)$, $g_1^{-1}(t) = -\sqrt{t}$, $g_2^{-1}(t) = \sqrt{t}$ (for $t > 0$) and (10) becomes

$$f_Y(t) = \frac{1}{2\sqrt{t}} f(-\sqrt{t}) + \frac{1}{2\sqrt{t}} f(\sqrt{t}) \qquad \text{if } t > 0. \tag{11}$$

It is instructive, however, to derive this simple result from scratch. For $t > 0$, $P[X^2 \leq t] = P[-\sqrt{t} \leq X \leq \sqrt{t}]$. Hence

$$F_Y(t) = F(\sqrt{t}) - F(-\sqrt{t}) \tag{12}$$

and differentiation of (12) yields (11).

Note that if X is symmetric, (11) becomes

$$f_Y(t) = \frac{1}{\sqrt{t}} f(\sqrt{t}) \qquad \text{if } t > 0. \tag{13}$$

Example 4. Let $g(x) = \cos x$ and let X be a $\mathcal{G}(1, 1)$ rv so that

$$f(x) = e^{-x} \qquad \text{if } x > 0.$$

It suffices to let $\mathcal{C}$ be a partition of $\mathcal{R}^+$. Let $A_i = ((i-1)\pi, i\pi]$ and

$$\begin{aligned} g_i^{-1}(t) &= (i-1)\pi + \cos^{-1} t && \text{if } i \text{ is odd} \\ &= i\pi - \cos^{-1} t && \text{if } i \text{ is even.} \end{aligned}$$

(Draw a picture.) Here the values of $\cos^{-1} t$ are in $(0, \pi]$. Then $|J_i(t)| = 1/\sqrt{1 - t^2}$ for $0 < t < 1$ and

$$
\begin{aligned}
f_Y(t) &= \frac{1}{\sqrt{1 - t^2}} \sum_{i=1}^{\infty} e^{-g_i^{-1}(t)} \\
&= \frac{1}{\sqrt{1 - t^2}} \left[\sum_{k=0}^{\infty} e^{-(2k\pi + \cos^{-1} t)} + \sum_{k=1}^{\infty} e^{-(2k\pi - \cos^{-1} t)} \right] \\
&= \frac{e^{-\cos^{-1} t} + e^{-(2\pi - \cos^{-1} t)}}{(1 - e^{-2\pi})\sqrt{1 - t^2}}.
\end{aligned}
$$

B. Higher Dimensions: Jacobians

Now assume that $S^+(\mathbf{X}) \subset \mathcal{R}^n$ and that $g(\mathbf{x}) \in \mathcal{R}^k$ with $k \leq n$. The cases with $k = 1$ and $k = n > 1$ are the most interesting; in the first case g is real valued and in the second g is a transformation on $\mathcal{R}^n$ to $\mathcal{R}^n$. We will in fact, begin with the latter and show how, by using auxiliary variables, this case may be usefully applied to the former or to intermediate values of k.

The following theorem (from multidimensional calculus), stated without proof, provides an n-dimensional analog of (7) and (8). The notation used below is:

$g = (g_1, \ldots, g_n)$.
g_i is a real-valued function, the ith component of g.
$g^{-1} = \gamma = (\gamma_1, \ldots, \gamma_n)$.
$g(B) = \{g(\mathbf{x}): \mathbf{x} \in B\}$.

The function g is *differentiable* if $\partial g_i(\mathbf{x})/\partial x_j$ exists for each $i, j = 1, \ldots, n$. In that case, γ is also differentiable; $J(\mathbf{t}) = \det(\partial \gamma_i(\mathbf{t})/\partial t_j)$ is the determinant of the $n \times n$ matrix with (i, j)th element $\partial \gamma_i(\mathbf{t})/\partial t_j$. Then, $J(\mathbf{t})$ is the *Jacobian* of the transformation g.

Theorem 2. *Let f be a real-valued function on $\mathcal{R}^n$ and g be a function on $\mathcal{R}^n$ taking values in $\mathcal{R}^n$. Assume that g is one-to-one and differentiable. Then*

$$
\int_A f(\mathbf{x}) \, d\mathbf{x} = \int_{g(A)} f(g^{-1}(\mathbf{t})) |J(\mathbf{t})| \, d\mathbf{t} \tag{14}
$$

for any $A \subset \mathcal{R}^n$ for which either side of (14) is defined.

Set $J_0(\mathbf{x}) = \det(\partial g_i(\mathbf{x})/\partial x_j)$, the Jacobian of g^{-1}. Then since $J(\mathbf{t}) = 1/J_0(g^{-1}(\mathbf{t}))$, it follows that $|J(\mathbf{t})| < \infty$ if, and only if, $J_0(g^{-1}(\mathbf{t})) \neq 0$. Since

g is one-to-one, replacing A by $A \cap \{\mathbf{x}; J_0(\mathbf{x}) \neq 0\}$ will not change the left side of (14) and will guarantee that $|J(\mathbf{x})| < \infty$ in the integral on the right side.

Theorem 2 will now be used to obtain the density of a function of an n-dimensional random vector.

Theorem 3. *Let* $\mathbf{X}$ *be an n-dimensional random vector with density* f. *Let* g *be a one-to-one function on* $\mathfrak{R}^n$ *taking values in* $\mathfrak{R}^n$ *and make the assumptions about g given above. Then the density of* $\mathbf{Y} = g(\mathbf{X})$ *is*

$$f_{\mathbf{Y}}(\mathbf{t}) = f(g^{-1}(\mathbf{t}))|J(\mathbf{t})|. \tag{15}$$

Proof. If

$$P[g(\mathbf{X}) \in B] = \int_B h(t) \, d\mathbf{t}, \tag{16}$$

for each n-dimensional rectangle B, then h is the density of $g(X)$. But

$$P[g(\mathbf{X}) \in B] = P[\mathbf{X} \in g^{-1}(B)] = \int_{g^{-1}(B)} f(\mathbf{x}) \, d\mathbf{x}.$$

Applying Theorem 2 yields

$$\begin{aligned} P[g(\mathbf{X}) \in B] &= \int_{g(g^{-1}(B))} f(g^{-1}(\mathbf{t}))|J(\mathbf{t})| \, d\mathbf{t} \\ &= \int_B f(g^{-1}(\mathbf{t}))|J(\mathbf{t})| \, d\mathbf{t} \end{aligned} \tag{17}$$

since $g(g^{-1}(B)) = B$. Since B is arbitrary, the integrand in (17) must be $f_{\mathbf{Y}}$, proving (15).

Example 5. One of the most important applications of Theorem 3 is to the case $g(\mathbf{x}) = \mathbf{Ax}$, where $\mathbf{A}$ is an $n \times n$ nonsingular matrix. Then $g^{-1}(\mathbf{t}) = \mathbf{A}^{-1}\mathbf{t}$ and, from (15),

$$f_{\mathbf{Y}}(\mathbf{t}) = |\det \mathbf{A}^{-1}|f(\mathbf{A}^{-1}\mathbf{t}) = \frac{f(\mathbf{A}^{-1}\mathbf{t})}{|\det \mathbf{A}|} \tag{18}$$

since $|J(\mathbf{t})| = |\det \mathbf{A}^{-1}|$, the absolute value of the determinant of $\mathbf{A}^{-1}$.

Suppose, for example, that $\mathbf{X}$ is $\mathfrak{N}_n(\mathbf{0}, \mathbf{I}_n)$, that is, that

$$f(\mathbf{x}) = \left(\frac{1}{\sqrt{2\pi}}\right)^n e^{-(1/2)\mathbf{x}'\mathbf{x}}.$$

Then $\mathbf{Y} = \mathbf{AX} + \boldsymbol{\mu}$ implies that $\mathbf{X} = \mathbf{A}^{-1}(\mathbf{Y} - \boldsymbol{\mu})$ and

$$f_{\mathbf{Y}}(\mathbf{t}) = |\det \mathbf{A}^{-1}| f(\mathbf{A}^{-1}(\mathbf{t} - \boldsymbol{\mu})) \tag{19}$$

$$= \frac{1}{(\sqrt{2\pi})^n |\det \mathbf{A}|} e^{-(1/2)(\mathbf{t}-\boldsymbol{\mu})'(\mathbf{AA'})^{-1}(\mathbf{t}-\boldsymbol{\mu})}.$$

Equation (19) was used earlier (Section 42) as part of a proof that the distribution of a normal random vector is determined by its mean vector and its covariance matrix.

We apply Theorem 3 in the following important example.

Example 6. Let $\mathbf{X} = (X_1, X_2, X_3)$ and assume that the X_i are independent, $\mathcal{G}(\alpha_i, \beta)$ rv's. Let $Y_1 = X_1 + X_2 + X_3$, $Y_2 = X_2/(X_1 + X_2)$, and $Y_3 = X_3/(X_1 + X_2 + X_3)$. Then $\gamma_1(\mathbf{t}) = t_1(1 - t_2)(1 - t_3)$, $\gamma_2(\mathbf{t}) = t_1 t_2 (1 - t_3)$, and $\gamma_3(\mathbf{t}) = t_1 t_3$. Hence

$$J(\mathbf{t}) = \det \begin{bmatrix} (1 - t_2)(1 - t_3) & -t_1(1 - t_3) & -t_1(1 - t_2) \\ t_2(1 - t_3) & t_1(1 - t_3) & -t_1 t_2 \\ t_3 & 0 & t_1 \end{bmatrix} = t_1^2(1 - t_3)$$

and for $t_1 > 0$, $0 < t_2, t_3 < 1$,

$$f_{\mathbf{Y}}(\mathbf{t}) = c[\gamma_1(\mathbf{t})]^{\alpha_1-1}[\gamma_2(\mathbf{t})]^{\alpha_2-1}[\gamma_3(\mathbf{t})]^{\alpha_3-1} e^{-t_1/\beta} |J(\mathbf{t})|$$

$$= c t_1^{\alpha_1+\alpha_2+\alpha_3-1} e^{-t_1/\beta} t_2^{\alpha_2-1}(1 - t_2)^{\alpha_1-1} t_3^{\alpha_3-1}(1 - t_3)^{\alpha_1+\alpha_2-1}.$$

Thus Y_1, Y_2, and Y_3 are independent, Y_1 is a $\mathcal{G}(\alpha_1 + \alpha_2 + \alpha_3, \beta)$ rv, Y_2 is a $\mathcal{B}e(\alpha_2, \alpha_1)$ rv, and Y_3 is a $\mathcal{B}e(\alpha_3, \alpha_1 + \alpha_2)$ rv.

Theorem 3 can be extended to differentiable functions g that are many-to-one. In this case, as illustrated for $n = 1$ in Subsection A, we define several one-to-one inverses, each of which is differentiable. The desired density of $g(\mathbf{X})$ is then the sum of terms, one from each inverse. Suppose, for example, that g is two-to-one and differentiable. We define one-to-one inverses g_1^{-1} and g_2^{-1} which are differentiable. Each inverse would have its own Jacobian, say $J_1(\mathbf{t})$ and $J_2(\mathbf{t})$, respectively. Then instead of (15), we have

$$h(\mathbf{t}) = f(g_1^{-1}(\mathbf{t}))|J_1(\mathbf{t})| + f(g_2^{-1}(\mathbf{t}))|J_2(\mathbf{t})|. \tag{20}$$

Although Theorem 3 requires g to be n-dimensional, it is often used to obtain the density of a lower-dimensional function of $\mathbf{X}$. Let $Y_1 = g_1(\mathbf{X})$ be the real-valued function whose density is to be determined. The idea is to invent $n - 1$ other variables $Y_i = g_i(\mathbf{X})$ (called *auxiliary* rv's) to produce a vector $\mathbf{Y} = g(\mathbf{X})$ which does satisfy the conditions of Theorem

3 and then to determine the marginal of Y_1 by suitable integration of (15). Success of this method depends on the appropriateness of the auxiliary rv's. Often $Y_i = X_i$ for $i = 2, \ldots, n$ is used. The integration is avoided if the choice of the auxiliary rv's results in Y_1 and $(Y_2, \ldots, Y_n)$ being independent as in Example 6.

Summary

Let **X** be a continuous n-dimensional random vector and $\mathbf{Y} = g(\mathbf{X})$ also be n-dimensional. Assuming that g is differentiable and one-to-one,

$$f_{\mathbf{Y}}(\mathbf{t}) = f_{\mathbf{X}}(g^{-1}(\mathbf{t}))|J(\mathbf{t})|, \tag{21}$$

where $J(\mathbf{t})$ is the Jacobian of the tranformation g. Extension of (21) to the many-to-one case is possible.

If $n = 1$, then $J(t) = \dfrac{d}{dt} g^{-1}(t)$.

Problems

1. Let the rv X have density f and g be a differentiable function that is symmetric about zero and one-to-one on $[0, \infty)$. Find the density of $g(X)$. Specialize your result to the case in which f is also symmetric about zero.
2. Let F be the distribution function of X.
    (a) Assuming that F is continuous, find the distribution function of $|X|$.
    (b) Modify your answer in part (a) when F need not be continuous.
    (c) If F has a density, f, find the density of $|X|$.
    (d) Specialize your answer in part (c) to the case of f symmetric.
3. Let X be a $\mathcal{U}(0, 1)$ rv. Find the densities of:
    (a) $-\log X$
    (b) X^α, where $\alpha \neq 0$
    (c) $g(X) = \log 2X$ if $X < 1/2$
              $= -\log 2(1 - X)$ if $X \geq 1/2$
4. Let X be a $\mathcal{B}e(\alpha, \beta)$ rv. Find the densities of:
    (a) $X/(1 - X)$
    (b) $(1 - X)/X$
5. Let X be a $\mathcal{G}(1, \beta)$ rv. Let $Y = e^{-X/\alpha}$.
    (a) Find the density of Y.
    (b) Specialize to the case $\alpha = \beta$. How is this related to a previous result?
6. Investigate the density of X^r as suggested in Example 2, case (II).

7. Assume that X is a $\mathcal{G}(\alpha, \beta)$ rv. Find the density of $Y = (X - 1)^2$.
8. Let $\mathbf{X}$ be an n-dimensional random vector with density $f(\mathbf{x}) = h(\|\mathbf{x}\|)$, where $\|\mathbf{x}\| = \sqrt{\sum_{i=1}^{n} x_i^2}$. Express the density of $\|\mathbf{X}\|$ in terms of h.
9. Let X_1 and X_2 be independent and assume that X_i is distributed $\mathcal{N}(0, \sigma_i^2)$. Find the density of the angle measured from the horizontal axis to $\mathbf{X} = (X_1, X_2)$.
10. Let X be a rv and $Y = X^2$. Find F_Y without making the continuity assumption.
11.† Assume that $\log X$ is distributed $\mathcal{N}(\mu, \sigma^2)$.
    (a) Find the density of X. This is called the *log normal density*.
    (b) Find $E(X)$ and $\mathrm{Var}(X)$.
12. Let X be a $\mathcal{G}(1, 1)$ rv and find the density of $\sin X$. *Hint*: Let $A_0 = (0, \pi/2]$.

48. MISCELLANEOUS SAMPLE STATISTICS

This section contains distributions of certain functions of random vectors that are useful in statistical applications. A *statistic* is any function of a sample.

A. Functions of Normal Samples

Distributions of certain functions of samples from normal distributions arise frequently in statistics. These will be named, but in most cases their densities will be left as problems.

Chi-Square Distribution

Let $Z_1, \ldots, Z_n$ be independent, $\mathcal{N}(0, 1)$ rv's. The distribution of $\sum_{i=1}^{n} Z_i^2$ is the *chi-square distribution with n degrees of freedom* (abbreviated χ_n^2).

Some discussion of the term "degrees of freedom" follows Theorem 3.

Lemma 1. *The χ_1^2 distribution is $\mathcal{G}(1/2, 2)$.*

Proof. Let Z be a $\mathcal{N}(0, 1)$ rv so that $W = Z^2$ has the χ_1^2 distribution. By (47.13),

$$f_W(t) = \frac{1}{\sqrt{t}} \varphi(\sqrt{t}) = \frac{1}{\sqrt{2\pi t}} e^{-t/2} \qquad \text{if } t > 0,$$

which is the $\mathcal{G}(1/2, 2)$ density.

Theorem 1.   *The χ_n^2 distribution is $\mathcal{G}(n/2, 2)$.*

Proof.   Immediate by Lemma 1 and (46.8)(e), the closure under convolution of subfamilies of $\mathcal{G}$ with common scale parameter.

$\mathcal{T}$-Distribution

Let Z be a $\mathfrak{N}(0, 1)$ rv and U be a χ_n^2 rv. Assume that U and Z are independent. Then

$$\frac{Z}{\sqrt{U/n}}$$

has the $\mathcal{T}$-distribution with n degrees of freedom (abbreviated $\mathcal{T}_n$).

$\mathcal{F}$-Distribution

Let U and V be independent rv's which are χ_m^2 and χ_n^2, respectively. Then $(U/m)/(V/n)$ has the *$\mathcal{F}$-distribution with m degrees of freedom in the numerator and n degrees of freedom in the denominator* (abbreviated $\mathcal{F}_{m,n}$).
Distribution properties of $\overline{X}_n$, the sample mean, and of $D_n^2 = \Sigma_{i=1}^n$ $(X_i - \overline{X}_n)^2$ or $V_n = D_n^2/n$, the sample variance, are of considerable interest in statistics. In the following theorems, the case of a sample from a normal distribution is considered. (Results in Section 43A that require the use of matrix algebra are invoked.)

Theorem 2.   *Let $X_1, \ldots, X_n$ be independent, $\mathfrak{N}(\mu, \sigma^2)$ rv's. Then $\overline{X}_n$ and D_n^2 are independent.*

Proof.   The rv's $X_1 - \overline{X}_n, \ldots, X_n - \overline{X}_n, \overline{X}_n$ are jointly normal. Furthermore,

$$\mathrm{Cov}(X_i, \overline{X}_n) = \frac{1}{n} \sum_{j=1}^n \mathrm{Cov}(X_i, X_j) = \frac{\sigma^2}{n}$$

and $\mathrm{Cov}(\overline{X}_n, \overline{X}_n) = \mathrm{Var}(\overline{X}_n) = \sigma^2/n$, so that $\mathrm{Cov}(X_i - \overline{X}_n, \overline{X}_n) = 0$. Thus, by Theorem 42.3, the random vector $(X_1 - \overline{X}_n, \ldots, X_n - \overline{X}_n)$ and the rv $\overline{X}_n$ are independent. This suffices to yield the result.

Theorem 3.   *Let $X_1, \ldots, X_n$ be indepedent, $\mathfrak{N}(\mu, \sigma^2)$ rv's. Then D_n^2/σ^2 is distributed χ_{n-1}^2.*

Proof. The proof is by induction. Without loss of generality, assume that $\sigma^2 = 1$. Now

$$D_2^2 = \left(X_1 - \frac{X_1 + X_2}{2}\right)^2 + \left(X_2 - \frac{X_1 + X_2}{2}\right)^2 = \frac{(X_1 - X_2)^2}{2}.$$

But $X_1 - X_2$ is a $\mathfrak{N}(0, 2)$ rv, so that D_2^2 is a χ_1^2 rv.

Suppose that the result is true for $n = m$. It suffices to show that D_{m+1}^2 is the sum of a χ_{m-1}^2 rv and the square of an independent $\mathfrak{N}(0, 1)$ rv. By elementary algebra,

$$D_{m+1}^2 = D_m^2 + \left[\sqrt{\frac{m}{m+1}}\,(X_{m+1} - \overline{X}_m)\right]^2. \tag{1}$$

Now D_m^2, $\overline{X}_m$, and X_{m+1} are independent so that the terms in (1) are independent. By the induction hypothesis, we have only to show that the second term is the square of a $\mathfrak{N}(0, 1)$ rv. Since X_{m+1} is a $\mathfrak{N}(\mu, 1)$ rv and $\overline{X}_m$ is a $\mathfrak{N}(\mu, 1/m)$ rv, $X_{m+1} - \overline{X}_m$ is a $\mathfrak{N}(0, (m+1)/m)$ rv, completing the proof.

In the definition of a chi-square rv, the phrase "degrees of freedom" was used for the number of terms in the sum. These terms are not subject to any linear restriction. However, the n summands in the definition of D_n^2 are subject to the linear restriction $\Sigma_{i=1}^n (X_i - \overline{X}_n) = 0$ and hence have only $n - 1$ degrees of freedom.

Theorem 4. *Let $X_1, \ldots, X_n$ be independent, $\mathfrak{N}(\mu, \sigma^2)$ rv's. Then*

$$T = \frac{\overline{X}_n - \mu}{D_n/\sqrt{n(n-1)}} = \frac{\overline{X}_n - \mu}{\sqrt{V_n/(n-1)}}$$

has the $\mathfrak{I}_{n-1}$ distribution.

Proof. By Theorem 3, the numerator and denominator of T are independent. Furthermore,

$$Z = \frac{\overline{X}_n - \mu}{\sigma/\sqrt{n}}$$

is a $\mathfrak{N}(0, 1)$ rv and $U = nV_n/\sigma^2$ is a χ_{n-1}^2 rv. By elementary algebra,

$$T = \frac{Z}{\sqrt{U/(n-1)}},$$

which is a $\mathfrak{I}_{n-1}$ rv.

Theorem 4 is applied in situations where $\overline{X}_n$ is used to estimate μ. It enables us to study $\overline{X}_n - \mu$, the error of estimation. If σ is known, then $(\overline{X}_n - \mu)/(\sigma/\sqrt{n})$ has the $\mathfrak{N}(0, 1)$ distribution and probabilities can be found from Table C. When σ is unknown, it must be replaced by an estimate. The usual choice of an estimate is $D_n/\sqrt{n-1}$, which yields the formula for T.

Theorem 5. *Let $X_1, \ldots, X_m$ be a sample from $\mathfrak{N}(\mu, \sigma^2)$ and $Y_1, \ldots, Y_n$ be a sample from $\mathfrak{N}(v, \tau^2)$. Assume that these samples are independent (i.e., all $m + n$ rv's are independent). Let $D_m^2(X) = \sum_{i=1}^n (X_i - X_m)^2$ and define $D_n^2(Y)$ similarly. Then*

$$\frac{\tau^2 D_m^2(X)/(m-1)}{\sigma^2 D_n^2(Y)/(n-1)} \tag{2}$$

has the $\mathfrak{F}_{m-1,n-1}$ distribution.

Proof. Immediate by Theorem 3 and the definition of $\mathfrak{F}$.

The quantity in (2) is used in statistical inference about σ^2/τ^2.

B. Order Statistics

Let $X_1, \ldots, X_n$ be iid rv's, each having density f. In certain statistical applications we are concerned with the n-dimensional random vector $\mathbf{X}^0 = (X_{(1)}, \ldots, X_{(n)})$ obtained by arranging the X_i in ascending order. Thus $X_{(1)} = \min X_i$, $X_{(2)}$ is the next smallest, $\ldots$, and $X_{(n)} = \max X_i$. Then $X_{(1)} \le X_{(2)} \le \cdots \le X_{(n)}$ and each $X_{(i)}$ is one of the X_i. By continuity of the X_i's $P[X_i = X_j$ for some $i \ne j] = 0$, so that all the equality signs may be deleted and $P[X_{(1)} < X_{(2)} < \cdots < X_{(n)}] = 1$.

Order Statistics

The random variable $X_{(i)}$ is called the *ith-order statistic.*

Theorem 6. *Under the conditions above, the density of $\mathbf{X}^0$ is*

$$g(\mathbf{t}) = n! \prod_{i=1}^n f(t_i) \qquad \text{if } t_1 < t_2 < \cdots < t_n. \tag{3}$$

Proof. Let F be the distribution function of an X_i. Then the joint distribution function of $X_1, \ldots, X_n$ is

$$H(\mathbf{u}) = \prod_{i=1}^n F(u_i)$$

for all $\mathbf{u} \in \mathcal{R}^n$. Let G be the distribution function of $\mathbf{X}^0$. For any $\mathbf{t} \in \mathcal{R}^n$ with $t_1 < t_2 < \cdots < t_n$ there are $n!$ points $\mathbf{u} \in \mathcal{R}^n$ whose coordinates are permutations of those of $\mathbf{t}$ and $H(\mathbf{u})$ is the same for each such point. Hence

$$G(\mathbf{t}) = n! \, H(\mathbf{t}) = n! \prod_{i=1}^{n} F(t_i) + \lambda(\mathbf{t}), \tag{4}$$

where $\lambda(\mathbf{t})$ does not depend on t_n.[1] Computing $\partial^n / \partial t_1 \cdots \partial t_n$ from (4) yields the expression in (3).

Since $P[X_{(1)} < X_{(2)} < \cdots < X_{(n)}$ is violated$] = 0$, it follows that $g(\mathbf{t}) = 0$ if $t_1 < t_2 < \cdots < t_n$ is violated.

The assumption that $X_1, \ldots, X_n$ have a density is not needed for the definition of order statistics. In fact, (4) is valid even if F is not continuous. Alternative expressions are needed if there are equalities among the t_i.

Marginal densities of the $X_{(i)}$ can be obtained from (3). We will, instead, prove the following theorem by an alternative method. The result was obtained in the special case of the X_i distributed $\mathcal{U}(0, 1)$ (Example 44.1) and partially extended (Problem 44.6). The proof here follows similar lines.

Theorem 7. *Under the conditions of Theorem 6, the density of $X_{(k)}$ is*

$$g_k(z) = k \binom{n}{k} f(z)[F(z)]^{k-1}[1 - F(z)]^{n-k}. \tag{5}$$

Proof. Let W_z be the number of $X_i \le z$. Then W_z is a $\beta(n, F(z))$ rv and $[X_{(k)} \le z] = [W_z \ge k]$. Hence the distribution function of $X_{(k)}$ is

$$G_k(z) = P[X_{(k)} \le z] = P[W_z \ge k]$$
$$= 1 - B(k - 1 \mid n, F(z)) = Be(F(z) \mid k, n - k + 1), \tag{6}$$

where Be is a beta distribution function. Differentiating in (6) with respect to z yields

$$g_k(z) = f(z)be(F(z) \mid k, n - k + 1),$$

which is the expression in (5).

Below we obtain the joint density of $X_{(i)}$ and $X_{(j)}$ for $i < j$ using a simple heuristic argument which, in principle, can be extended to yield the joint density of any collection of the $X_{(i)}$. A rigorous proof follows along the lines of the proof of Theorem 7.

---

[1] When $n = 2$, for example, $\lambda(\mathbf{t}) = -F^2(t_1)$. See Problem 10 for general n.

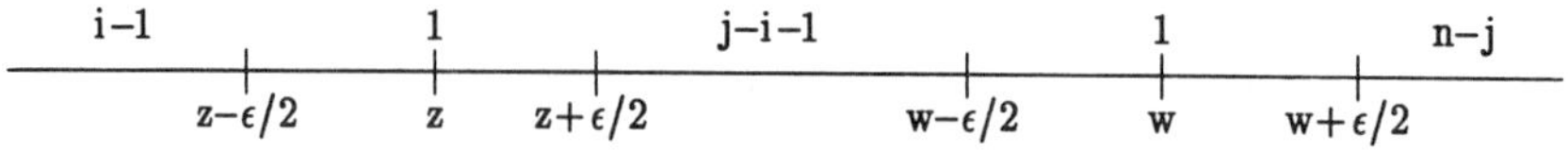

$$\underset{z-\epsilon/2}{\overset{i-1}{\rule{0pt}{0pt}}} \qquad \underset{z}{\overset{1}{\rule{0pt}{0pt}}} \qquad \underset{z+\epsilon/2}{\overset{j-i-1}{\rule{0pt}{0pt}}} \qquad \underset{w-\epsilon/2}{\overset{\rule{0pt}{0pt}}{\rule{0pt}{0pt}}} \qquad \underset{w}{\overset{1}{\rule{0pt}{0pt}}} \qquad \underset{w+\epsilon/2}{\overset{n-j}{\rule{0pt}{0pt}}}$$

Figure 1

In order that $z - \epsilon/2 < X_{(i)} < z + \epsilon/2$ and $w - \epsilon/2 < X_{(j)} < w + \epsilon/2$, it is (roughly) necessary that $(i - 1)$ X_i's be below $z - \epsilon/2$, $(n - j)$ be above $w + \epsilon/2$, and the remaining $(j - i - 1)$ be between $z + \epsilon/2$ and $w - \epsilon/2$ (see Figure 1). Using $f(x) \Delta x$ to approximate the probability that $X \in (x - \Delta x/2, x + \Delta x/2)$, letting g_{ij} denote the joint density of $X_{(i)}$ and $X_{(j)}$, and using the multinomial probability function gives

$$P\left[z - \frac{\epsilon}{2} < X_{(i)} < z + \frac{\epsilon}{2}, w - \frac{\epsilon}{2} < X_{(j)} < w + \frac{\epsilon}{2} \right] \approx \epsilon^2 g_{ij}(z, w)$$

$$\approx \frac{n!}{(i-1)!(j-i-1)!(n-j)!} [F(z)]^{i-1} \epsilon f(z)$$

$$\times [F(w) - F(z)]^{j-i-1} \epsilon f(w)[1 - F(w)]^{n-j},$$

provided that $z + \epsilon/2 < w - \epsilon/2$. Dividing by ϵ^2 and letting $\epsilon \to 0$,

$$g_{ij}(z, w) = \frac{n!}{(i-1)!(j-i-1)!(n-j)!} f(z)f(w)$$

$$\times [F(z)]^{i-1}[F(w) - F(z)]^{j-i-1}[1 - F(w)]^{n-j} \quad (7)$$

for $z < w$.

Range and Midrange

For a sample $X_1, \ldots, X_n$, the *range* is $R = X_{(n)} - X_{(1)}$ and the *midrange* is $M = (X_{(1)} + X_{(n)})/2$.

Theorem 8. *Let $X_1, \ldots, X_n$ be a sample from the density f and distribution function F. Then*

$$f_{R,M}(u, v) = n(n - 1)f\left(v - \frac{u}{2}\right) f\left(v + \frac{u}{2}\right)$$

$$\times \left[F\left(v + \frac{u}{2}\right) - F\left(v - \frac{u}{2}\right) \right]^{n-2} \qquad \text{if } u > 0. \quad (8)$$

Proof. From (7), the joint density of $X_{(1)}$ and $X_{(n)}$ is

$$g_{1n}(z, w) = n(n - 1)f(z)f(w)[F(w) - F(z)]^{n-2} \qquad \text{if } z < w.$$

The inverse of the tranformation from $(X_{(1)}, X_{(n)})$ to (R, M) is $\psi_1(u, v) = v - u/2$ and $\psi_2(u, v) = v + u/2$. Thus $|J(u, v)| = 1$ and (8) follows.

Example 1. If $X_1, \ldots, X_n$ is a sample from $\mathfrak{U}(0, 1)$, then

$$f_{R,M}(u, v) = n(n - 1)u^{n-2} \quad \text{if } 0 < u < 1, \frac{u}{2} < v < 1 - \frac{u}{2},$$

Hence

$$f_R(u) = n(n - 1)u^{n-2}(1 - u) \quad \text{if } 0 < u < 1,$$

the $\mathfrak{Be}(n - 1, 2)$ density. Also

$$f_M(v) = n(n - 1) \int_0^{2v} u^{n-2}\, du = n2^{n-1}v^{n-1} \qquad \text{if } 0 < v < 1/2$$

$$= n(n - 1) \int_0^{2(1-v)} u^{n-2}\, du = n2^{n-1}(1 - v)^{n-1} \qquad \text{if } 1/2 \le v < 1.$$

Summary

For a sample from a normal distribution,

1. $\overline{X}_n$ and D_n^2 are independent.
2. D_n^2/σ^2 has the χ_{n-1}^2 distribution.
3. $(\overline{X}_n - \mu)/(D_n/\sqrt{n(n - 1)})$ has the $\mathfrak{I}_{n-1}$ distribution.
4. The $\mathfrak{I}$ distributions are related to ratios of independent sample variances.

The joint density of the order statistics

$$X_{(1)} < X_{(2)} < \cdots < X_{(n)}$$

is

$$g(t) = n! \prod_{i=1}^{n} f(t_i) \quad \text{if } t_1 < t_2 < \cdots < t_n,$$

where f is the density of an X_i. The marginal density of $X_{(k)}$ is

$$g_k(z) = k \binom{n}{k} f(z)[F(z)]^{k-1}[1 - F(z)]^{n-k}.$$

A heuristic method is given for obtaining higher-dimensional marginals.

Problems

1. Let W be an $\mathfrak{I}_{m,n}$ rv. Show that $mW/(n + mW)$ is a $\mathfrak{Be}(m/2, n/2)$ rv.
2. Find the $\mathfrak{I}_{m,n}$ density.
3. Find the $\mathfrak{I}_n$ density.

4. Let $X_1, \ldots, X_m$ and $Y_1, \ldots, Y_n$ be samples from $\mathfrak{N}(\mu, \sigma^2)$. Assume that these samples are independent. Let $D_m^2(X) = \sum_{i=1}^m (X_i - \overline{X}_m)$ and define $D_n^2(Y)$ similarly. Show that

$$\frac{\overline{X}_m - \overline{Y}_n}{\sqrt{\dfrac{D_m^2(X) + D_n^2(Y)}{m + n - 2}\left(\dfrac{1}{m} + \dfrac{1}{n}\right)}}$$

has the $\mathfrak{I}_{m+n-2}$ distribution.

5. Let U be a χ_n^2 rv. Find the density of $\sqrt{U}$ (the χ_n density).

6. Find the mean and variance for the following distributions.
   (a) χ_n
   (b) χ_n^2
   (c) $\mathfrak{I}_{m,n}$
   (d) $\mathfrak{I}_n$

7. Find the density of the range of a sample from $\mathcal{G}(1, \beta)$.

8. Let $X_1, \ldots, X$ be iid rv's with continuous distribution function F. Identify the distribution of $Y_k = F(X_{(k)})$ by name and parameters.

9. Let $X_1, \ldots, X_n$ be iid rv's with density f and distribution function F. For $i < j$, let $R_{ij} = X_{(} - X_{(i)}$ and $M_{ij} = (X_{(i)} + X_{()})/2$.
   (a) Find the joint density of R_{ij} and M_{ij}.
   (b) Specialize your answer in part (a) to the $\mathcal{G}(1, \beta)$ case.
   (c) In part (b), find the marginal density of R_{ij}.

10. Show that $\lambda(t)$ in (4) does not depend on t_n. Hint: What are the intersections of events such as $[X_1 \leq t_1, \ldots, X_{n-2} \leq t_{n-2}, X_{n-1} \leq t_{n-1}, X_n \leq t_n]$ and $[X_1 \leq t_1, \ldots, X_{n-2} \leq t_{n-2}, X_{n-1} \leq t_n, X_n \leq t_{n-1}]$?

GENERAL PROBLEMS

In each of Problems 1 to 6, let X_i be the lifetime of component C_i. Assume that each X_i is a $\mathcal{G}(1, 1/\lambda)$ rv and the X_i are independent. For each problem, give the density of the lifetime of the system.

1. Components C_1 and C_2 are wired in series.

2. Components C_1 and C_2 are wired in parallel.

3. Components C_1, C_2, and C_3 are wired as in Figure 1. The system works if C_1 works and either C_2 or C_3 works.

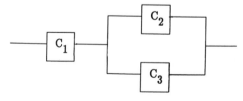

Figure 1

4. Component C_2 is a backup to C_1 so that C_2 begins to operate when C_1 fails and the system fails when C_2 fails. Thus the life of the system is the sum of the lives of C_1 and C_2. This will be pictured as in Figure 2.

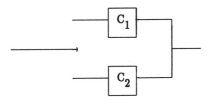

Figure 2

5. Components C_1, C_2, and C_3 are wired as in Figure 3.

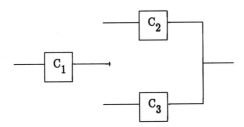

Figure 3

6. Components C_1, C_2, and C_3 are wired in Figure 4.

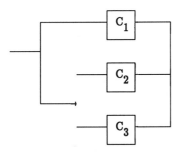

Figure 4

7. Let $X_1, \ldots, X_n$ be independent $\mathcal{U}(0, 1)$ rv's. Find the density of $X_{(i)} - X_{(i-1)}$ $(i = 2, \ldots, n)$.

8. Let $X_1, \ldots, X_n$ be independent, $\mathcal{G}(1, \beta)$. Set $X_{(0)} = 0$.
   (a) Find the density of the n-dimensional random vector $\mathbf{U}$, where
   $$U_i = (n - i + 1)(X_{(i)} - X_{(i-1)})(i = 1, \ldots, n).$$
   (b) Verify that the U_i are independent. What are the distributions of the U_i?
   (c) Find $E(X_{(i)} - X_{(i-1)})$ and $\text{Var}(X_{(i)} - X_{(i-1)})$.
9. (a) Let X be a $\mathcal{N}(\mu, \sigma^2)$ rv and find the density of X^2.
   (b) Express the density in part (a) in the form $A(y)[B_1(y) + B_2(y)]$.
   (c) Obtain the power series expansions of $B_1(y)$ and $B_2(y)$ and simplify the result.
   (d) After algebraic manipulations, the expression in part (c) should be of the form of a mixture $\Sigma_{k=0}^\infty g(k)h_k(y)$, where g is a probability function and the h_k are densities. What models do g and the h_k represent?
10. Devise general formulas [analogous to (47.9) and (47.10)] for the density of a product of continuous rv's. Apply the results to $Y = X_1 X_2$ where
    (a) X_1 and X_2 are iid $\mathcal{U}(-1, 1)$.
    (b) X_1 and X_2 are independent, X_1 is $\mathcal{U}(0, 1)$, and X_2 is $\mathcal{G}(2, 1)$.
11. Let $Z = X + Y$, where X and Y are independent. Derive an expression for $f_X(x \mid Z = z)$ and obtain specific results when:
    (a) X and Y are Poisson
    (b) X and Y are normal
    (c) X and Y are Cauchy
12. Let $f_X(x \mid \theta) = g(x)h(t(x) \mid \theta)$ and $Y = t(X)$. In each of the cases in (a), (b), and (c), show that $f_Y(y \mid \theta) = g_1(y)h(y, \theta)$ and identify g_1.
    (a) X is discrete.
    (b) X is continuous, t is one-to-one and differentiable.
    (c) As in part (b), but t is n-to-one.
13. Let X and Y be iid rv's whose family of distributions is an exponential family (defined in GP 7.10 with notation used there).
    (a) Under each set of conditions in Problem 12 show that the family of distributions of $t(X)$ is an exponential family.
    (b) Show that the family of distributions of $t(X) + t(Y)$ is an exponential family.

REFERENCES

DeGroot, Morris H. (1986). *Probability and Statistics*, 2nd ed. Reading, Mass.: Addison-Wesley Publishing Co., Inc.

Neuts, Marcel F. (1973). *Probability*. Boston: Allyn and Bacon, Inc.

Woodroofe, Michael (1975). *Probability with Applications*. New York: McGraw-Hill Book Company.

9.

Generating Functions

49. DEFINITIONS AND PROPERTIES

Generating functions provide an additional useful tool for some probability problems. Many of the problems already solved in previous sections have shorter, but less direct solutions through generating functions. Some of the results stated here have proofs beyond the scope of this book. Such proofs will be omitted without comment.

Generating Function of a Sequence

If $A(s) = \sum_{k=0}^{\infty} a_k s^k$ converges for $|s| < s_0$, then A is the *generating function of the sequence* $\{a_k\}$. When $A(s)$ is specified, the values of the a_k can be recovered by differentiation: $a_k = A^{(k)}(0)/k!$.

The names given to generating functions useful in probability are not always consistent with this definition. However, in each case differentiation can be used to generate the indicated characteristic.

1. Let X be a nonnegative, integer-valued rv and set $a_k = f_X(k)$. In this case denote A by ψ_X. We call ψ_X the *probability generating function of* X. Note that $\psi_X(s) = E(s^X)$, that $\psi_X(s)$ exists at least for $|s| \leq 1$, and that $\psi_X(1) = 1$. Using $\psi_X(s) = E(s^X)$ for $s \geq 0$ extends the definition to more general rv's subject to existence of the expectation. Another name for ψ_X is the *factorial moment generating function*.

371

2. When the expectation exists, set $L_X(s) = E(s^{-X})$. We call L_X the *reverse factorial moment generating function.*[1]
3. The *moment generating function of* X is $M_X(s) = E(e^{sX})$, so that $M_X(0) = 1$ and $M_X(s)$ may only exist for $s = 0$. If $M_X(s)$ exists for $|s| < s_0$, then M_X is the generating function of the sequence $a_k = E(X^k)/k!$.

The following relations exist between these functions:

$$M_X(s) = \psi_X(e^s);$$

$$L_X(s) = \psi_X\left(\frac{1}{s}\right).$$

Since some results apply equally well to ψ_X, to M_X, and to L_X, let T_X stand for any of them when this is the case. In each case, $T_X(s) = E(\theta^X(s))$ for a suitably chosen function θ. In any statement about $T_X(s)$, no matter which generating function it is, assume that s is such that $T_X(s)$ exists.

From Theorem 5, when $X \geq 0$ is integer valued, its probability generating function ψ_X represents its model. The theorem below gives a sufficient condition under which M_X represents the model for X.

Theorem 1. *If* $M_X(s) = M_Y(s)$ *for every* s *in some open interval, then* $F_X = F_Y$.

Theorem 2. *If* X *and* Y *are iid rv's, then* $T_{X+Y} = T_X T_Y$.

Proof. Since $\theta^{X+Y} = \theta^X \theta^Y$, the result follows from properties of expectations of products of independent rv's.

Theorem 2 and its corollaries are the principal reasons for interest in generating functions.

Corollary 1. *Let* $X, X_1, X_2, \ldots$ *be iid rv's.*

(a) *If* $S_n = \sum_{i=1}^n X_i$, *then* $T_{S_n}(s) = T_X^n(s)$.
(b) *If* N *is a nonnegative, integer-valued rv independent of* $X, X_1, X_2, \ldots$ *and* $S_N = \sum_{i=1}^N X_i$, *then* $T_{S_N}(s) = \psi_N(T_X(s))$.

Proof. (a) Left to the reader.
  (b) First,

$$T_{S_N}(s) = E(\theta^{S_N}(s)) = E(E(\theta^{S_N}(s) \mid N)).$$

---

[1]This is not a standard term but one we find useful in some cases.

From (a), $E(\theta^{S_N}(s) \mid N) = T_X^N(s)$. The remainder of the proof follows from the definition of ψ_N.

Example 1. Let X be a $\mathscr{P}o(\mu)$ rv. Then

$$\psi_X(s) = e^{-\mu} \sum_{x=0}^{\infty} s^x \frac{\mu^x}{x!} = e^{\mu(s-1)}$$

so that

$$M_X(s) = e^{\mu(e^s-1)}$$

and

$$L_X(s) = e^{\mu(1/s-1)}.$$

Example 2. Let $P(A) = p$ and $q = 1 - p$. Then,

$$\psi_{I_A}(s) = q + ps;$$
$$M_{I_A}(s) = q + pe^s;$$
$$L_{I_A}(s) = q + \frac{p}{s}.$$

Hence if S_n is a $\mathscr{B}(n, p)$ rv, then

$$\psi_{S_n}(s) = (q + ps)^n;$$
$$M_{S_n}(s) = (q + pe^s)^n;$$
$$L_{S_n}(s) = \left(q + \frac{p}{s}\right)^n.$$

Let N be a $\mathscr{P}o(\mu)$ rv and let the conditional distribution of S_N given $N = n$ be $\mathscr{B}(n, p)$. Then

$$\psi_{S_N}(s) = e^{\mu(q+ps-1)} = e^{p\mu(s-1)};$$
$$M_{S_N}(s) = e^{\mu(q+pe^s-1)} = e^{p\mu(e^s-1)};$$
$$L_{S_N}(s) = e^{\mu(q+p/s-1)} = e^{p\mu(1/s-1)}$$

This proves that S_N has the $\mathscr{P}o(p\mu)$ distribution (the result of Problem 40.25).

Example 3. If X is degenerate at a, then $M_X(s) = e^{as}$, while in the other two cases, $T_X(s) = \theta^a(s)$.

Corollary 2. *Let X be a rv, a and b be constants, and $Y = a + bX$. Then $M_Y(s) = e^{as}M_X(bs)$ and, in the other cases, $T_Y(s) = \theta^a(s)T_X(s^b)$.*

Proof. Left as a problem.

Example 4. Let Z be a $\mathfrak{N}(0, 1)$ rv. Then

$$M_Z(s) = \frac{1}{\sqrt{2\pi}} \int_{-\infty}^{\infty} e^{sz}e^{-z^2/2} \, dz = \frac{1}{\sqrt{2\pi}} e^{s^2/2} \int_{-\infty}^{\infty} e^{-(z-s)^2/2} \, dy = e^{s^2/2}.$$

If X is a $\mathfrak{N}(\mu, \sigma^2)$ rv, then by Theorem 2,

$$M_X(s) = e^{\mu s + \sigma^2 s^2/2}.$$

Theorem 3. *If the moment generating function M_X exists in an open interval containing 0, then $M_X^{(k)}(0)$, $\psi_X^{(k)}(1)$, and $L_X^{(k)}(1)$ exists for all $k = 1, 2, \ldots$. Furthermore, in that case,*

$$E(X^k) = M_X^{(k)}(0);$$
$$E(X(X - 1) \cdots (X - k + 1)) = \psi_X^{(k)}(1);$$

and

$$E(X(X + 1) \cdots (X + k - 1)) = (-1)^k L_X^{(k)}(1).$$

Proof. (Partial) The proof requires a verification that in each case derivatives and sums or integrals can be interchanged. The remainder of the proof then follows from the relationships

$$M_X^{(k)}(s) = E\left[\frac{d^k}{ds^k}\left(e^{sX}\right)\right];$$
$$\psi_X^{(k)}(s) = E\left[\frac{d^k}{ds^k}\left(s^X\right)\right];$$
$$L_X^{(k)}(s) = E\left[\frac{d^k}{ds^k}\left(s^{-X}\right)\right].$$

Corollary 3. *Let $g(s) = \log T_X(s)$. Then*

(a) $T_X = M_X$ *implies that* $g'(0) = E(X)$ *and* $g''(0) = Var(X)$.
(b) $T_X = \psi_X$ *implies that* $g'(1) = E(X)$ *and* $g''(1) = Var(X) - E(X)$.
(c) $T_X = L_X$ *implies that* $g'(1) = -E(X)$ *and* $g''(1) = Var(X) + E(X)$.

Proof. In any case, $g'(s) = T'_X(s)/T_X(s)$ and $g''(s) = [T''_X(s)T_X(s) - T'^2_X(s)]/T^2_X(s)$. For $T_X = \psi_X$, it follows that $g'(1) = T'_X(1) = E(X)$ and

$$g''(1) = T''_X(1) - [T'_X(1)]^2 = E(X(X - 1)) = (E(X))^2.$$

The remainder of the proof is left as a problem.

Theorem 4. *Let X be a nonnegative, integer-valued rv with probability function f_X. Then $\psi_X^{(k)}(0)$ exists for $k = 1, 2, \ldots,$ and*

$$f_X(k) = \frac{\psi_X^{(k)}(0)}{k!}.$$

Proof. Left as a problem.

Example 5. In a sequence of Bernoulli trials, let T_2 be the waiting time for the pattern SS. For example, if the first two trials are S's, then $T_2 = 2$, while if the first is F and the next two are S's, then $T_2 = 3$. Let $P_k = f_{T_2}(k)$, so that

$$P_2 = p^2;$$
$$P_3 = p^2q = qP_2; \tag{1}$$
$$P_k = qP_{k-1} + pqP_{k-2} \quad \text{for } k = 4, 5, \ldots.$$

In each equation in (1) multiply both sides by s^k and sum over k to obtain

$$\psi_{T_2}(s) = p^2s^2 + qs\psi_{T_2}(s) + pqs^2\psi_{T_2}(s).$$

Hence for s sufficiently close to 0,

$$\psi_{T_2}(s) = \frac{p^2s^2}{1 - qs - pqs^2} = p^2s^2 \sum_{n=0}^{\infty} [qs(1 + ps)]^n.$$

The coefficient of s^k is P_k, so that

$$P_k = p^2 \sum_{n=[(k-1)/2]}^{k-2} q^n p^{k-n-2} \binom{n}{k - n - 2} \tag{2}$$

for $k = 2, 3, \ldots$. The expression in (2) would be more difficult to obtain without the use of a generating function.
   Let

$$g(s) = \log \psi_{T_2}(s) = 2(\log p + \log s) - \log(1 - qs - pqs^2).$$

Then

$$g'(s) = \frac{2}{s} + \frac{q + 2pqs}{1 - qs - pqs^2};$$

$$g''(s) = -\frac{2}{s^2} + \frac{2pq(1 - qs - pqs^2) + (q + 2pqs)^2}{(1 - qs - pqs^2)^2}.$$

Hence, after some simplifying algebra,

$$E(T_2) = \frac{1 + p}{p^2}$$

and

$$\mathrm{Var}(T_2) = \frac{q(1 + 3p + p^2)}{p^4}.$$

The evaluation of $E(T_2)$ was the subject of Problem 34.10(b).

From the definition of M_X,

$$M_X(s) = \sum_{k=0}^{\infty} \frac{E(X^k)}{k!} s^k = \sum_{k=0}^{\infty} \frac{M_X^{(k)}(0)}{k!} s^k.$$

The last expression is the Taylor series expansion of $M_X(s)$ about 0 and can be truncated at the mth term. In this form,

$$M_X(s) = \sum_{k=0}^{m} \frac{E(X^k)}{k!} s^k + R_m(s), \tag{3}$$

where $\lim_{s \to 0} R_m(s)/s^m = 0$. In the next section (3) will be used with $m = 2$.

Example 6. Assume that buses arrive at a station according to a Poisson process with rate λ and that the numbers $X_1, X_2, \ldots$ of passengers leaving these buses are $\mathcal{P}_o(\mu)$ rv's. Let N_t be the number of buses arriving during $(0, t)$ and assume that $N_t, X_1, X_2, \ldots$ are independent. The probability generating function $\psi(s, t)$ of Y_t, the total number of passengers leaving the buses during $(0, t)$, will be found.

Let $P_n(t) = P[Y_t = n]$. Then from Poisson process assumptions (Section 39),

$$P_n(t + h) = (1 - \lambda h)P_n(t) + \lambda h \sum_{k=0}^{n} \frac{\mu^k}{k!} e^{-\mu} p_{n-k}(t) + o(h), \tag{4}$$

where $o(h)$ stands for a quantity which, after division by h, converges to 0 as $h \to 0$.

Set $\psi(s, t) = \psi_{Y_t}(s)$. Multiply both sides of (4) by s^n, where $|s| < 1$ and sum over n to obtain

$$\psi(s, t + h) = (1 - \lambda h)\psi(s, t) + \lambda h e^{-\mu}A(s, t) + o(h), \qquad (5)$$

where

$$A(s, t) = \sum_{n=0}^{\infty} s^n \sum_{k=0}^{n} \frac{\mu^k}{k!} P_{n-k}(t) = \sum_{k=0}^{\infty} \frac{(s\mu)^k}{k!} \sum_{n=0}^{\infty} s^n P_n(t) = e^{s\mu}\psi(s, t).$$

Subtract $\psi(s, t)$ from both sides of (5), divide by h, and let $h \to 0$. This yields

$$\frac{\partial}{\partial t} \psi(s, t) = -\lambda[1 - e^{-\mu(1-s)}]\psi(s, t),$$

which has the general solution

$$\psi(s, t) = C(s)e^{-\lambda t[1 - e^{-\mu(1-s)}]}.$$

Since $P_0(0) = 1$, it follows that $\psi(s, 0) \equiv 1$, so that $C(s) \equiv 1$ and

$$\psi(s, t) = e^{-\lambda t[1 - e^{-\mu(1-s)}]}. \qquad (6)$$

A useful technique for finding generating functions has been illustrated. In this example, a simpler method exists. Observe that $Y_t = \sum_{i=1}^{N_t} X_i$, where $N_t, X_1, X_2, \ldots$ are independent, N_t is $\mathscr{P}_o(\lambda t)$, and the X_i are $\mathscr{P}_o(\mu)$. Then apply Corollary 1 and Example 1. This demonstrates that (6) is valid for all s.

Finally,

$$P_0(t) = \psi(0, t) = e^{-\lambda t(1 - e^{\mu})};$$

$$\frac{\partial}{\partial s} \psi(s, t) = \lambda\mu t e^{-\mu(1-s)}e^{-\lambda t[1 - e^{-\mu(1-s)}]}$$

so that

$$P_1(t) = \lambda\mu t e^{-\mu}e^{-\lambda t(1 - e^{-\mu})}.$$

Similarly,

$$P_2(t) = \frac{1}{2} \lambda\mu^2 t e^{-\mu}(1 + \lambda t e^{-\mu})e^{-\lambda t(1 - e^{-\mu})}$$

and, in principle, any $P_n(t)$ can be found.

The next three examples illustrate moment generating functions that exist only on proper subsets of $\mathscr{R}$.

Example 7. Let X have the exponential density $f(x) = e^{-x}$ for $x > 0$. Then $M_X(s) = 1/(1 - s)$ for $s < 1$ and fails to exist elsewhere. All moments exist and $M_X(s)$ is defined on an open interval that includes 0.

Example 8. Let X have the density

$$f(x) = \frac{1}{\alpha} x^{-(\alpha+1)} \qquad \text{if } x > 1,$$

where $\alpha > 0$. For $s \leq 0$ the existence of $M_X(s)$ follows from $e^{Xs} \leq 1$. For $s > 0$, let $k > \alpha$ and $x > 1/s$. Then

$$e^{xs} > \frac{(xs)^{\alpha+2}}{(k + 2)!} > x^{\alpha+1},$$

the last inequality being valid for $x > (k + 2)!/s^{\alpha+2}$. Hence $M_X(s)$ does not exist for $s > 0$. The βth moment exists only for $\beta < \alpha$. The nonexistence of higher moments implies that there is no open interval containing 0 on which $M_X(s)$ exists.

Example 9. Let X be a $\mathfrak{N}(0, 1)$ rv and $Y = e^X$, so that Y is a lognormal rv. Where it exists,

$$M_Y(s) = E[e^{se^X}] = \frac{1}{\sqrt{2\pi}} \int_{-\infty}^{\infty} e^{se^x} e^{-x^2/2} \, dx.$$

Clearly, $M_Y(s)$ exists for $s \leq 0$. As in Example 7, for $s > 0$ and x sufficiently large, $se^x > x^2/2$, so that $M_Y(s)$ does not exist. All moments exist, but there is another distribution with the same moments. In this case the moments do not *determine the distribution*.

As a consequence of Theorems 1 and 4, all the moments exist and they determine the distribution if, and only if, the moment generating function exists on an open interval containing 0.

The definitions can be extended to generating functions of an n-dimensional random vector $\mathbf{X}$. The vector $\mathbf{s}$ is also n-dimensional. These extensions are

$$\psi_{\mathbf{X}}(\mathbf{s}) = E\left[\prod_{i=1}^{n} s_i^{X_i}\right];$$

$$L_{\mathbf{X}}(\mathbf{s}) = E\left[\prod_{i=1}^{n} s_i^{-X_i}\right];$$

and

$$M_{\mathbf{X}}(\mathbf{s}) = E(e^{\mathbf{s}'\mathbf{X}}).$$

The results given previously, with the exception of Corollary 3, have simple n-dimensional analogs.

Summary

Various generating functions are introduced and used to obtain moments and probability functions.

Problems

1. Prove Theorem 3.
2. Complete the proof of Theorem 4.
3. In a Bernoulli sequence, find the distribution of the waiting time for the pattern SF. Also find the expectation and variance of the waiting time.
4. Complete the proof of Corollary 2.
5. Use generating function arguments to show closure under convolution of the following families:
    (a) $\mathcal{B}$ for fixed p
    (b) $\mathcal{P}_a$
    (c) $\overline{\mathcal{B}}$ for fixed p
    (d) $\mathcal{G}$ for fixed scale parameter
    (e) $\mathcal{N}$
6. Why can't the generating functions of this section be used to verify closure under convolution of the Cauchy family? See Problem 47.19.
7. Prove Theorem 5.
8. (a) Let N be a $\overline{\mathcal{B}}(r, p_1)$ rv and let the conditional distribution of S_N given $N = n$ be $\mathcal{B}(n, p_2)$. Find ψ_{S_N}.
    (b) In the special case $r = 1$, $p_1 = p_2 = p$, show that

$$f_{S_N}(k) = \frac{q}{1 + q} \qquad \text{if } k = 0$$

$$= \frac{q^{k-1}}{(1 + q)^{k+1}} \qquad \text{if } k = 1, 2, \ldots .$$

9. Find $E(Y_t)$ and $\text{Var}(Y_t)$ in Example 5.
10. Use generating function techniques to prove Theorem 40.6, the theorem on the Poisson process.
11. (a) Let X be a lognormal rv (see Problem 48.11). Find all moments (including those of noninteger order) of X.

    (b) Generalize the solution of part (a) to a rv X for which log X has a prescribed moment generating function.

12. Let X be a nonnegative, integer-valued rv and $Q(s) = \Sigma_{k=0}^{\infty} s^k P[X > k]$.
    (a) Find the relationship between Q and ψ_X.
    (b) Express $E(X)$ and $Var(X)$ in terms of Q.

13. Let the family of distributions of X be an exponential family (defined in GP 7.10) and use the natural parameter. Find the moment generating function of $t(X)$.

14. Let X be uniform on $\{1, \ldots, N\}$. Use ψ_x to find $E(X)$ and $Var(X)$.

15. Generalize Problem 14 by letting X be any symmetric rv for which the mean does not exist.

16. Find $E(Y_t)$ and $Var(Y_t)$ in the model of Example 6.

17. Modify the model of Example 6 by assuming that each $X_i = W_i - 1$, where $W_1, W_2, \ldots$ are iid $\overline{\mathcal{B}}(1, p)$ rv's.
    (a) Find $\psi(s, t)$.
    (b) Find $P_n(t)$ for $n = 0, 1, 2, 3$.
    (c) Find $E(Y_t)$ and $Var(Y_t)$.

18. Let $\{Y_t\}$ be a Poisson process with rate λ. Find $\psi_{Y_t}(s)$ and use it to verify that Y_t is a $\mathcal{P}o(\lambda t)$ rv.

19. Show that $\psi_X(1 + s)$ and $L_X(1 - s)$ are generating functions in the sense of the general definition. What are the sequences $\{a_k\}$?

50. A PROOF OF THE CENTRAL LIMIT THEOREM

Assume that $X, X_1, X_2 \ldots$ are iid rv's with moment generating function M. In this case we will prove the central limit theorem. That theorem, stated as Theorem 42.1, requires only the weaker condition of existence of the second moment. Recall that the central limit theorem states that

$$\lim_{n\to\infty} P[S_n^* \le z] = \phi(z),$$

where $S_n = \Sigma_{i=1}^{n} X_i$ and S_n^* is the standardization of S_n.
    The following preliminary theorem is stated without proof.

Theorem 1.   *For $n = 1, 2, \ldots$, let X_n have the moment generating function M_n and suppose that $\lim_{n\to\infty} M_n(s) = M(s)$ on some open interval. If M is the moment generating function of the distribution function F, then*

$$\lim_{n\to\infty} P[X_n \le x] = F(x), \tag{1}$$

at all x where F is continuous.

The example that follows shows that the convergence of $F_n(x)$ to $F(x)$ in (1) need not be valid if x is a discontinuity point of F.

Example 1. Let X_n be a $\mathcal{U}(-1/(2n), 1/(2n))$ rv so that

$$F_n(x) = n\left(x + \frac{1}{2n}\right) \qquad \text{if } |x| \leq \frac{1}{2n}$$

$$= 1 \qquad \text{if } x > \frac{1}{2n}$$

and

$$M_n(s) = \frac{n}{s}\left(e^{s/(2n)} - e^{-s/(2n)}\right).$$

Then $M_n(s) \to 1$ for all s as $n \to \infty$. The limit is the moment generating function of a rv degenerate at 0 [with distribution function $F(x) = 1$ for $x \geq 0$]. However, $F_n(x) \to G(x)$ as $n \to \infty$, where

$$G(x) = \frac{1}{2} \qquad \text{if } x = 0$$

$$= 1 \qquad \text{if } x > 0.$$

Note that $F(x) = G(x)$ for $x \neq 0$ and that 0 is a discontinuity point of F.

Proof of the Central Limit Theorem. Since $S_n^* = \sum_{i=1}^n X_i^*/\sqrt{n}$, we can, without loss of generality, assume that $E(X_i) = 0$ and $\text{Var}(X_i) = 1$. Then

$$M_{S_n^*}(s) = \left[M_X\left(\frac{s}{\sqrt{n}}\right)\right]^n.$$

Using (49.3) yields

$$M_{S_n^*}(s) = \left[1 + \frac{s^2}{2n} + R_2\left(\frac{s}{\sqrt{n}}\right)\right]^n.$$

Since $\lim_{n\to\infty} nR_2(t/\sqrt{n}) = 0$, it follows that $M_{S_n^*}(s) \to e^{s^2/2}$ as $n \to \infty$. This limit is the $\mathcal{N}(0, 1)$ moment generating function (Example 49.4), so that applying Theorem 1 completes the proof.

Problems

1. Use a generating function argument to obtain the Poisson approximation to the binomial.
2. Use a generating function argument to prove a version of the weak law

of large numbers (Theorem 37.2). Without loss of generality, assume that $\mu = 0$. What additional assumptions did you make for this proof?

REFERENCES

Billingsley, Patrick (1986). *Probability and Measure*, 2nd ed. New York: John Wiley and Sons, Inc. Contains proofs of Theorems 49.1 and 50.1 in the case in which the open interval contains the origin. The extensions stated here are problems. The proofs require mathematics beyond the level assumed here.

Feller, William (1968). *An Introduction to Probability Theory and Its Applications*, Vol. I, 3rd ed. New York: John Wiley and Sons, Inc. Extensive discussion of probability generating functions and generating functions for tail probabilities of nonnegative, integer-valued rv's. Includes extensive applications.

Appendix

DESCRIPTION OF THE TABLES AND THEIR COMPUTATIONS

The following pages contain tables of $B(r \mid n, p)$ and $\Pi(r \mid \mu)$ for various values of the parameters, of $\Phi(z)$, and of random digits. These tables were prepared using an IBM PC-AT computer.

Table A: Table of $B(r \mid n, p)$

The table of $B(r \mid n, p)$ is given for $n = 5, 10, 15, 20, 25$ and $p = .1, .2, .25, .3, .4, .5$. For each n and p, we first compute $b(0 \mid n, p) = B(0 \mid n, p)$. Then the recursion relation (40.12) with $r = k + 1$ is used to compute $b(r \mid n, p)$ for $r = 1, \ldots, n$. Finally, $B(r \mid n, p) = b(r \mid n, p) + B(r - 1 \mid n, p)$ completes the computation.

If $B(r \mid n, p)$ is 0 to three decimal places or $B(r - 1 \mid n, p)$ is 1 to three decimal places, then the $B(r \mid n, p)$ entry is left blank.

The computations were done in BASIC.

Table B: Table of $\Pi(r \mid \mu)$

The table of $\Pi(r \mid \mu)$ is given for μ from .05 to 1 in steps of .05, to 8 in steps of .5 and for $\mu = 9, 10$. For each μ, we first compute $\pi(0 \mid \mu) = \Pi(0 \mid \mu)$. The recursion relation (40.22) with $r = k + 1$ is used to

383

compute $\pi(r \mid \mu)$ for $r = 1, 2, \ldots$. Finally, $\Pi(r \mid \mu) = \pi(r \mid \mu) + \Pi(r - 1 \mid \mu)$ completes the computation.

The convention of Table A for leaving entries blank is followed. The computations were done in BASIC.

Table C: Table of $\Phi(z)$

The table of $\Phi(z)$ is given for z from 0 to 3.09 in steps of .01. Let

$$P_0 = 18.3983860429584 \qquad Q_0 = 18.3983859792919$$
$$P_1 = 22.40039515948224 \qquad Q_1 = 43.16075727248846$$
$$P_2 = 13.0613856633851 \qquad Q_2 = 43.36348705913165$$
$$P_3 = 4.028389617759693 \qquad Q_3 = 23.640303763581146$$
$$P_4 = .56420137667455481 \qquad Q_4 = 7.14083719979940819$$
$$Q_5 = 1.$$

For z in our range, set

$$A(y) = P_0 + P_1 y + P_2 y^2 + P_3 y^3 + P_4 y^4$$

and

$$B(y) = Q_0 + Q_1 y + Q_2 y^2 + Q_3 y^3 + Q_4 y^4 + Q_5 y^5,$$

where $y = z/\sqrt{2}$. Then

$$\Phi(z) \approx 1 - \frac{A(y)e^{-y^2}}{2B(y)}. \tag{1}$$

This gives relative accuracy of $1 - \Phi(z)$ to within 10^{-8} as long as $y < 8$, that is, $z < 8\sqrt{2}$. For larger z, we can still use (1), but the P_i and Q_i have different values and the relative accuracy of $1 - \Phi(z)$ is only to within 10^{-4}.[1]

In fact, we omitted the fifth-degree term from $B(y)$, rounded the coefficients to 14 significant figures, and set $Q_0 = P_0$. This still yields accuracy to well within the three decimal places given.

The inverse table, following the table of $\Phi(z)$, is computed by successively bracketing values. Having specified the value Φ_0 for $\Phi(z)$, fix z_L and z_U so that $\Phi(z_L) < \Phi_0 < \Phi(z_U)$. Such choices are easy with the help of the table of $\Phi(z)$. Set $z = (z_L + z_U)/2$. If $\Phi(z) < \Phi_0$, redefine $z_L = z$, and

---

[1] See John F. Hart et al., *Computer Approximations* (New York: John Wiley and Sons, 1968).

if $\Phi(z) > \Phi_0$, redefine $z_U = z$. This is continued until either $\Phi(z) = \Phi_0$ or $z_U - z_L < .0001$.

Computations were all done in BASIC.

Table D: Table of Random Numbers

The table provides 5000 digits obtained from the APL pseudorandom-number generator. The one- and two-digit numbers on the left of the rows are row identification numbers (not random).

Table A Binomial Distribution Function, Table of $B(r \mid n, p)$

					p		
n	r	.10	.20	.25	.30	.40	.50
5	0	0.590	0.328	0.237	0.168	0.078	0.031
	1	0.919	0.737	0.633	0.528	0.337	0.188
	2	0.991	0.942	0.896	0.837	0.683	0.500
	3	1.000	0.993	0.984	0.969	0.913	0.813
	4		1.000	0.999	0.998	0.990	0.969
	5			1.000	1.000	1.000	1.000
10	0	0.349	0.107	0.056	0.028	0.006	0.001
	1	0.736	0.376	0.244	0.149	0.046	0.011
	2	0.930	0.678	0.526	0.383	0.167	0.055
	3	0.987	0.879	0.776	0.650	0.382	0.172
	4	0.998	0.967	0.922	0.850	0.633	0.377
	5	1.000	0.994	0.980	0.953	0.834	0.623
	6		0.999	0.996	0.989	0.945	0.828
	7		1.000	1.000	0.998	0.988	0.945
	8				1.000	0.998	0.989
	9					1.000	0.999
	10						1.000

Table A   (*Continued*)

n	r	.10	.20	.25	.30	.40	.50
					p		
15	0	0.206	0.035	0.013	0.005		
	1	0.549	0.167	0.080	0.035	0.005	
	2	0.816	0.398	0.236	0.127	0.027	0.004
	3	0.944	0.648	0.461	0.297	0.091	0.018
	4	0.987	0.836	0.686	0.515	0.217	0.059
	5	0.998	0.939	0.852	0.722	0.403	0.151
	6	1.000	0.982	0.943	0.869	0.610	0.304
	7		0.996	0.983	0.950	0.787	0.500
	8		0.999	0.996	0.985	0.905	0.696
	9		1.000	0.999	0.996	0.966	0.849
	10			1.000	0.999	0.991	0.941
	11				1.000	0.998	0.982
	12					1.000	0.996
	13						1.000
20	0	0.122	0.012	0.003	0.001		
	1	0.392	0.069	0.024	0.008	0.001	
	2	0.677	0.206	0.091	0.035	0.004	
	3	0.867	0.411	0.225	0.107	0.016	0.001
	4	0.957	0.630	0.415	0.238	0.051	0.006
	5	0.989	0.804	0.617	0.416	0.126	0.021
	6	0.998	0.913	0.786	0.608	0.250	0.058
	7	1.000	0.968	0.898	0.772	0.416	0.132
	8		0.990	0.959	0.887	0.596	0.252

Table A (*Continued*)

				p			
n	r	.10	.20	.25	.30	.40	.50
20	9		0.997	0.986	0.952	0.755	0.412
	10		0.999	0.996	0.983	0.872	0.588
	11		1.000	0.999	0.995	0.943	0.748
	12			1.000	0.999	0.979	0.868
	13				1.000	0.994	0.942
	14					0.998	0.979
	15					1.000	0.994
	16						0.999
	17						1.000
25	0	0.072	0.004	0.001			
	1	0.271	0.027	0.007	0.002		
	2	0.537	0.098	0.032	0.009		
	3	0.764	0.234	0.096	0.033	0.002	
	4	0.902	0.421	0.214	0.090	0.009	
	5	0.967	0.617	0.378	0.193	0.029	0.002
	6	0.991	0.780	0.561	0.341	0.074	0.007
	7	0.998	0.891	0.727	0.512	0.154	0.022
	8	1.000	0.953	0.851	0.677	0.274	0.054
	9		0.983	0.929	0.811	0.425	0.115
	10		0.994	0.970	0.902	0.586	0.212
	11		0.998	0.989	0.956	0.732	0.345
	12		1.000	0.997	0.983	0.846	0.500
	13			0.999	0.994	0.922	0.655
	14			1.000	0.998	0.966	0.788

Table A (*Continued*)

n	r	.10	.20	.25	.30	.40	.50
25	15				1.000	0.987	0.885
	16					0.996	0.946
	17					0.999	0.978
	18					1.000	0.993
	19						0.998
	20						1.000

p

Table B   Poisson Distribution Function, Table of $\Pi(r \mid \mu)$

r	.05	.10	.15	.20	.25	.30	.35	.40
				μ				
0	0.951	0.905	0.861	0.819	0.779	0.741	0.705	0.670
1	0.999	0.995	0.990	0.982	0.974	0.963	0.951	0.938
2	1.000	1.000	0.999	0.999	0.998	0.996	0.994	0.992
3			1.000	1.000	1.000	1.000	1.000	0.999
4								1.000

r	.45	.50	.55	.60	.65	.70	.75	.80
				μ				
0	0.638	0.607	0.577	0.549	0.522	0.497	0.472	0.449
1	0.925	0.910	0.894	0.878	0.861	0.844	0.827	0.809
2	0.989	0.986	0.982	0.977	0.972	0.966	0.959	0.953
3	0.999	0.998	0.998	0.997	0.996	0.994	0.993	0.991
4	1.000	1.000	1.000	1.000	0.999	0.999	0.999	0.999
5					1.000	1.000	1.000	1.000

r	.85	.90	.95	1.0	1.5	2.0	2.5	3.0
				μ				
0	0.427	0.407	0.387	0.368	0.223	0.135	0.082	0.050
1	0.791	0.772	0.754	0.736	0.558	0.406	0.287	0.199
2	0.945	0.937	0.929	0.920	0.809	0.677	0.544	0.423

Table B   (*Continued*)

r	.85	.90	.95	1.0	1.5	2.0	2.5	3.0
				μ				
3	0.989	0.987	0.984	0.981	0.934	0.857	0.758	0.647
4	0.998	0.998	0.997	0.996	0.981	0.947	0.891	0.815
5	1.000	1.000	1.000	0.999	0.996	0.983	0.958	0.916
6				1.000	0.999	0.995	0.986	0.966
7					1.000	0.999	0.996	0.988
8						1.000	0.999	0.996
9							1.000	0.999
10								1.000

r	3.5	4.0	4.5	5.0	5.5	6.0
			μ			
0	0.030	0.018	0.011	0.007	0.004	0.002
1	0.136	0.092	0.061	0.040	0.027	0.017
2	0.321	0.238	0.174	0.125	0.088	0.062
3	0.537	0.433	0.342	0.265	0.202	0.151
4	0.725	0.629	0.532	0.440	0.358	0.285
5	0.858	0.785	0.703	0.616	0.529	0.446
6	0.935	0.889	0.831	0.762	0.686	0.606
7	0.973	0.949	0.913	0.867	0.809	0.744
8	0.990	0.979	0.960	0.932	0.894	0.847
9	0.997	0.992	0.983	0.968	0.946	0.916
10	0.999	0.997	0.993	0.986	0.975	0.957
11	1.000	0.999	0.998	0.995	0.989	0.980
12		1.000	0.999	0.998	0.996	0.991

Table B   (*Continued*)

			μ			
r	3.5	4.0	4.5	5.0	5.5	6.0
13			1.000	0.999	0.998	0.996
14				1.000	0.999	0.999
15					1.000	0.999
16						1.000

			μ			
r	6.5	7.0	7.5	8.0	9.0	10
0	0.002	0.001	0.001			
1	0.011	0.007	0.005	0.003	0.001	
2	0.043	0.030	0.020	0.014	0.006	0.003
3	0.112	0.082	0.059	0.042	0.021	0.010
4	0.224	0.173	0.132	0.100	0.055	0.029
5	0.369	0.301	0.241	0.191	0.116	0.067
6	0.527	0.450	0.378	0.313	0.207	0.130
7	0.673	0.599	0.525	0.453	0.324	0.220
8	0.792	0.729	0.662	0.593	0.456	0.333
9	0.877	0.830	0.776	0.717	0.587	0.458
10	0.933	0.901	0.862	0.816	0.706	0.583
11	0.966	0.947	0.921	0.888	0.803	0.697
12	0.984	0.973	0.957	0.936	0.876	0.792
13	0.993	0.987	0.978	0.966	0.926	0.864
14	0.997	0.994	0.990	0.983	0.959	0.917
15	0.999	0.998	0.995	0.992	0.978	0.951
16	1.000	0.999	0.998	0.996	0.989	0.973
17		1.000	0.999	0.998	0.995	0.986

Table B (*Continued*)

r	6.5	7.0	7.5	8.0	9.0	10
				μ		
18			1.000	0.999	0.998	0.993
19				1.000	0.999	0.997
20					1.000	0.998
21						0.999
22						1.000

Table C Standard Normal Distribution Function, Table of $\Phi(z)$

z	.00	.01	.02	.03	.04	.05	.06	.07	.08	.09
0.0	0.500	0.504	0.508	0.512	0.516	0.520	0.524	0.528	0.532	0.536
0.1	0.540	0.544	0.548	0.552	0.556	0.560	0.564	0.567	0.571	0.575
0.2	0.579	0.583	0.587	0.591	0.595	0.599	0.603	0.606	0.610	0.614
0.3	0.618	0.622	0.626	0.629	0.633	0.637	0.641	0.644	0.648	0.652
0.4	0.655	0.659	0.663	0.666	0.670	0.674	0.677	0.681	0.684	0.688
0.5	0.691	0.695	0.698	0.702	0.705	0.709	0.712	0.716	0.719	0.722
0.6	0.726	0.729	0.732	0.736	0.739	0.742	0.745	0.749	0.752	0.755
0.7	0.758	0.761	0.764	0.767	0.770	0.773	0.776	0.779	0.782	0.785
0.8	0.788	0.791	0.794	0.797	0.800	0.802	0.805	0.808	0.811	0.813
0.9	0.816	0.819	0.821	0.824	0.826	0.829	0.831	0.834	0.836	0.839
1.0	0.841	0.844	0.846	0.848	0.851	0.853	0.855	0.858	0.860	0.862
1.1	0.864	0.866	0.869	0.871	0.873	0.875	0.877	0.879	0.881	0.883
1.2	0.885	0.887	0.889	0.891	0.893	0.894	0.896	0.898	0.900	0.901
1.3	0.903	0.905	0.907	0.908	0.910	0.911	0.913	0.915	0.916	0.918
1.4	0.919	0.921	0.922	0.924	0.925	0.926	0.928	0.929	0.931	0.932
1.5	0.933	0.934	0.936	0.937	0.938	0.939	0.941	0.942	0.943	0.944
1.6	0.945	0.946	0.947	0.948	0.949	0.951	0.952	0.953	0.954	0.954
1.7	0.955	0.956	0.957	0.958	0.959	0.960	0.961	0.962	0.962	0.963
1.8	0.964	0.965	0.966	0.966	0.967	0.968	0.969	0.969	0.970	0.971
1.9	0.971	0.972	0.973	0.973	0.974	0.974	0.975	0.976	0.976	0.977
2.0	0.977	0.978	0.978	0.979	0.979	0.980	0.980	0.981	0.981	0.982
2.1	0.982	0.983	0.983	0.983	0.984	0.984	0.985	0.985	0.985	0.986

Table C   (*Continued*)

z	.00	.01	.02	.03	.04	.05	.06	.07	.08	.09
2.2	0.986	0.986	0.987	0.987	0.987	0.988	0.988	0.988	0.989	0.989
2.3	0.989	0.990	0.990	0.990	0.990	0.991	0.991	0.991	0.991	0.992
2.4	0.992	0.992	0.992	0.992	0.993	0.993	0.993	0.993	0.993	0.994
2.5	0.994	0.994	0.994	0.994	0.994	0.995	0.995	0.995	0.995	0.995
2.6	0.995	0.995	0.996	0.996	0.996	0.996	0.996	0.996	0.996	0.996
2.7	0.997	0.997	0.997	0.997	0.997	0.997	0.997	0.997	0.997	0.997
2.8	0.997	0.998	0.998	0.998	0.998	0.998	0.998	0.998	0.998	0.998
2.9	0.998	0.998	0.998	0.998	0.998	0.998	0.998	0.999	0.999	0.999
3.0	0.999	0.999	0.999	0.999	0.999	0.999	0.999	0.999	0.999	0.999

Continuation of the table yields $\Phi(3.29) = .999$ and $\Phi(3.30) = 1.000$.

Inverse Table.

p	.750	.800	.850	.875	.900	.920	.950
$\Phi^{-1}(p)$	0.675	0.842	1.036	1.150	1.282	1.405	1.645

p	.960	.975	.980	.990	.995	.998	.999
$\Phi^{-1}(p)$	1.751	1.960	2.054	2.326	2.576	2.878	3.090

Table D Table of Random Numbers

	1–5	6–10	11–15	16–20	21–25	26–30	31–35	36–40	41–45	46–50
1	18462	07246	22302	18041	19667	46693	05971	30060	97021	96067
2	16019	39265	20161	67253	28872	33894	41644	03113	91729	72201
3	23577	06766	06269	37668	09622	85239	60354	26220	57137	68231
4	59715	17924	51416	67894	94142	67810	77746	10404	27874	79795
5	24681	34552	18370	20718	13793	11548	70202	40862	52657	40449
6	32297	18679	00367	46765	42260	00996	09812	55200	12696	49932
7	83962	19887	97101	91059	91104	20022	72230	33924	92616	16957
8	89921	71963	64804	48446	44764	17494	56698	30710	09088	56606
9	56299	40085	87462	02308	11992	88104	26064	78438	49395	80413
10	95172	77643	05223	55440	19115	84004	79656	53315	89278	85314
11	17836	81954	56346	14583	57045	64631	34971	71975	03560	13783
12	01260	18669	30056	95721	05518	80003	20881	57431	34949	40792
13	26771	93300	37953	09051	88405	02175	62543	30411	41717	35374
14	22989	72774	74412	23007	04311	90244	72554	48920	55266	62888
15	90527	70682	58525	19258	16774	41469	59140	08800	72461	81387
16	11777	06581	71995	38991	95238	49334	50400	18616	31064	23719
17	19368	93799	16262	04479	09556	19986	59509	83170	87232	61223
18	67303	35495	35390	65188	68335	32953	80675	11230	32129	38592
19	52460	27491	70789	42052	26914	59993	76886	47622	94392	37878
20	28241	85297	52404	09474	74742	54082	79898	88229	47985	58173
21	16271	98487	75323	00203	60295	65669	40931	09726	16199	01061
22	42581	69428	12630	18376	08504	23115	85124	70456	76017	00442
23	14966	83129	01445	15122	70327	25999	89920	48147	71806	87807
24	04372	09493	35077	86023	32798	92164	74444	84866	36366	54999
25	82331	57841	17077	36818	55170	05121	36088	74098	74453	22255

Table D   (*Continued*)

26	61174	48672	31811	16788	59455	73877	13097	46086	08643	00954
27	49412	28102	28739	27706	12911	35210	45871	14881	12841	84737
28	22233	09526	54321	78067	43610	56997	90590	87531	77235	79631
29	83535	35656	39097	16999	73101	27932	64318	29015	85957	11754
30	65156	27392	63498	14874	91762	56529	90628	36963	33737	46287
31	16104	53974	60035	18680	20420	19882	54212	83668	71350	33110
32	64653	31390	51429	43016	80397	60287	11676	50339	08033	59148
33	27281	38847	79383	47183	02709	47316	29194	86437	04956	36554
34	51678	48432	34666	84575	20652	17878	89769	56256	84292	90822
35	46106	07908	15043	27521	48709	37052	17520	06876	82155	48969
36	00652	90560	75839	20127	44845	76996	28658	26340	27814	08302
37	07151	97469	82607	20874	30660	21810	79054	68310	04432	03037
38	40144	25428	95002	00801	85800	52043	73394	65214	30096	84918
39	48723	67316	02436	89920	55981	70510	44044	82934	02144	69739
40	01010	14825	04801	65624	07472	38486	02611	98237	15356	30821
41	15697	23158	92203	87468	39739	78958	18914	52140	62743	22822
42	65806	95327	38126	19029	18936	20319	57237	04947	35332	91227
43	82920	32257	46271	71081	67749	79523	76709	13750	57032	59299
44	72818	74293	63402	59498	91378	23768	84862	51796	85615	73385
45	26235	96759	71240	23370	35846	03633	42761	81157	78746	38234
46	02019	00862	83253	40802	88086	38783	65344	04082	29354	04668
47	91287	17364	26073	11585	37006	79168	45032	20286	89515	77776
48	73229	25883	42671	74088	22578	27976	26709	77765	05569	81222
49	00607	13176	16816	70993	77420	08374	66507	12246	53135	31782
50	47665	13455	84800	06488	42861	80230	47707	87397	75043	02634

Table D (*Continued*)

51	94487	58720	63142	67201	72680	78949	91448	82598	79324	07781
52	55056	78011	41745	91334	14156	17453	30328	43952	40090	21253
53	04771	13167	35668	91480	13525	45351	16064	99865	77915	19113
54	91341	04424	35619	42392	48482	05483	73406	56123	31190	75860
55	77404	89807	86893	32337	15115	33430	98021	68293	05507	29201
56	52764	61490	40025	05022	41890	03582	36736	27061	98442	66402
57	37995	98539	07421	62841	60503	30272	64862	90649	49356	79177
58	26293	52773	48530	73679	17892	90357	78355	31393	96410	48533
59	46920	26436	90972	65838	76892	22925	61313	72054	64693	10521
60	14113	44419	33169	21006	13391	83333	86480	23494	72799	84594
61	14333	00618	50746	41567	44786	29036	53614	98584	43757	69308
62	57126	40327	50552	34550	66322	09625	10618	25143	35464	07186
63	16363	12715	04506	32801	95693	72715	90479	50413	43313	12333
64	35603	31948	21093	74935	66686	57180	57446	95082	42074	11771
65	56994	64934	26444	20449	78112	14534	73506	99140	06001	37330
66	39275	36047	84470	04279	46007	23428	78773	01367	97521	73545
67	55087	85123	14771	38976	18575	09357	86074	99272	54069	74982
68	34982	62478	49965	96323	35214	33109	47269	44169	59972	85366
69	48531	16125	60364	88113	67494	71235	19141	29968	14377	98969
70	04279	39704	11584	79585	57139	05534	71299	29683	98283	24618
71	62197	49911	47575	29700	96491	20971	09327	68971	17988	27556
72	38758	83600	38616	45374	49030	61629	82533	61314	46672	32436
73	68600	95259	33645	44563	24497	39423	31290	37780	59433	84988
74	89670	74716	53107	30815	52794	81796	55827	30426	36377	76837
75	36804	94237	96866	58218	67276	64727	23649	84014	49069	88382

Table D   (*Continued*)

76	54710	99988	79341	22261	73322	82744	29929	37392	27165	30872
77	98205	88376	49076	27089	42953	28092	26393	99541	89613	97943
78	17987	83023	39723	31828	93602	49990	35901	54504	66875	65235
79	54853	81229	22064	75602	56133	67709	69623	91321	08292	57881
80	16214	43716	60285	95553	69344	78918	62279	44927	27471	58317
81	41220	54362	79628	37684	37645	43761	79162	62413	23774	51682
82	31065	01054	79012	82496	86052	25524	69183	50599	35797	08691
83	76692	04142	34197	24201	98412	07669	26632	32882	35571	47325
84	01128	69492	04204	68048	09169	71429	39510	17421	62078	32011
85	72580	09693	49112	95979	28183	05269	12300	61799	35034	77065
86	44098	46102	17512	51133	65615	14777	94588	64072	71665	35016
87	28470	43349	62763	30342	77260	54757	01715	77579	86501	68053
88	18209	02881	15289	24342	39457	49368	81402	24377	79062	34309
89	94203	57977	73649	92838	57395	00482	83728	58159	37918	66878
90	52476	87240	12557	97300	38735	95140	29514	51919	35823	18788
91	38527	47824	96449	95031	10985	37965	79525	56866	51411	56678
92	99531	83502	98087	66156	97147	72450	30175	11864	00352	63734
93	82025	03430	36288	50714	93948	70836	49089	89122	75555	64944
94	16106	75557	94106	47758	91983	35622	51496	17261	97048	99778
95	75073	82079	99239	04489	25754	59993	66764	45433	45525	67559
96	84612	41971	35164	48933	21137	16554	64841	24591	99365	84433
97	52025	11490	45824	75760	93409	12419	99722	46408	37622	07936
98	00241	51521	98587	40958	60562	84905	67292	30664	59825	52457
99	67486	18261	74608	96599	15134	59976	49785	00253	25544	50855
100	05245	61794	08047	37489	17930	86823	86709	04990	47961	42009

Index